Towards More Sustainable Food Systems

Towards More Sustainable Food Systems

Editor

Sigrid Kusch-Brandt

MDPI

Basel • Beijing • Wuhan • Barcelona • Belgrade • Novi Sad • Cluj • Manchester

Editor
Sigrid Kusch-Brandt
Faculty of Mathematics,
Natural Sciences and Management
University of Applied Sciences Ulm
Ulm
Germany

Editorial Office
MDPI
St. Alban-Anlage 66
4052 Basel, Switzerland

This is a reprint of articles from the Special Issue published online in the open access journal *International Journal of Environmental Research and Public Health* (ISSN 1660-4601) (available at: www.mdpi.com/journal/ijerph/special_issues/TMSFS).

For citation purposes, cite each article independently as indicated on the article page online and as indicated below:

Lastname, A.A.; Lastname, B.B. Article Title. *Journal Name* **Year**, *Volume Number*, Page Range.

ISBN 978-3-0365-9049-3 (Hbk)
ISBN 978-3-0365-9048-6 (PDF)
doi.org/10.3390/books978-3-0365-9048-6

Contents

About the Editor

Sigrid Kusch-Brandt

Dr. Sigrid Kusch-Brandt is a Lecturer for Circular Economy and Sustainable Management of Resources at the University of Applied Sciences Ulm, a Visiting Research Fellow within Engineering and Physical Sciences at the University of Southampton, and a consultant to international organisations, research institutions, public authorities, and various companies. In earlier academic positions, she served as a Substitute Professor for waste management at the Technical University of Dresden and as a Visiting Professor for 'Sustainable and Renewable Resources' at the University of Padua for six years. With her work, she aims to contribute to the more sustainable management of environmental resources, including through a more effective integration of circular economy principles. She has a specific interest in the environmentally sound usage of biogenic resources.

Preface

Food systems are under increasing pressure. They must meet the food requirements of a growing world population, and socio-economic changes also influence the type of foods which are under demand. At the same time, food systems are a major contributor to global environmental change, and environmental changes adversely impact agricultural productivity. This Special Issue explores opportunities and challenges towards achieving more sustainable food systems. Essential changes required in food systems are highlighted, such as more effective food distribution, the avoidance or valorisation of food waste, and less meat consumption. How to actually achieve these required changes across food value chains is also presented. This Special Issue supports solution-oriented approaches towards addressing one of most complex challenges of this century. The reader is invited to study the publications included in this Special Issue in detail.

Sigrid Kusch-Brandt
Editor

International Journal of
Environmental Research and Public Health

MDPI

Editorial

Towards More Sustainable Food Systems—14 Lessons Learned

Sigrid Kusch-Brandt [1,2]

[1] Civil, Environmental and Architectural Engineering, University of Padua, 35131 Padua, Italy; mail@sigrid-kusch.eu
[2] Water and Environmental Engineering Group, University of Southampton, Southampton SO16 7QF, UK

Received: 28 May 2020; Accepted: 3 June 2020; Published: 4 June 2020

Abstract: Food production, processing, distribution and consumption are among the major contributors to global environmental change. At the same time, food systems need to effectively respond to the demands of a growing world population, and already today many communities and individuals are affected by food insecurity. Moving towards sustainable food value chains is one of the greatest and most complex challenges of this century. To explore promising solutions and specific problems in this context, and to discuss achieved progress, this Special Issue of the International Journal of Environmental Research and Public Health was initiated. The publications enrich our knowledge about essential changes required in the food systems, such as more effective food distribution, avoidance or valorisation of food waste and less meat consumption. Knowing what to change and knowing how to actually achieve such change are two different themes. It becomes evident that there is still an incomplete picture regarding how innovations in the food system can be strengthened to catalyse transformations at a larger scale. Grassroot initiatives require more supporting efforts to effectively influence policies, and the lack of coordination among civil society initiates must be overcome. Sustainability-oriented companies in food supply chains also have a major role to play.

Keywords: sustainability in the food sector; food supply chains; food insecurity; food waste and loss; innovation and change; food governance

1. Introduction

Sustainable food systems are a key challenge for all future development. This Special Issue of the International Journal of Environmental Research and Public Health was initiated as a platform to present approaches and findings that contribute towards more sustainable food systems, and to discuss challenges that need to be addressed with priority. The Special Issue's information website [1] and the invitation to submit a paper asked the following question:

"How can the value chain of food production, processing, distribution, consumption, and waste management become more sustainable?"

A total of 14 submissions were positively evaluated and included in the Special Issue. While certainly these publications do not cover all topics relevant in the field of sustainable food production and consumption, each of these contributions adds substance to the large mosaic of changes and efforts required to achieve food systems that are fit to sustain a prosperous and healthy world in the future, with the human population living in dignity and with responsible stewardship over natural and economic systems. The aim of this editorial is to highlight one key insight from each publication, thus compiling 14 lessons learned from the authors of this Special Issue and their work. These 14 lessons learned are a selection only, and a broader spectrum of insights and findings is available in the publications.

2. Overview of Publications in the Special Issue 'Towards More Sustainable Food Systems'

Table 1 compiles an overview of publications included in the Special Issue, and provides information regarding whether they fall into four core thematic clusters identified.

Table 1. Publications in the Special Issue 'Towards More Sustainable Food Systems', listed in the order as they are discussed in this editorial (+++: main focus area of publication; +: integrated topic).

	Title of Publication	Local Food Security	Changed Environments	Reducing Environmental Impacts	Supporting Desirable Changes
1	Exploring food access and sociodemographic correlates of food consumption and food insecurity in Zanzibari households (Nyangasa et al. [2])	+++			+
2	Why tenure responsive land-use planning matters: insights for land use consolidation for food security in Rwanda (Chigbu et al. [3])	+++			+
3	The environmental impact and formation of meals from the pilot year of a Las Vegas convention food rescue program (To et al. [4])	+++		+++	+
4	Impact of organic manure on growth, nutrient content and yield of chilli pepper under various temperature environments (Khaitov et al. [5])	+	+++	+	
5	Handmade comal tortillas in Michoacán: traditional practices along the rural–urban gradient (Arnés and Astier [6])	+	+++		
6	Smart approaches to food waste final disposal (Cecchi and Cavinato [7])			+++	
7	Climate change and consumer's attitude toward insect food (Chang et al. [8])	+		+++	+
8	Consumer attitudes towards environmental concerns of meat consumption: a systematic review (Sanchez-Sabate and Sabaté [9])			+++	+
9	Consumers' willingness to pay for organic foods in China: bibliometric review for an emerging literature (Li R. et al. [10])			+++	+
10	Food system transformation: integrating a political–economy and social–ecological approach to regime shifts (Pereira et al. [11])	+	+		+++
11	Rome, a policy without politics: the participatory process for a metropolitan scale food policy (Mazzocchi and Marino [12])	+			+++
12	Toward livestock supply chain sustainability: a case study on supply chain coordination and sustainable development in the pig sector in China (Zhuo and Ji [13])			+	+++
13	Moderating effect of dynamic environment in the relationship between guanxi, trust, and repurchase intention of agricultural materials (Li L. et al. [14])	+		+	+++
14	Evaluation of policies on inappropriate treatment of dead hogs from the perspective of loss aversion (Yang and Wang [15])			+	+++

The publications concentrate on four main themes: (1) food security among the local population, (2) potential impacts of changing environments on food production and already observed manifestations of changes, (3) specific measures for reducing adverse environmental impacts related to food consumption, and (4) understanding how to effectively support desirable changes in the food supply systems.

Table 1 shows that most publications address more than one of these four main themes, which is not surprising when considering the complexity of the food systems. Each publication is rich in insights, and only one insight is selected from each of the 14 publications to be highlighted in this editorial.

3. Fourteen Lessons Learned

Table 2 provides an overview of the 14 insights extracted from the publications to be presented in this editorial. Drawing from the 14 publications of the Special Issue, and considering the information compiled in Table 1, the 14 insights discussed in the following mainly, although not exclusively, address three main challenges of food sustainability:

1. Food security, including under changing environments (lessons 1 to 5)

2. Spotlights on specific measures that may reduce adverse environmental impacts related to food consumption (lessons 6 to 9)
3. How to effectively support desirable changes in the food supply systems (lessons 10 to 14)

Table 2. Overview of 14 insights extracted from the publications of the Special Issue.

	Lesson Learned	Reference
	Food security, including under changing environments	
Lesson 1	Improvement of infrastructure to facilitate distribution of food continues to be of urgent need in overcoming food insecurity in developing countries.	Nyangasa et al. [2]
Lesson 2	Food insecurity in local communities and land tenure insecurity are interlinked phenomena; tenure responsive land-use planning is an essential mechanism in improving food security.	Chigbu et al. [3]
Lesson 3	In high-income countries, food insecurity continues to be a challenge of significant scale, and it can be alleviated by rescuing uneaten food from large resorts such as convention centres.	To et al. [4]
Lesson 4	While the positive impacts of organic fertilisers are generally well known, more efforts are required to understand in detail how climatic change will impact the performance of organic fertilisers.	Khaitov et al. [5]
Lesson 5	To halt erosion of traditional culinary knowledge and preserve traditional agro-alimentary systems, as a means to reduce local food insecurity risks in a further globalising and urbanising world, it is essential to understand the multifaceted differences in food production and consumption patterns among rural and urban populations.	Arnés and Astier [6]
	Spotlights on specific measures that may reduce adverse environmental impacts	
Lesson 6	Where food waste cannot be avoided, the coupling of food waste valorisation and wastewater treatment, in an integrated system, creates important synergies to reduce greenhouse gas emissions and deliver additional marketable outputs in a biorefinery approach.	Cecchi and Cavinato [7]
Lesson 7	Foods with insect ingredients have high environmental advantages compared to common meat-based foods, but environmental concerns do not have an impact on consumer's purchase intention, while food neophobia has a significant role in limiting the purchase of insect food.	Chang et al. [8]
Lesson 8	Consuming less meat has the potential to make a significant contribution to reducing greenhouse gas emissions, but environmental concerns do not have a major impact on the decision to consume meat; however, there might be regionally differing cultural and economic determinants, which merit more attention in order to understand changing food consumption practices.	Sanchez-Sabate and Sabaté [9]
Lesson 9	Purchasing power is the strongest factor in influencing the decision to buy organic food, and thus the price is most important when deciding in favour of organic produce, although individual norms, knowledge about health implications and consideration of animal welfare may also play a role.	Li R. et al. [10]
	How to effectively support desirable changes in the food supply systems	
Lesson 10	Transformative change towards a more sustainable and just food regime needs to disrupt from the bottom up, where local innovations act within their contextual frame, addressing the challenges that their specific communities are facing and looking at scaling in different ways to effectively challenge conventional top-down solutions.	Pereira et al. [11]
Lesson 11	Insufficient connections between food-system-oriented initiatives from civil society and small market actors result in the low impact of such initiatives on strategic policymaking; to effectively intervene in food systems, it is essential to engage in participatory coordination, in order to identify common interests among the different types of initiatives, while respecting that the diversity of models brings benefits to the community.	Mazzocchi and Marino [12]
Lesson 12	In supply chains, sustainability-oriented practices are initially transferred from single focal companies to other actors in the supply chain; a close and stable cooperation functions as learning environment.	Zhuo and Ji [13]
Lesson 13	Trust between farmers and agricultural retailers plays an important role in the farmers' decisions regarding what agricultural materials to buy, including new types of materials, such as fertilisers derived from circular economy schemes.	Li L. et al. [14]
Lesson 14	Integrating people's aversion to loss into the design of environmental policies can create more effective environmental protection schemes.	Yang and Wang [15]

3.1. Food Security, Including under Changing Environments (Lessons 1 to 5)

Lesson 1: Improvement of infrastructure to facilitate distribution of food continues to be of urgent need in overcoming food insecurity in developing countries.

By looking at the nutrition situation of the urban and rural population in Zanzibar, the exploratory study of Nyangasa et al. [2] contributes to a better understanding of the factors that influence patterns of poor food consumption and food insecurity among different types of households. Severe food insecurity is more likely to occur in larger households with poor food access, and this is not limited to urban areas but also applies to rural households. Interestingly, Nyangasa et al. observed that in Zanzibar, urban households with good food access have a higher chance of acceptable food consumption than rural households with poor food access. According to the researchers, this may be explained by better infrastructure in urban areas, enabling good the accessibility of foods. The authors conclude that improvement of infrastructure is a key priority in enhancing the distribution of food within the rural–urban areas. In addition, different forms of coping strategies are required, particularly in rural areas, such as efficient food storage techniques and home gardening.

Lesson 2: Food insecurity in local communities and land tenure insecurity are interlinked phenomena; tenure responsive land-use planning is an essential mechanism in improving food security.

The work of Chigbu et al. [3], conducted in Rwanda, illustrates that there exists a close connection between land-use decisions and food security outcomes. Access to agricultural land is not enough; rather it is the capacity of the land-user to make critical, household-specific decisions that affects the food security of local communities. Such decisions include what to plant, and how to use the land and harvested agricultural products. Imposition of priority crops on farmers is identified as a form of tenure insecurity in the context of land-use, and, according to Chigbu et al., such an approach failed to be food security responsive in Rwanda. The authors introduce the concept of tenure responsive land-use planning, and propose it as a method for food security improvement.

Lesson 3: In high-income countries, food insecurity continues to be a challenge of significant scale, and it can be alleviated by rescuing uneaten food from large resorts such as convention centres.

To et al. [4] present a successful food rescue pilot programme, operated as a cooperation between a convention centre in Las Vegas, USA, and a local food bank. In this context, food safety requires high attention, and only surplus food that has never been served to a guest, has not left the kitchen and has maintained a safe temperature gets donated in the programme. Temperature control is required throughout all further transport and distribution through a network of charitable organisations. In the US, where more than 12% of households are food insecure, while at the same time around 30% of all food sold and prepared at retail and consumer levels is wasted, such food rescue efforts of edible food make an important contribution to more sustainable food systems. Positive impacts on health and wellbeing, higher financial independence and thus more opportunity to avoid or end damaging personal living situations, and the reduction of greenhouse gas emissions are some of the positive impacts of successful food re-distribution initiatives.

Lesson 4: While the positive impacts of organic fertilisers are generally well known, more efforts are required to understand in detail how climatic change will impact the performance of organic fertilisers.

Application rates of chemical fertilisers are of high environmental concern, and the use of organic fertilisers, such as livestock manure, is an effective strategy to ensure more responsible nutrient management for agricultural crops and long-term productivity of agricultural soils. However, the experimental results of Khaitov et al. [5] suggest that temperature variations, in particular an increase of the average temperature during the agricultural plants' growth season (as must be expected in many regions as a result of global climatic change), can significantly impact the performance of organic manure. Soil-climatic conditions can accelerate the mineralisation processes of the applied organic manures. For chilli pepper fruit, Khaitov et al. found a substantially improved fruit yield under elevated temperatures, but the authors highlight the need to consider specific biological properties of different crops, in order to better understand how changing environmental conditions impact the effectiveness of organic amendments to agricultural land.

Lesson 5: To halt erosion of traditional culinary knowledge and preserve traditional agro-alimentary systems, as a means to reduce local food insecurity risks in a further globalising and urbanising world, it is essential to understand the multifaceted differences in food production and consumption patterns among rural and urban populations.

A loss of traditional gastronomical traditions and agro-alimentary systems implies a risk of higher vulnerability to food insecurity for local communities, along with an increased risk of negative environmental impacts such as biodiversity loss. Arnés and Astier [6] look at tortilla production in

Mexico, where the traditional cuisine is listed as cultural heritage of humanity by UNESCO, and several institutions work to preserve the traditional food systems. Nevertheless, the field study of Arnés and Astier reveals that ancestral knowledge of tortilla making is eroding. This is most evident in urban areas, but it also applies to rural populations. Even where tortillas continue to be referred to as handmade, their ingredients and making might differ. One reason is the substitution of traditional maize with hybrid improved maize, introduced by large seed companies, causing a lack of availability of native grain, especially in urban centres. Another is the practice of mixing native maize with other components in both urban and rural households and gastronomies. The authors also highlight that in rural areas, the consumption of the traditional type of tortillas is associated with poverty and is perceived as a sign of backwardness, while in urban areas it is a luxury. This demand for traditional tortillas in the peri-urban and urban areas might offer opportunities to strengthen traditional practices.

3.2. Spotlights on Specific Measures that May Reduce Adverse Environmental Impacts Related to Food Consumption (Lessons 6 to 9)

Lesson 6: Where food waste cannot be avoided, the coupling of food waste valorisation and wastewater treatment, in an integrated system, creates important synergies to reduce greenhouse gas emissions and deliver additional marketable outputs in a biorefinery approach.

Around one third of food is never consumed, which wastes resources and generates greenhouse gas emissions. Where food waste cannot be avoided, its valorisation is of high priority. An established food waste valorisation pathway is anaerobic digestion with biogas production. A more environmentally beneficial food waste valorisation pathway is studied by Cecchi and Cavinato [7], namely the simultaneous treatment of food waste and wastewater, which offers several benefits, such as the more effective purification of wastewater in the biological treatment step, a higher biogas yield in the anaerobic digestion step, the recovery of phosphorous, and the option to recover biodegradable polymers in a biorefinery approach as additional marketable outputs. The authors report results from two full scale plants operated in Italy with success.

Lesson 7: Foods with insect ingredients have high environmental advantages compared to common meat-based foods, but environmental concerns do not have an impact on consumer's purchase intention, while food neophobia has a significant role in limiting the purchase of insect food.

Compared to common livestock breeding, raising edible insects emits far less greenhouse gas and consumes significantly fewer resources, such as water and land. Thus, insect food is considered an environmentally favourable choice. However, by conducting a questionnaire survey, Chang et al. [8] found that environmental concerns do not have an influence on a consumer's decision in favour or against buying food which contains insect ingredients. Arguments that focus on environmental implications therefore are not likely to increase market uptake or support market success. The researchers identified that food neophobia has significant effects on purchase intentions. Consequently, facilitating positive feelings about experiences with edible insect foods can be considered a promising strategy for encouraging more widespread uptake of insect-based food.

Lesson 8: Consuming less meat has the potential to make a significant contribution to reducing greenhouse gas emissions, but environmental concerns do not have a major impact on the decision to consume meat; however, there might be regionally differing cultural and economic determinants, which merit more attention in order to understand changing food consumption practices.

Sanchez-Sabate and Sabaté [9] also explore whether environmental concerns impact the food consumption decisions of consumers, in particular the decision to consume meat, which represents a major contributor to global warming. Sanchez-Sabate and Sabaté implemented a systematic literature review and focused on consumers' intention to consume less meat, and not necessarily no meat at all. It is interesting to note that the latter, i.e., fully turning to vegetarian or vegan food consumption practices, has actually been studied more frequently than the simple reduction of meat quantities consumed (meat-reducers or flexitarians). The findings of Sanchez-Sabate and Sabaté suggest that environmental concerns are relevant for a small minority of consumers only. Even among people who are environmentally aware, reduction of meat consumption is among the least preferred strategies for alleviating climate change. There might, however, be regionally differing cultural and economic determinants which merit more attention.

Lesson 9: Purchasing power is the strongest factor in influencing the decision to buy organic food, and thus the price is most important when deciding in favour of organic produce, although individual norms, knowledge about health implications and consideration of animal welfare may also play a role.

Li R. et al. [10] reviewed the literature regarding consumers' willingness to pay for organic food in China. They found that several factors can influence the decision to buy organic produce, but the consumer's purchasing power is of decisive impact. It is therefore most essential that marketers of organic foods consider the price. Other factors, such as health benefits or environmental concerns, can play a role in the consumers' decision, but even when factoring in such considerations, most consumers will not purchase the organic foods if they are expensive compared to conventional foods that are cheaper and readily available. It should, however, be noted that the authors reviewed only a small number of publications, because research on organic food purchasing in China is at an emerging level, with few results available so far.

3.3. How to Effectively Support Desirable Changes in the Food Supply Systems (Lessons 10 to 14)

Lesson 10: Transformative change towards a more sustainable and just food regime needs to disrupt from the bottom up, where local innovations act within their contextual frame, addressing the challenges that their specific communities are facing and looking at scaling in different ways to effectively challenge conventional top-down solutions.

To understand how desirable changes can potentially be supported, Pereira et al. [11] use the concept of regime shifts to understand the key drivers and innovation processes of past structural shifts in the food system. The dynamics of past transformations are complex, and, according to the authors, previous regimes manifested as a result of powerful actors. Pereira et al. argue that transformative change towards a more sustainable and just regime needs to disrupt from the bottom up, where local innovations act within their contextual frame, addressing the challenges that specific communities are facing and looking at scaling in different ways, thus disrupting conventional top-down solution.

Lesson 11: Insufficient connections between food-system-oriented initiatives from civil society and small market actors result in the low impact of such initiatives on strategic policymaking; to effectively intervene in food systems, it is essential to engage in participatory coordination, in order to identify common interests among the different types of initiatives, while respecting that the diversity of models brings benefits to the community.

But how can initiatives from civil society and small local market actors be strengthened to achieve innovation in the food system? Looking at the numerous initiatives across Rome, Italy, Mazzocchi and Marino [12] highlight that initiatives tend to be insufficiently connected, and their experiences lack integration both horizontally (i.e., among them) and vertically (i.e., with respect to strategic policymaking and influence on governance levels that intervene in food systems). To be effective in achieving change, cooperation between initiatives and stakeholders must be improved and integrated into policymaking. Institutionalisation is an important step, but the authors mention that it is a risk to be too closely associated with one political personality; the duration of electoral mandates may be too short compared to the time required for an effective transition of the food system. To coordinate various initiatives to become more effective in influencing food policies, the researchers report success from a participatory approach, which first grouped the various initiatives into clusters and then targeted the identification of common interests among the clusters, while still respecting that each model brings different benefits to the community.

Lesson 12: In supply chains, sustainability-oriented practices are initially transferred from single focal companies to other actors in the supply chain; a close and stable cooperation functions as learning environment.

The case study of Zhuo and Ji [13] analyses supply chain coordination in the pig sector in China under a sustainability lens, in particular looking for pathways that foster sustainability along the whole supply chain. The findings illustrate that good practices are initially transferred from single focal companies to other actors in the supply chain, including to upstream farmers with whom the companies cooperate. Provided that the coordinative relationship is close and stable, the cooperation functions as a learning environment for the gaining of sustainable development capabilities, and thus the initial performance of a few supply chain actors catalyses a process where the entire supply chain becomes more sustainable. The authors therefore recommend the establishment of governmental schemes, that provide financial and technological support to core companies for adopting sustainable practices and embracing innovation and learning processes. In addition, effective supply chain coordination methods can facilitate the adoption of sustainability practices across the supply chain.

Lesson 13: Trust between farmers and agricultural retailers plays an important role in the farmers' decisions regarding what agricultural materials to buy, including new types of materials, such as fertilisers derived from circular economy schemes.

Li L. et al. [14] address cooperation at farm level and report the results of a questionnaire survey conducted in China. According to the observations made, good personal relations and trust contribute to reducing information asymmetry between farmers and retailers, and strongly influence a farmer's decisions regarding what kind of materials to use. Therefore, the establishment and maintenance of personal relations is critical to promoting the security and sustainability of food systems, including through the acceptance of new types of nutrient inputs produced with emerging technologies under circular economy schemes (such as phosphorous recovered from sewage sludge or food waste).

Lesson 14: Integrating people's aversion to loss into the design of environmental policies can create more effective environmental protection schemes.

Finally, Yang and Wang [15] evaluate the effectiveness of different policies that target the disposal of dead animals on farms in China, and whether they discourage farmers from implementing inappropriate measures which endanger public health and environmental protection, but might be economically advantageous. By means of a questionnaire survey, the researchers explored how dead hogs were handled on farms, and found that punishment policies had a significant impact on inappropriate treatment of dead hogs. Farmers feared police detention time more than financial penalties. Financial loss occurring alongside the loss of an animal, i.e., a dead hog, is highlighted as a major concern for a farmer. Interestingly, the loss aversion theory was fruitful to explain the effectiveness of policies in terms of whether an individual engages in inappropriate solutions. The loss aversion theory postulates that individuals are more sensitive to losses than to returns, thus perceiving a positive and negative deviation from the decision point as being different: one unit of loss is considered more important than one unit of gain (the weight assigned to the loss is higher). Thus, evasion of loss is more important for individuals. The authors suggest integrating people's aversion to loss into the creation of more effective environmental policy schemes.

4. Concluding Remarks

The studies presented in this Special Issue contain much more than the 14 insights highlighted in this editorial. The publications reflect a broad research agenda in the field of food system sustainability, and they document both the progress being made in identifying suitable pathways, and the high complexity of the challenges. This enriches our knowledge of the essential changes required in the food systems. Knowing what a desirable future looks like is an important but insufficient achievement. Initiating and effectively supporting progress towards such a desirable future requires additional knowledge and skills. A richness in insights, regarding how to facilitate desirable changes in order to see them happen in the future, is a strength of this Special Issue. The reader will also find information about further research needs.

Funding: This research received no external funding.

Acknowledgments: I thank the editorial team of International Journal of Environmental Research and Public Health for the continuous support in compiling this Special Issue; special thanks go to Lin Li and Holly Zha. Furthermore, the reviewers are gratefully acknowledged for their efforts in evaluating the submitted manuscripts and for providing feedback to the authors. I also thank the authors for submitting their work and for enriching this Special Issue with new perspectives.

Conflicts of Interest: The authors declare no conflict of interest.

References

1. Special Issue "Towards More Sustainable Food Systems". Available online: https://www.mdpi.com/journal/ijerph/special_issues/TMSFS (accessed on 26 May 2020).
2. Nyangasa, M.A.; Buck, C.; Kelm, S.; Sheikh, M.; Hebestreit, A. Exploring Food Access and Sociodemographic Correlates of Food Consumption and Food Insecurity in Zanzibari Households. *Int. J. Environ. Res. Public Health* **2019**, *16*, 1557. [CrossRef]
3. Chigbu, U.E.; Ntihinyurwa, P.D.; de Vries, W.T.; Ngenzi, E.I. Why Tenure Responsive Land-Use Planning Matters: Insights for Land Use Consolidation for Food Security in Rwanda. *Int. J. Environ. Res. Public Health* **2019**, *16*, 1354. [CrossRef] [PubMed]
4. To, S.; Coughenour, C.; Pharr, J. The Environmental Impact and Formation of Meals from the Pilot Year of a Las Vegas Convention Food Rescue Program. *Int. J. Environ. Res. Public Health* **2019**, *16*, 1718. [CrossRef] [PubMed]
5. Khaitov, B.; Yun, H.J.; Lee, Y.; Ruziev, F.; Le, T.H.; Umurzokov, M.; Bo Bo, A.; Cho, K.M.; Park, K.W. Impact of Organic Manure on Growth, Nutrient Content and Yield of Chilli Pepper under Various Temperature Environments. *Int. J. Environ. Res. Public Health* **2019**, *16*, 3031. [CrossRef] [PubMed]
6. Arnés, E.; Astier, M. Handmade Comal Tortillas in Michoacán: Traditional Practices along the Rural-Urban Gradient. *Int. J. Environ. Res. Public Health* **2019**, *16*, 3211. [CrossRef] [PubMed]
7. Cecchi, F.; Cavinato, C. Smart Approaches to Food Waste Final Disposal. *Int. J. Environ. Res. Public Health* **2019**, *16*, 2860. [CrossRef] [PubMed]
8. Chang, H.-P.; Ma, C.-C.; Chen, H.-S. Climate Change and Consumer's Attitude toward Insect Food. *Int. J. Environ. Res. Public Health* **2019**, *16*, 1606. [CrossRef] [PubMed]
9. Sanchez-Sabate, R.; Sabaté, J. Consumer Attitudes Towards Environmental Concerns of Meat Consumption: A Systematic Review. *Int. J. Environ. Res. Public Health* **2019**, *16*, 1220. [CrossRef] [PubMed]
10. Li, R.; Lee, H.-Y.; Lin, Y.-T.; Liu, C.-W.; Tsai, P.F. Consumers' Willingness to Pay for Organic Foods in China: Bibliometric Review for an Emerging Literature. *Int. J. Environ. Res. Public Health* **2019**, *16*, 1713. [CrossRef] [PubMed]
11. Pereira, L.M.; Drimie, S.; Maciejewski, K.; Tonissen, P.B.; Biggs, R.O. Food System Transformation: Integrating a Political–Economy and Social–Ecological Approach to Regime Shifts. *Int. J. Environ. Res. Public Health* **2020**, *17*, 1313. [CrossRef] [PubMed]
12. Mazzocchi, G.; Marino, D. Rome, a Policy without Politics: The Participatory Process for a Metropolitan Scale Food Policy. *Int. J. Environ. Res. Public Health* **2020**, *17*, 479. [CrossRef] [PubMed]
13. Zhuo, N.; Ji, C. Toward Livestock Supply Chain Sustainability: A Case Study on Supply Chain Coordination and Sustainable Development in the Pig Sector in China. *Int. J. Environ. Res. Public Health* **2019**, *16*, 3241. [CrossRef]
14. Li, L.; Li, G.; Feng, X.; Liu, Z.; Tsai, F.-S. Moderating Effect of Dynamic Environment in the Relationship between Guanxi, Trust, and Repurchase Intention of Agricultural Materials. *Int. J. Environ. Res. Public Health* **2019**, *16*, 3773. [CrossRef] [PubMed]
15. Yang, C.; Wang, J. Evaluation of Policies on Inappropriate Treatment of Dead Hogs from the Perspective of Loss Aversion. *Int. J. Environ. Res. Public Health* **2019**, *16*, 2938. [CrossRef] [PubMed]

International Journal of
Environmental Research and Public Health

MDPI

Article

Exploring Food Access and Sociodemographic Correlates of Food Consumption and Food Insecurity in Zanzibari Households

Maria Adam Nyangasa [1,†], Christoph Buck [1,†], Soerge Kelm [2], Mohammed Sheikh [3] and Antje Hebestreit [1,*]

1 Leibniz Institute for Prevention Research and Epidemiology—BIPS, 28359 Bremen, Germany; nyangasa@leibniz-bips.de (M.A.N.); buck@leibniz-bips.de (C.B.)
2 Center for Biomolecular Interactions Bremen, University Bremen, 28334 Bremen, Germany; kelm@uni-bremen.de
3 Environmental Analytical Chemistry and Eco-toxicology, Zanzibar State University, P.O. Box 146 Unguja, Zanzibar; m.sheikh@suza.ac.tz
* Correspondence: hebestr@leibniz-bips.de; Tel.: +49-421-218-56849; Fax: +49-421-218-56821
† These authors contributed equally to the work.

Received: 25 February 2019; Accepted: 1 May 2019; Published: 4 May 2019

check for updates

Abstract: Rapid growth of the Zanzibari population and urbanization are expected to impact food insecurity and malnutrition in Zanzibar. This study explored the relationship between food access (FA) and sociodemographic correlates with food consumption score and food insecurity experience scale. Based on cross-sectional data of 196 randomly selected households, we first investigated the association between sociodemographic correlates and Food Consumption Score (FCS) and Food Insecurity Experience Scale using multilevel Poisson regression. Secondly, the role of FA in these associations was investigated by interaction with the respective correlates. About 65% of households had poor food consumption, and 32% were severely food-insecure. Poor FA was more prevalent in households with poor food consumption (71%). Polygamous households and larger households had a higher chance for severe food insecurity. In the interaction with FA, only larger households with poor FA showed a higher chance for severe food insecurity. In households having no vehicle, good FA increased the chance of having acceptable FCS compared to poor FA. By contrast, urban households with good FA had a twofold chance of acceptable FCS compared to rural household with poor FA. Poor FA, poor food consumption and food insecurity are challenging; hence, facilitating households' FA may improve the population's nutrition situation.

Keywords: demographic correlates; food access; household; food insecurity experience scale; Zanzibar; sub-Saharan Africa

1. Introduction

The world population is expected to increase by 2.5 billion between 2007 and 2050, with most of the growth foreseen to occur in urban areas of developing countries [1]. This rapid growth and urbanization are expected to increase poverty and negatively impact the food security environment of urban dwellers, leading to food insecurity and malnutrition [2]. According to the World Bank report from 2015 [3], 29% of Tanzanians could not meet their basic consumption needs, and about 10% of the population could not afford to buy basic food stuff. In Tanzania and other developing countries, the leading factor in household food insecurity in urban areas is the dependency on food purchase [2,4,5]. Therefore, a slight increase in food prices has a major impact on vulnerable households, pushing them into hunger and poverty [6]. About 80% of the household food requirement in peri-urban areas of

Unguja Island, Zanzibar is purchased [7,8], while about 60% of food consumed in rural areas is obtained through home gardening and farming [9]. Home gardening and farming contribute substantially to households' own food production [10] and may thus enhance household food availability and food security in rural areas. Own food production increases purchasing power, due to savings on food bills [4,11] and income from selling of the produce. It also provides a diversity of nutritious food that helps to improve the health status of the household [11] as well as serves as a means of food provision during food shortage.

Studies conducted in Nigeria [12] and Ethiopia [13] have shown that male heads of household (HH) play a substantial role in determining household food security, while others reported that female HH are more likely to spend most of their income on food, thus guaranteeing food security for their households [14,15]. Household size has also been shown to influence food security and acceptable food consumption, with smaller household size being associated with household food security [16] and acceptable food consumption, and large household size with poor food consumption and perceived food insecurity [17]. Food insecurity has also been found to be associated with low monthly income [18]. Income earned from any source improves the food situation of a household [19]; thus, households with more employed adult members are likely to have a better food situation compared to households with more unemployed adult members [20].

Several studies conducted in low-income countries, especially in Africa [4,5,16,18], have investigated the determinants of either household food security or food consumption behavior in households. However, this is the first study in Zanzibar that recruited randomly selected households and enrolled all members of a household for further insights on correlates of food insecurity and food consumption. The present study aimed to explore the relationship between food access (FA) and sociodemographic household factors with the Food Consumption Score (FCS) and Food Insecurity Experience Scale (FIES). Findings from this study can provide baseline information on the interaction between food access and household factors, and the collected data can be used for further research on health interventions to improve food consumption and food security in Zanzibar.

2. Materials and Methods

2.1. Study Area, Population, and Sampling

A population-based cross-sectional study was conducted in 2013 in Unguja Island, Zanzibar, whereby entire households were enrolled as sampling units. For the purposes of this study, a household is defined according to Beaman and Dillon [21], with emphasis on eating from the same pot, i.e., having the same food provider. This is particularly important as our study population consisted not only of monogamous but also of polygamous families, who do not necessarily live together in the same house or compound.

Household aspects were reported by the head of household in a questionnaire-administered personal interview. In total, 244 randomly selected households were contacted from 80 Shehias (wards), and 239 (97.9%) participated in the survey. Due to missing information on socioeconomic status, demographic correlates, and responses from FCS and FIES instruments, 43 households were excluded from the analysis, resulting in a final sample of 196 (82%) households. Further details on sampling procedures, data collection, and quality management are provided elsewhere [22].

Prior to the data collection process, all participants gave written, informed consent. All procedures applied in this study were approved by the Ethics Committees of the University of Bremen (in September 26, 2013) as well as the Zanzibar Ministry of Health and the Zanzibar Medical Research and Ethics Committee (ZAMREC/0001/AUGUST/013) in accordance with the ethical standards according to the 1964 Declaration of Helsinki and its later amendments.

2.2. Questionnaires

A structured household questionnaire was used to collect general household information. The information included data on socioeconomic and demographic indicators of the household and of the head of household, such as area of residence, number of animals owned, number of vehicles belonging to the household, household size, marital status, education level of the HH, occupation of the HH, etc. In cases of polygamy, household information was also collected from the households of the other wife or wives. Information on indicators of household food consumption was collected using a standardized questionnaire (Food Consumption Score, FCS), adapted from the United Nations World Food Programme (UNWFP) [23]. Household food insecurity was measured at the household level using a standardized questionnaire for the Food Insecurity Experience Scale (FIES), which was adapted from the Food and Agriculture Organization (FAO) [24]. All questionnaires were developed in English, translated into Swahili and back-translated to check for translation errors.

2.2.1. Food Consumption Score (FCS)

FCS is a composite score constructed from (1) household dietary diversity based on nine food groups (staples, pulses, fruits, vegetables, meat and fish, dairies, sugar, oil and fat, condiments) consumed during the 7 days preceding the survey, (2) food frequency, counted as the number of days each food group was consumed during the 7 days preceding the survey, and (3) relative nutritional importance of different food groups, applying a weighting system [23], thus reflecting the quality and quantity of food consumed in the household.

Higher weights were given to energy-dense foods with proteins of high quality and a range of bioavailable micronutrients, while lower weights were given to oil and sugar, which are energy-dense but contain—if any—proteins of low quality and low levels of micronutrients [25]. Cut-off points established by the UNWFP [23] were used to classify FCS. They were computed by summing up the weighted frequencies of the different food groups consumed in the household. FCS $\leq$ 28 was categorized as "poor", FCS > 28 and < 42 as "borderline", and FCS $\geq$ 42 as "acceptable" [23]. For the regression analysis in this study, poor consumption and borderline were merged to a new category, "poor", resulting in two categories of food consumption score, i.e., "poor" and "acceptable".

2.2.2. Food Insecurity Experience Scale (FIES)

Food insecurity was assessed using the Food Insecurity Experience Scale (FIES), a standardized set of questions developed by the FAO [24] that has been applied in a large number of countries following a standardized procedure. The scale, which is an experience-based metric of severity of food insecurity that relies on people's direct/actual responses, includes components of uncertainty and worry about food, inadequate food quality, and insufficient food quantity. It consists of a set of eight items that assess food-related behaviors associated with difficulties in accessing food due to limited resources. The instrument measures the degree of food insecurity/hunger experienced by individuals during the 12 months preceding the survey.

Household scores of food insecurity on the eight items were scaled based on a Rasch model as an application of the item–response theory (IRT) [26]. In IRT, the response to each item is modelled as a function of item and household parameters to measure the position of households on a latent trait, independently of the item difficulty. We conducted the Rasch model using the eRm package 0.16–2 [27] in R 3.5.1 to derive household scores for the following regression analyses. Households who responded 'no' to any of the eight items received the lowest value. Item characteristic curves indicated item reliability, and the person separation index was found to be high (0.84). Household scores ranged from −3.57 to 3.76 and clustered either to the lowest negative scores (<−2), around 0 (−2 to 2) or to the highest scores. Based on these cut-off points, we categorized household scores into mild, moderate, and severe (hunger), which are the three categories used by the FAO [24] to define levels of food insecurity/hunger. For the Poisson regression analysis, food security, mild and moderate food insecurity were dichotomized into "mild to moderate" (0) and "severe food insecurity" (1). A household is considered as food-secure if members have always had enough food and no hunger worries [24].

2.3. Correlates

Correlates of FCS and FIES were assessed for the HH (gender, education level, number of jobs, marital status) and at household level (household size, number of types of animals kept in the household, vehicles owned by any of the household members, area of residence, and food access). The highest education level of the HH was assessed using the International Standard Classification of Education (ISCED) [28] and was categorized for the analysis as low education level (primary school and below) and high education level (secondary school and above). Number of jobs of HH for assessing main household income was defined as "no job" and "≥1 job". Marital status of HH was calculated in three categories: Married monogamous, married polygamous, and other (single, widowed, cohabitating or divorced). To facilitate interpretation, two categories for marital status were derived: Married (monogamous or polygamous) and not married (single, widow, cohabitation, divorced). Cohabitation was categorized as "not married" since it is characterized by a different socioeconomic status compared to those households with married HH. Using the mean household size of 6 members in this study population as a cut-off, household size was classified as large (six or more members) and small (less than six members). To assess household wealth, the number of types of animals kept in the household from a list of six items including ducks, goats, sheep, cows, fish, and chicken, and the number of assets from a list of eight items including electricity, radio, mobile phone, iron, kerosene lamps, television, refrigerator, and non-mobile phone was summed up. The median number of animals and assets was five, and this figure was used to categorize wealth into wealthy (≥5) and poor (<5). The number of vehicles per household was assessed as type of transportation owned by any member of the household from a list of 5 categories (bicycle, car/truck, boat, motorcycle, none). Two categories were built: "none" and "at least one type of vehicle". Area of residence was assessed as rural or urban.

An important component of food consumption and food insecurity included in the analysis was FA. In this study, FA was defined as the ability of a household to acquire adequate amount of food through mixed strategies. Indicators for FA were assessed, and derived variables were combined to a composite score (see Table 1 below). The derived variables were: (a) Food source; main source of food consumed during the last seven days (purchased, borrowed, own production, traded food/barter, received as gift, food aid, other), (b) food purchased; types of food (cereals, starchy vegetables/ tubers, vegetables, fruits, legumes, meat, egg, milk, fish, oils and fats, any kind of beverage, other), frequently bought from shop/market during the last seven days, (c) own food; types of foods (from those listed above) of own household production, and (d) market distance; distance to the nearest market or shop (<30 min, 30 to 60 min, 1 h to 2 h, >2 h).

To classify households as having poor or good FA, a composite score for FA was computed as the mean of the values of the derived variables multiplied by 4 (number of all derived variables). The FA-score ranged from 4–8, and FA was then categorized as "poor food access" (≤6) and "good food access" (>6). The cut-off was set according to the distribution of FA-status in the study population, as half of the population included in the study had FA score of 6 and below.

Table 1. Overview of measured indicators for FA and derivation of variables for the composite score.

Indicator of Food Access	Derived Variable	Categories
Main source of food consumed One main source to be selected from 6 categories: purchased, borrowed, traded food/barter, received as gift, food aid, own production	Food source = one main source per household	1: borrowed, received as gift, food aid, other 2: own production, traded food/barter, purchased
Types of food groups frequently bought from shop/market Food group (e.g., cereals) out of 11 food groups was 0: not bought, 1: bought	Food purchased = sum of all food groups purchased	1: ≤4 food groups, 2: >4 food groups
Types of food groups of own household production Food groups out of 11 food groups 0: not produced (purchased, borrowed, traded food/barter, received as gift, food aid), 1: own production	Own food = sum of all food groups with own production	1: ≤2 food groups, 2: >2 food groups
Distance to the nearest market/shop far (>30 min walking distance); near (<30 min walking distance)	Market distance	1: far, 2: near

2.4. Statistical Analysis

Study characteristics such as socioeconomic and demographic variables were calculated for categories of food consumption and food insecurity. First, the associations between the exposure variables (food access, socioeconomic and demographic correlates) with either food consumption (FCS; Model 1) or food insecurity (FIES; Model 2) as outcome variables were explored. To explore these associations, linear, Poisson, and logistic regression models were considered and evaluated with regard to model fit and Pearson residuals as well as quantile-quantile (Q–Q) plots. Eventually, we conducted multilevel logistic regression models to calculate odds ratios (OR) and 95% confidence limits (CI) and to account for clustering within Shehia level using a random intercept. Secondly, statistical interactions between each correlate and food access (correlate*FA) were investigated. Hence, in Models 3a–h and 4a–h, the predictive power of each socioeconomic and demographic factor with FA on FCS and FIES was tested separately. Each model was again adjusted for the remaining correlates, including a random intercept for the Shehia level. All statistical analyses were performed using SAS 9.3 (SAS Institute, Cary, NC, USA). Due to the exploratory design of our study, we only considered confidence limits as a precision measure of the point estimates but did not apply a level of significance. Moreover, we did not adjust for multiple testing. Noteworthy associations are presented considering higher (OR > 1.5) or lower (OR < 0.66) chances for the modelled response category.

3. Results

3.1. Household Characteristics

The sample data were based on the responses of the HH. The majority of the households were headed by men (63%, 123/196), and about 55% of the HH (107/196) were in a monogamous marriage. More than half of the HH had one or more sources of income that he/she contributed to the household. Most of the households were in rural areas, and the overall mean household size of the participating households was 6 persons. More than 60% of the households had a good socioeconomic status, with one or more than one animal kept and at least one vehicle owned by a member belonging to the household. Overall, about 65% of the households had poor food consumption, and about 32% were severely food-insecure (Table 2). Acceptable food consumption was more prevalent in households with higher-educated HH (40%), in monogamous households (38%), and in larger households (six or more members) (38%). Severe food insecurity was more prevalent in polygamous households (40%), in households with low-educated HH (40%) and in larger households (six or more members) (40%). Looking at each question of the FIES, the majority of the households (73.5%) indicated having eaten few kinds of food in the last 12 months due to lack of money, and 26% went without eating for a whole day due to lack of money (Table 3). About 54% (106/196) of the study population had poor FA, of which 71% had poor food consumption and 35% experienced severe food insecurity.

Table 2. Proportion of food consumption and food insecurity experience scale according to demographic and socioeconomic factors.

	Food Consumption Score				Food Insecurity Experience Scale				Total
	Poor		Acceptable		Mild to Moderate		Severe		
	N	%	*N*	%	*N*	%	*N*	%	*N*
All	128	65.3	68	34.7	134	68.4	62	31.6	196
Household Demographics									
Gender									
Male	72	58.5	51	41.5	80	65.0	43	35.0	123
Female	56	76.7	17	23.3	54	74.0	19	26.0	73
Marital status of HH									
Not married [a]	27	79.4	7	20.6	28	82.4	6	17.6	34
Married monogamous	66	61.7	41	38.3	73	68.2	34	31.8	107
Married polygamous	35	63.6	20	36.4	33	60.0	22	40.0	55
Education level									
Low	69	71.1	28	28.9	59	60.8	38	39.2	97
High	59	59.6	40	40.4	75	75.8	24	24.2	99
Number of jobs									
No job	61	70.1	26	29.9	57	65.5	30	34.5	87
One or more jobs	67	61.5	42	38.5	75	68.8	24	22.2	109
Area									
Rural	105	68.6	48	31.4	102	66.7	51	33.3	153
Urban	23	53.5	20	46.5	32	74.4	11	25.6	43
Household size [b]									
Small	68	68.7	31	31.3	75	75.8	24	24.2	99
Large	60	61.9	37	38.1	59	60.8	38	39.2	97
Socioeconomic Factors									
Wealth (household assets and animals) [c]									
Poor	61	73.5	22	26.5	51	61.4	32	38.6	83
Wealthy	67	59.3	46	40.7	83	73.5	30	26.5	113
Number of vehicles									
None	31	72.1	12	27.9	28	65.1	15	34.9	43
At least one vehicle	97	63.4	56	36.6	106	69.3	47	30.7	153
Food access [d]									
Poor	75	70.8	31	29.2	69	65.1	37	34.9	106
Good	53	58.9	37	41.1	65	72.2	25	27.8	90

[a] Not married includes single, divorced, widow, and cohabitation; [b] cut-off was derived from the mean number of household members in this study, small (≤6) and large (>6); [c] wealth (poor <5 and wealthy ≥5, calculated as the median number of animals and assets in the household); [d] cut-off (poor ≤6, good >6).

Table 3. Questions of the Food Insecurity Experience Scale and affirmatively answered questions by the study population in Zanzibar ($N = 196$). During the last 12 months, was there a time when …

No	Food Insecurity Experience Scale Questions	*N*	%
1	You were worried you would run out of food because of a lack of money?	112	57.1
2	You were unable to eat healthy and nutritious food because of a lack of money?	134	68.4
3	You ate only a few kinds of foods because of a lack of money?	144	73.5
4	You had to skip a meal because there was not enough money to get food?	100	51.0
5	You ate less than you thought you should because of a lack of money?	117	59.7
6	Your household ran out of food because of a lack of money?	103	52.6
7	You were hungry but did not eat because there was not enough money for food?	76	38.8
8	You went without eating for a whole day because of a lack of money?	51	26.0

3.2. Correlates of Food Consumption and Food Insecurity

Households with a higher-educated HH had a lower chance of reporting severe food insecurity (OR 0.53; 95% CI 0.28–1.08) compared to those with lower-educated HH (Table 4). Those HH married in monogamy had a higher chance of reporting acceptable food consumption (OR 1.71; 95% CI 0.55–5.38) but at the same time a higher chance of severe food insecurity (OR 1.83; 95%CI 0.55; 6.08) compared to those HH not married (single, widowed, cohabitating or divorced) (Table 4). Polygamous households had a higher chance of severe food insecurity (OR 3.95; 95% CI 1.17–13.4) and also reported higher chance for acceptable food consumption (OR 1.78; 95% CI 0.54–5.83) compared to those not married. Larger households had a higher chance of severe food insecurity (OR 2.44; 95% CI 1.16–5.13) than smaller households, while wealthy households had a lower chance of severe food insecurity (OR 0.52; 95% CI 0.25–1.11) than poor households.

Table 4. Associations of socioeconomic and demographic correlates of 196 households with food consumption (Model 1) and food insecurity (Model 2) in terms of odds ratios (OR) and 95% confidence intervals (CI) as well as model fit (generalized chi-square/degrees of freedom), respectively.

	Model 1: Food Consumption (Ref: Poor)		**Model 2: Food Insecurity (Ref: Mild to Moderate)**	
Model fit	χ^2/DF	0.87	χ^2/DF	0.86
Between Shehia variance (SE)		0.48 (0.38)		0.37 (0.38)
	OR	**(95% CI)**	**RR**	**(95% CI)**
Gender (ref: female)	1.76	(0.75–4.11)	1.65	(0.68–4.01)
Marital status of HH (ref: not married)				
monogamous	1.71	(0.55–5.38)	1.83	(0.055–6.08)
polygamous	1.78	(0.54–5.83)	3.95	(1.17–13.4)
Education (ref: low)	1.36	(0.69–2.70)	0.53	(0.26–1.08)
Number of jobs (ref: no job)	1.22	(0.59–2.50)	0.58	(0.28–1.22)
Area of residence (ref: rural)	2.08	(0.85–5.10)	0.64	(0.24–1.70)
Household size (ref: small)	1.02	(0.51–2.05)	2.44	(1.16–5.13)
Wealth (ref: poor ≥ 5)	1.35	(0.64–2.83)	0.52	(0.25–1.11)
Number of vehicles (ref: none)	1.04	(0.42–2.57)	0.78	(0.32–1.89)
Food access (ref: poor ≤ 6)	1.56	(0.79–3.10)	0.69	(0.33–1.43)

3.3. Role of Food Access on the Correlates, Food Consumption and Food Insecurity

There were few relevant changes in the chances observed in the interaction of food access with gender, marital status of HH, number of jobs, and wealth on both food consumption and food insecurity. However, urban households with good FA showed a higher chance of acceptable food consumption compared to rural households with poor FA (Table 5, Model 3e). Households with no vehicle had a higher chance of acceptable food consumption if they had good FA compared to those with poor FA (OR 6.21; 95% CI 1.20–32.3). However, having at least one vehicle tentatively increased the chance of having acceptable food consumption for both good and poor FA (OR 2.17; 95% CI 0.68–6.87; OR 1.83; 95% CI 0.59–5.71, respectively) (Model 3h).

In comparison to households with low-educated HH and poor FA, we observed a lower chance of severe food insecurity in households with higher-educated HH either with good FA (OR 0.42; 95% CI 0.15–1.17) or poor FA (OR 0.40; 95% CI 0.16; 1.02) and in households with lower-educated HH and good FA (OR 0.49; 95% CI 0.18–1.30) (Model 4c). Considering poor FA, larger households (six or more members) had a higher chance of severe food insecurity (OR 3.42; 95% CI 1.29–9.10) compared to smaller households (Model 4f).

Table 5. Results of the multilevel logistic regressions in terms of odds ratios (OR) and 95% confidence limits as well as model fit (generalized chi-square/degrees of freedom) to investigate the interaction of food access with socioeconomic and demographic correlates on food consumption (Model 3a–h) and food insecurity (Model 4a–h), each adjusted for the remaining correlates.

Model	Covariate	Food Access	Model 3: Food Consumption (Ref: Poor)					Model 4: Food Insecurity (Ref: Mild to Moderate)				
			Ref: *N* (%)	OR	(95% CI)	Between Shehia Variance (SE)	χ^2/DF	Ref: *N* (%)	OR	(95%CI)	Between Shehia Variance (SE)	χ^2/DF
a	**Gender**											
	Male	Good access	34 (55.7)	3.03	(0.96–9.62)	0.48 (0.38)	0.88	41 (67.2)	0.85	(0.29–2.53)	0.41 (0.39)	0.87
	Male	Poor access	38 (61.3)	2.26	(0.73–7.00)			39 (62.9)	0.99	(0.36–2.78)		
	Female	Good access	19 (65.5)	2.24	(0.66–7.66)			24 (82.8)	0.44	(0.12–1.61)		
	Female	Poor access	37 (84.1)	1.00				30 (68.2)	1.00			
b	**Marital status** [a]											
	Married	Good access	43 (55.8)	1.96	(0.49–7.88)	0.50 (0.39)	0.85	53 (68.8)	1.67	(0.43–6.39)	0.39 (0.39)	0.87
	Married	Poor access	58 (68.2)	1.09	(0.28–4.25)			53 (62.4)	2.28	(0.63–8.20)		
	Not married	Good access	10 (76.9)	0.62	(0.09–4.20)			12 (92.3)	0.44	(0.04–5.01)		
	Not married	Poor access	17 (81.0)	1.00				16 (76.2)	1.00			
c	**Education**											
	High	Good access	28 (58.3)	2.21	(0.82–5.98)	0.49 (0.39)	0.87	37 (77.1)	0.42	(0.15–1.17)	0.38 (0.38)	0.88
	High	Poor access	31 (60.8)	2.09	(0.80–5.48)			38 (74.5)	0.40	(0.16–1.02)		
	Low	Good access	25 (59.5)	2.53	(0.91–7.06)			28 (66.7)	0.49	(0.18–1.30)		
	Low	Poor access	44 (80.0)	1.00				31 (56.4)	1.00			
d	**Number of jobs**											
	One or more	Good access	36 (60.0)	2.05	(0.80–5.24)	0.55 (0.40)	0.86	42 (70.0)	0.44	(0.17–1.13)	0.50 (0.42)	0.83
	One or more	Poor access	31 (63.3)	1.95	(0.73–5.17)			35 (71.4)	0.36	(0.14–0.95)		
	No Job	Good access	17 (56.7)	2.81	(0.96–8.25)			23 (76.7)	0.32	(0.10–1.01)		
	No Job	Poor access	44 (77.2)	1.00				34 (59.6)	1.00			
e	**Area of residence**											
	Urban	Good access	6 (35.3)	5.48	(1.42–21.2)	0.46 (0.38)	0.88	14 (82.4)	0.44	(0.09–2.08)	0.39 (0.38)	0.87
	Urban	Poor access	17 (65.4)	1.21	(0.39–3.71)			18 (69.2)	0.62	(0.19–1.96)		
	Rural	Good access	47 (64.4)	1.16	(0.53–2.51)			51 (69.9)	0.70	(0.32–1.53)		
	Rural	Poor access	58 (72.5)	1.00				51 (63.8)	1.00			
f	**Household size**											
	Large	Good access	23 (52.3)	1.56	(0.59–4.12)	0.46 (0.38)	0.88	30 (68.2)	1.56	(0.54–4.48)	0.38 (0.38)	0.87
	Large	Poor access	37 (69.8)	0.70	(0.26–1.86)			29 (54.7)	3.42	(1.29–9.10)		
	Small	Good access	30 (65.2)	1.08	(0.41–2.84)			35 (76.1)	1.15	(0.41–3.22)		
	Small	Poor access	38 (71.1)	1.00				40 (75.5)	1.00			
g	**Wealth**											
	Wealthy	Good access	28 (53.8)	2.21	(0.78–6.28)	0.48 (0.38)	0.87	40 (76.9)	0.38	(0.13–1.09)	0.43 (0.40)	0.86
	Wealthy	Poor access	39 (63.9)	1.59	(0.57–4.48)			43 (70.5)	0.42	(0.16–1.14)		
	Poor	Good access	25 (65.8)	1.90	(0.63–5.71)			25 (65.8)	0.51	(0.18–1.47)		
	Poor	Poor access	36 (80.0)	1.00				26 (57.8)	1.00			

Table 5. *Cont.*

Model	Covariate	Food Access	Model 3: Food Consumption (Ref: Poor)					Model 4: Food Insecurity (Ref: Mild to Moderate)				
			Ref: *N* (%)	OR	(95% CI)	Between Shehia Variance (SE)	χ^2/DF	Ref: *N* (%)	OR	(95%CI)	Between Shehia Variance (SE)	χ^2/DF
h	**Number of Vehicles**											
	At least one	Good access	48 (60.8)	2.17	(0.68–6.87)	0.44 (0.38)	0.89	56 (70.9)	0.51	(0.18–1.43)	0.38 (0.38)	0.86
	At least one	Poor access	49 (66.2)	1.83	(0.59–5.71)			50 (67.6	0.63	(0.24–1.69)		
	None	Good access	5 (45.5)	6.21	(1.20–32.3)			9 (81.8)	0.31	(0.05–2.06)		
	None	Poor access	26 (81.3)	1.00				19 (59.4)	1.00			

[a] Two categories for marital status of HH were used (1 = not married (single, widow, divorce, cohabitation) 2 = married (monogamous or polygamous)).

4. Discussion

This exploratory study aimed at adding to the on-going debate on how good FA may help to improve the nutrition situation of the Zanzibar population. We observed that poor FA, poor food consumption, and food insecurity are a problem in many Zanzibari households. Poor FA was more prevalent in households with poor food consumption and severe food insecurity. In particular, polygamous households and larger households had a higher chance of severe food insecurity. Good FA increased the chance of acceptable food consumption for urban households and households with no vehicle, whereas poor FA increased the chance for severe food insecurity for larger households.

4.1. Proportions of FCS and FIES in the Study Population

The proportion of households with acceptable food consumption was about 35% in the present study. This is lower than reported in Ethiopia (73%) [29] and in the Nyarugusa refugee camp in Tanzania (86%) with a sample of 343 households in a WFP/United Nations High Commissioner for Refugees (UNHCR) study [30]. Food consumption was assessed using comparable instruments in the present study and in the UN World Food Programme Study [23]; thus, the higher proportions in the Nyarugusa camp could have been due to the fact that 83% of the households' main source of food was from food aid, unlike in Unguja Island, where households relied mostly on purchase and on their own production. The overall proportion of mild to moderate food insecurity in our study population was about 68%, slightly higher than the 49% observed in Burkina Faso among 330 surveyed households [31], and doubles that in Ethiopia (34%) [32]. This survey was conducted during the short-rainy season (October–December 2013), which is a time for sowing and growing of food products and when most household stocks are depleted [33]. This may have contributed to the relatively low proportions of households with acceptable food consumption. Still, the proportion of households with experienced mild to moderate food insecurity was higher compared to other household surveys, which were also conducted during the lean seasons [29,31,32,34]. More than three quarters of the households in our study resided in the rural area and hence depended mostly on own food production, and food products grown in the rainy season are harvested between January and February. A survey conducted in Burkina Faso confirmed the seasonal effects on food security of 1056 households with a lower food security during the lean season compared to the post-harvest season [34]. Hence, data collection should also be conducted during the post-harvest season in order to gain insight into FCS and FIES when food availability improves on Unguja Island.

4.2. Role of Food Access on the Correlates, Food Consumption and Food Insecurity

In the present study, higher education level of the HH seemed to influence the level of food consumption and food insecurity of the households with poor FA, agreeing with findings from Rwanda [35] and Uganda [36], reflecting that "some level of education is important to household food security". It is assumed that a literate HH has a greater capacity of adapting to improved technologies and coping strategies, thus increasing production/food supply in the household [16]. The lower chance for severe food insecurity in households with higher-educated HH and good FA, higher-educated HH and poor FA, and lower-educated HH and good FA indicated that education level of the HH as well as FA both played a comparable role in impacting the food insecurity status of the household.

The fact that in this study, larger households had higher chances for severe food insecurity than smaller households may be explained by the fact that smaller household sizes are generally better manageable in terms of food demand and supply. The latter can be improved through own production, which increases food consumption and decreases food insecurity as also reported in other studies [18,37]. In contrast to our findings, studies from Niger and Nigeria observed a decreasing likelihood of a household being food-insecure with an increasing household size [38]. However, to estimate the effect of household size on food insecurity is challenging, as this would require verifying numerous other factors such as the number of all active members in the household (contributors of income or food), income sources (salary), and expenditure on food. While the number of jobs of the HH showed no effect on food insecurity, wealthier households, on the other hand, revealed a lower chance of having severe food insecurity compared to poor households. Interestingly, our results showed that larger households with poor FA had a higher chance for severe food insecurity compared to smaller households, while those with good FA only had a smaller chance for severe food insecurity compared to small households with poor FA, this further confirms the role FA played in impacting food insecurity in the study population. Studies investigating the role of FA on food consumption and food insecurity in Sub-Saharan Africa are scarce, and more research is advisable.

In our study, households with polygamous HH had a higher chance for of severe food insecurity than households with not married (e.g., widowed or divorced) HH. While this is in line with previous findings from Tanzania mainland [39], other studies reported contrasting findings [40,41]. The latter postulated that the large number of individuals in polygamous households means that a great number of individuals can provide financial and labor support among each other, thus reducing the chance of having food insecurity. However, these studies compared polygamous households against monogamous households. When we compared monogamous against unmarried households, we observed a relatively lower chance of experiencing severe food insecurity. This is supporting our findings that larger households were more likely to be food-insecure compared to smaller households.

The finding that urban households with good FA experienced a higher chance of acceptable food consumption than rural households with poor FA may be explained by the good infrastructure in urban areas, which enables good accessibility for foods and may potentially enhance the food diversity of households. The good infrastructure includes factors such as the quality of roads or market density, which facilitate food distribution and transport into the communities. Studies in Malawi and Kenya [42,43] also reported that food insecurity for rural households was affected by long distance to the market and poor market access. Furthermore, in Malawi, the distance to the market was reported to affect food consumption for both rural and urban households [42]. The fact that in our study, even households without a vehicle had a higher chance of acceptable food consumption when they had good FA compared to those with poor FA indicates the importance of other aspects of the FA—in addition to infrastructure—such as borrowing foods, receiving foods as a gift, and own food production or animal rearing. The interplay of multiple factors facilitating food accessibility should be considered in future intervention studies.

4.3. Strengths and Limitations

This exploratory study provides valuable data on the interplay of food access, socioeconomic and demographic correlates with food consumption and food insecurity of Zanzibari households. Even though our study was conducted during the short-rainy season, which made it difficult to have access to some of the villages, an important strength was the overall high proportion of households that participated in the study. Further, the highly standardized study protocol using partly validated and pretested methods and instruments is a clear strength of the study. When comparing the study characteristics such as HH demography (gender, marital status, education level, occupation), household demography (rural/urban, household size), and the socioeconomic status of the household of the full survey sample and the sample presented in this study sample, no substantial differences were observed (results not shown). Thus, a selection bias can be ruled out.

The study, however, has limitations. Firstly, the FIES was explicitly developed for cross-cultural comparability and assesses correlates of food insecurity across different areas that have the same climatic or agricultural calendar [44]. We, however, conducted our study only in Unguja Island and hence have no basis for a comparison with other Zanzibari islands such as Pemba Islands, or with the population of Mainland Tanzania. As household information was based on self-reports, social desirability could have influenced the responses given. The survey was conducted during October–December, which is the time of the year when most household stocks are depleted (lean season) [33]; this may have affected our results on the food consumption and food insecurity situation of the households and must be acknowledged as a limitation. Further, the collection of cross-sectional data means that effects of seasonal variations could not be investigated. To overcome this limitation, it would be advisable to collect longitudinal data.

One major limitation we like to address is that the overall study was planned, powered, and conducted to estimate the prevalence of malnutrition in the Zanzibari population [22]. However, with this study, we intended to explore important household survey data on a broader level even though for this particular approach, the sample size was underpowered. Nevertheless, the data are a useful source for exploring the role of FA in the association between sociodemographic household factors with food consumption and food insecurity. Findings from this study will add knowledge and inform the development of intervention strategies and policies aiming at improving food consumption and food security in Zanzibar. Still, more research with a larger sample size is advisable.

5. Conclusions

Based on our findings, poor access to food may be seen as a modifiable factor for food consumption and perceived food insecurity in Zanzibari households, in particular for the association with educational level and household size. To improve food and nutrition security in Zanzibar, implementation of policies and programs that address education activities and different forms of practical coping strategies, such as efficient food storage techniques and home gardening, in their agendas are needed, particularly in rural areas. In parallel, strategies should consider improvement of infrastructure to facilitate distribution of produce within the rural–urban areas, as well as education campaigns on food quality and utilization, emphasizing on the importance of food group and balanced diets.

Ethics Approval and Consent to Participate: Ethical approval was obtained from the Ethics Committees of the University of Bremen in Germany with a reference number 06-3 and of the Zanzibar Ministry of Health and the Zanzibar Medical Research and Ethics Committee in Zanzibar, Tanzania with a reference number ZAMREC/0001/AUGUST/013. Written informed consents were taken from all participants and parents/guardians gave a written informed consent for their children. The consent forms were approved by the Institutional Ethics Committee.

Author Contributions: This manuscript represents original work that has not been published previously and is currently not considered by another journal. The authors' responsibilities were as follows: A.H. and M.A.N. had the idea of the analysis; A.H., M.A.N., S.K., and M.S. were responsible for data collection. M.A.N. and C.B. conducted statistical analyses; M.A.N. and C.B. did the analysis and data interpretation; M.N. wrote the manuscript and had primary responsibility for final content and submitting the manuscript for publication; M.A.N., S.K., M.S., C.B., and A.H. were responsible for critical revisions and final approval of the manuscript.

Funding: This research was funded by Leibniz-Gemeinschaft: SAW-2012-ZMT-4.

Acknowledgments: This work was done as part of the Leibniz Graduate School SUTAS (Sustainable Use of Tropical Aquatic Systems; http://www.zmt-bremen.de/SUTAS.html). This study would not have been possible without the voluntary collaboration of the Zanzibari families who participated in the extensive examinations. We are grateful for the support from regional and local community leaders and municipalities. The authors gratefully acknowledge the assistance from all the fieldworkers.

Conflicts of Interest: The authors declare no conflict of interest.

Abbreviations

FA	Food Access
FANTA	Food and Nutrition Technical Assistance
FAO	Food and Agriculture Organization
FCS	Food Consumption Score
FIES	Food Insecurity Experience Scale
GLIMMIX	Generalized Linear Mixed Models
HH	Head of Household
ISCED	International Standard Classification of Education
SNNPR	South Nation's, Nationalities and Peoples Region
SSA	Sub-Saharan Africa

SUTAS	Sustainable Use of Tropical Aquatic Systems
TDHS	Tanzanian Demographic and Health Survey
UNHCR	United Nations High Commissioner for Refugees
WFP	World Food Programme
WHO	World Health Organization

References

1. United Nations (UN). *United Nations Population Prospects. The 2007 Revision Population Database*; Department of Economic and Social Affairs Population Division: New York, NY, USA, 2008; p. 244.
2. United Nations World Food Programme (UNWFP). *Comprehensive Food Security and Vulnerability Analysis (CFSVA) and Nutrition Assessment 2010. Kenya High Density Urban Areas*; United Nations World Food Programme Headquarters: Rome, Italy, 2012; p. 91.
3. World Bank. *Main Report: Tanzania Mainland Poverty Assessment*; World Bank: Washington, DC, USA, 2015; p. 180.
4. Kimani-Murage, E.W.; Schofield, L.; Wekesah, F.; Mohamed, S.; Mberu, B.; Ettarh, R.; Egondi, T.; Kyobutungi, C.; Ezeh, A. Vulnerability to food insecurity in Urban slums: Experiences from Nairobi, Kenya. *J. Urban Health* **2014**, *91*, 1098–1113. [CrossRef]
5. Leduka, R.; Crush, J.; Frayne, B.; Mccordic, C.; Matobo, T.; Makoa, T.E.; Mphale, M.; Phaila, M.; Letsie, M. *The State of Poverty and Food Insecurity in Maseru, Lesotho. No.21 ed.*; African Food Security Urban Network (AFSUN): Cape Town South Africa, 2015; pp. 1–79.
6. Gustafson, D.J. Rising food costs & global food security: Key issues & relevance for India. *Indian J. Med. Res.* **2013**, *138*, 398–410.
7. The Revolutionary Government of Zanzibar (RGoZ). *Zanzibar food security & nutrition situational analysis*; Ministry of Agriculture, Livestock and Environment and Ministry of Health Social Welfare: Unguja, Zanzibar, 2006.
8. The United Republic of Tanzania. Comprehensive Food Security and Nutrition Assessment Report. Available online: www.ipcinfo.org/fileadmin/user_upload/ipcinfo/docs/IPC_Tanzania_AFI_Situation_2018Feb.pdf (accessed on 1 February 2017).
9. Revolutionary Government of Zanzibar (RGoZ). *The Zanzibar Strategy for Growth and Reduction of Poverty*; The Revolutionary Government of Zanzibar: Unguja, Zanzibar, March 2007.
10. Musotsi, A.A.; Sigot, A.J.; Onyango, M.O.A. The role of home gardening in household food security in Butere division of Western Kenya. *Afr. J. Food Agric. Nutr. Dev.* **2008**, *8*, 4. [CrossRef]
11. Koyenikan, M.J. Perception of home garden potentials among women in Edo South ecological zone, Nigeria. *Gend. Behav.* **2007**, *5*, 1042–1052. [CrossRef]
12. Obayelu, A. Comparative analysis of households' socioeconomic and demographic characteristics and food security status in urban and rural areas of Kwara and Kogi states of North-central Nigeria. *Afr. J. Food Agric. Nutr. Dev.* **2012**, *12*, 6027–6054.
13. Kebede, M. The gender perspective of household food security in Meskan district of the Gurage zone, Southern Ethiopia. *Afr. Res. Rev.* **2009**, *3*. [CrossRef]
14. Masinde, G.V. Food security coping strategies in female and male headed households in Kenyan slums: The case of Kawangware, Nairobi. *Int. J. Soc. Sci. Entrep.* **2014**, *1*, 36–54.
15. Food and Agriculture Organization of the United Nations (FAO). Women Play a Decisive Role in Household Food Security, Dietary Diversity and Children's Health. Available online: http://www.fao.org/gender/gender-home/gender-programme/gender-food/en/ (accessed on 1 July 2016).
16. Olayemi, A.O. Effects of family size on household food security in Osun State, Nigeria. *Asian J. Agric. Rural Dev.* **2012**, *2*, 136.
17. Olson, C.M.; Rauschenbach, B.S.; Frongillo, E.A., Jr.; Kendall, A. Factors Contributing to Housefold Food Insecurity in a Rural Upstate New York County. Available online: https://www.irp.wisc.edu/publications/dps/pdfs/dp110796.pdf (accessed on 1 September 1996).
18. Aidoo, R.; Mensah, J.O.; Tuffour, T. Determinants of household food security in the Sekyere-Afram plains district of Ghana. *Eur. Sci. J.* **2013**. [CrossRef]

19. Bogale, A.; Shimelis, A. Household level determinants of food insecurity in rural areas of Dire Dawa, Eastern Ethiopia. *Afr. J. Food Agric. Nutr. Dev.* **2009**, *9*, 1914–1926.
20. Ali Naser, I.; Jalil, R.; Muda, W.; Manan, W.; Nik, W.; Suriati, W.; Mohd Shariff, Z.; Abdullah, M.R. Association between household food insecurity and nutritional outcomes among children in Northeastern of peninsular Malaysia. *Nutr. Res. Pract.* **2014**, *8*, 304–311. [CrossRef] [PubMed]
21. Beaman, L.; Dillon, A. Do Household Definitions Matter in Survey Design. Available online: http://www.fao.org/fileadmin/templates/ess/documents/meetings_and_workshops/ICAS5/PDF/ICASV_1.2_109_Paper_Beaman.pdf (accessed on 14 July 2016).
22. Nyangasa, M.A.; Kelm, S.; Sheikh, M.A.; Hebestreit, A. Design, response rates, and population characteristics of a cross-sectional study in Zanzibar, Tanzania. *JMIR Res. Protoc.* **2016**, *5*, e235. [CrossRef] [PubMed]
23. United Nations World Food Programme (WFP). Technical Guidance Sheet—Food Consumption Analysis: Calculation and Use of the Food Consumption Score in Food Security Analysis. Available online: https://www.wfp.org/content/technical-guidance-sheet-food-consumption-analysis-calculation-and-use-food-consumption-score-food-s (accessed on 1 February 2018).
24. Food and Agriculture Organization of the United Nations (FAO). The Food Insecurity Experience Scale—Development of a Global Standard for Monitoring Hunger Worldwide. Available online: http://www.fao.org/fileadmin/templates/ess/voh/FIES_Technical_Paper_v1.1.pdf (accessed on 1 October 2013).
25. Leroy, J.L.; Ruel, M.; Frongillo, E.A.; Harris, J.; Ballard, T.J. Measuring the food access dimension of food security: A critical review and mapping of indicators. *Food Nutr. Bull.* **2015**, *36*, 167–195. [CrossRef]
26. Cafiero, C.; Viviani, S.; Nord, M. Food security measurement in a global context: The food insecurity experience scale. *Measurement* **2018**, *116*, 146–152. [CrossRef]
27. Maire, P.; Hatzinger, R. Extended rasch modeling: The erm package for the application of IRT models in R. *J. Stat. Softw.* **2007**, *20*, 1–20.
28. United Nations Educational Scientific and Cultural Organization (UNESCO). *International Standard Classification of Education*; UNESCO Institute for Statistics: Montreal, QC, Canada, 2011.
29. Ethiopia—Comprehensive Food Security and Vulnerability Analysis (CFSVA). Available online: https://www.wfp.org/content/ethiopia-comprehensive-food-security-and-vulnerability-analysis-2014 (accessed on 25 February 2019).
30. Tanzania—Community and Household Surveillance in North Western Tanzania: Programme Outcome Monitoring in Nyarugusu Refugee Camp. Available online: https://www.wfp.org/content/tanzania-community-and-household-surveillance-north-western-june-2011 (accessed on 25 February 2019).
31. Melgar-Quinonez, H.R.; Zubieta, A.C.; MkNelly, B.; Nteziyaremye, A.; Gerardo, M.F.; Dunford, C. Household food insecurity and food expenditure in Bolivia, Burkina Faso, and the Philippines. *J. Nutr.* **2006**, *136*, 1431S–1437S. [CrossRef]
32. Tadesse Tantu, A.; Demissie Gamebo, T.; Kuma Sheno, B.; Yohannis Kabalo, M. Household food insecurity and associated factors among households in Wolaita Sodo Town, 2015. *Agric. Food Secur.* **2017**, *6*, 19. [CrossRef]
33. Food and Agriculture Organization of the United Nations (FAO). Food security snapshot. United Republic of Tanzania. Available online: http://www.fao.org/giews/countrybrief/country.jsp?code=TZA (accessed on 11 April 2019).
34. Becquey, E.; Delpeuch, F.; Konate, A.M.; Delsol, H.; Lange, M.; Zoungrana, M.; Martin-Prevel, Y. Seasonality of the dietary dimension of household food security in Urban Burkina Faso. *Br. J. Nutr.* **2012**, *107*, 1860–1870. [CrossRef] [PubMed]
35. Habyarimana, J.B. Determinants of household food insecurity in developing countries evidences from a probit model for the case of rural households in Rwanda. *Sustain. Agric. Res.* **2015**, *4*, 78. [CrossRef]
36. Bahiigwa, G. *Household food security in Uganda: An empirical analysis*; Economic Policy Research Center: Kampala, Uganda, 1999.
37. Deaton, A.; Paxson, C. Economies of scale, household size, and the demand for food. *J. Polit. Econ.* **1998**, *106*, 897–930. [CrossRef]
38. Zakari, S.; Ying, L.; Song, B. Factors influencing household food security in West Africa: The case of Southern Niger. *Sustainability* **2014**, *6*, 1191–1202. [CrossRef]

39. Lawson, D.W.; James, S.; Ngadaya, E.; Ngowi, B.; Mfinanga, S.G.; Borgerhoff Mulder, M. No evidence that polygynous marriage is a harmful cultural practice in Northern Tanzania. *Proc. Natl. Acad. Sci. USA* **2015**, *112*, 13827–13832. [CrossRef] [PubMed]
40. Owoo, N.S. Food insecurity and family structure in Nigeria. *SSM Popul. Health* **2018**, *4*, 117–125. [CrossRef] [PubMed]
41. Meludu, N.T.; Ifie, P.A.; Akinbile, L.A.; Adekoya, E.A. The role of women in sustainable food security in Nigeria: A case of Udu local government area of Delta state. *J. Sustain. Agric.* **1999**, *15*, 87–97. [CrossRef]
42. Tembo, D.; Simtowe, F. The Effects of Market Accessibility on Household Food Security: Evidence from Malawi. In Proceedings of the Conference on International Research on Food Security, Natural Resource Management and Rural Development, Hamburg, Germany, 6–8 October 2009; German Institute for Agriculture in the Tropics and Subtropics: Witzenhausen, Germany.
43. Ndegwa, P.B. Assessment of Factors Influencing Food Security in Wenje Division, Tana River County-Kenya. *Food Sci. Qual. Manag.* **2015**, *44*.
44. Food and Agriculture Organization of the United Nations (FAO). *Methods for Estimating Comparable Rates of Food Insecurity Experienced by Adults throughout the World*; Food and Agriculture Organization (FAO): Rome, Italy, 2016.

International Journal of
Environmental Research and Public Health

MDPI

Article

Why Tenure Responsive Land-Use Planning Matters: Insights for Land Use Consolidation for Food Security in Rwanda

Uchendu Eugene Chigbu [1,*], Pierre Damien Ntihinyurwa [1], Walter Timo de Vries [1] and Edith Ishimwe Ngenzi [2]

1 Chair of Land Management, Technical University of Munich, Arcisstrasse 21, 80333 Munich, Germany; pdamien.ntihinyurwa@tum.de (P.D.N.); wt.de-vries@tum.de (W.T.d.V.)

2 Department of Land Administration and Management, Institut D'enseignement Supérieur de Ruhengeri, Musanze Street, Ruhengeri NM 155, Rwanda; ishimweedith@gmail.com

* Correspondence: ue.chigbu@tum.de

Received: 31 January 2019; Accepted: 12 April 2019; Published: 15 April 2019

check for updates

Abstract: Land use consolidation aims to address food insecurity challenges in Rwanda. However, there is contradictory evidence on whether this tool has met food security objectives or not. This study addresses two questions: How has the land use consolidation improved (or not improved) food security at the local level? How can food security challenges be addressed using a renewed approach to land use consolidation that adopts a tenure responsive land use planning procedure? We investigate these questions in Nyange Sector (in the Musanze District) of Rwanda using mixed research methods. The study generates theoretical and policy relevant outcomes. Theoretically, it links the concept of tenure responsive land-use planning to food security improvements. Policy wise, it provides an operational framework for implementing land use consolidation to make it more responsive to food security (based on tenure responsive land-use planning measures) in Rwanda.

Keywords: food security; land; land use consolidation; land-use planning; land tenure; rural development; Rwanda; tenure responsive; tenure responsive land-use planning; tenure security

1. Introduction

In most sub-Saharan African (SSA) societies, land use practices are rooted in ensuring improved livelihood options, community cohesion and food security [1]. Even though various approaches are being applied towards achieving this aim—both in scientific and cultural terms—the region continues to face various land-related challenges. Some of these challenges include land degradation [2], water pollution [3] rural and urban poverty [4], poor environmental sanitation, climate change related problems [5], gender inequality (especially women's lack of access to land) [6], inefficient land administration [7], loss of culture [1], land fragmentation [8], to mention a few. Out of the many problems bedeviling SSA countries, two critical challenges the region faces are how to deal with the widespread development of urban areas [9] and how to improve livelihoods development of the rural areas in a "responsible" manner [10,11]. Compounding the severity of these problems are the persistent threats posed by climate change, poverty, disaster risk, and lack of access to land (and other natural resources) for use for different aspects of human development. This study takes a rural dimension to investigate one of these problems.

The challenge of improving food security (or food insecurity as a treat to improved rural livelihoods) is increasingly understood to have a relationship to the types of vulnerabilities its users face on land, as well as to the type of privileges (and rights) owners (and users) enjoy on land [12,13]. For example, reduced food security could be caused by environmental degradation, poor energy usage, natural disasters or the way people use land [14–17]. Also, the level of tenure security land users and landowners have on a parcel of land can influence its productivity in both positive and negative ways [8,18–20]. One of the many countries in Africa where the relationship between land use and food security (among many other land-related problems) has become a policy concern is Rwanda. The land challenge in Rwanda is a complex one despite that the country has surveyed, adjudicated and registered all its land. The country-wide registration of land in Rwanda has led many scholars to assume that Rwandan rural landowners have high land tenure security that will enable them to participate fully in the development process of the country [6,7,10,18,21]. This assumption is however not completely valid due to a number of additional land related problems. First of all, Rwanda is the most densely populated country in Africa. Its dense population is one of the factors that makes land a critical resource in the country. The majority of the country's working population earn their livelihood from land (especially through agriculture) and related employment [21,22]. Rwanda's food security challenges have been blamed on the scarcity of land, and land policy related strategies (that are not adequately responding to the nature of challenges being faced by the country) [23]. Secondly, land tenure practices aside, the unique topography and high population densities of Rwanda are some of the reasons the country has a highly fragmented landscape which is in dire need of consolidation [8,24–28]. The consequence of these particular characteristics is that land use decisions remain contested and that food insecurity persists.

In recognition of the persistence of food insecurity in the country—and from the rural perspective—the government has made policy efforts towards improving the situation. The government has put improved agricultural production and food self-sufficiency at the top of its development agenda [29,30]. It introduced land use consolidation (LUC) measures since 2008 as a component of the Crop Intensification Program (CIP) to broadly meet food security by boosting the national agricultural production. However, the outcome of LUC remains questionable and debatable. There is contradictory evidence from literature on whether the program has met food security objectives or not. On the positive side, some studies (especially Rwandan government reports) claim that LUC has doubled (and even tripled) agriculture yields of the priority crops, and thereby increased food availability and food quantity at the national level [24,31–35]. On the negative side, Pritchard [29] argues that due to the rapid and forceful implementation of the program, "tenure and agricultural policies are unnecessarily undermining the livelihood stability of rural subsistence farmers". Brown and Hughes [23], note that "land use consolidation and limitations on land subdivisions have produced emerging threats to tenure security", as well as food security. Others argue that the program leads to monoculture farming since in its implementation all farmers with closed parcels are asked to grow only eight priority food crops [8,24,26,31–34]. This position is emphasized by the Rwanda National Food and Nutrition Plan 2013–2018 which recognizes that despite significant economic and poverty reduction progress brought by different governmental programs including LUC regarding tremendous increases in national agriculture production of the priority crops, improvements in nutrition and household food security remains a foundational issue [36,37] Irrespective of what side of the argument these scholars stand, they all agree that there are still worrying levels of inappropriate land use practices and livelihood options that cause food insecurity at the national level [2,24,26,30–40].

This study deals with an important question in Rwanda's land policy and land reform implementation at the local level. It takes a critical look at the impact of the LUC program on food security at the local level. It uses the FAO Food Security definition-oriented approach (which puts focus on the households and individual's dietary needs and food preferences) to assess food security status and driving factors before and after the LUC in the rural sector of Nyange, in Rwanda. This is done with the objective of discerning a renewed approach of LUC through tenure responsive land-use planning. The study approaches this objective by answering two key question that focus on the extent to which LUC impacts food security and what should be the recommendations to increase the effectiveness of LUC towards food security. The questions are:

- How (much) has the program improved (or not improved food security) in at the local level local in Rwanda?
- How can food security challenges be addressed using a renewed approach to land use consolidation that adopts tenure responsive land use planning procedure.

By answering these "how-to" questions, this study departs from existing literature in two ways. Firstly, it answers the question of whether LUC has an impact on food security at the local level. Considering that many of the contradictory literature (arguing for and against) on the food security performance of LUC at the national level, investigating the impact of LUC at the local level is necessary because it is possible that LUC may not have improved food security nationally but may have had positive impacts in pockets of administrative sectors in the country. Secondly, by focusing on how a tenure responsive land-use planning approach can be used to make LUC respond to the food security needs of the people, the study introduces a new approach to food security improvements.

In order to answer these questions, the study begins by framing the tenure responsive land-use planning as the conceptual basis for ensuring food security improvement at the local level. This is followed by description of the methodology, outcomes and discussion leading to a proposal for making Rwanda's LUC to generate food security through a responsive land-use planning approach. Finally, the study concludes.

2. Literature Review

2.1. *Framing Tenure Responsive Land-Use Planning in Response to Food Security: Delineation of Intertwined Concepts, Definitions and Relationships*

To understand tenure responsive land-use planning in response to food security, six major terms are worth understanding. They are land-use planning, tenure responsive land-use planning, land tenure, land tenure security, food systems and food security. All of these concepts share a relationship with land. Individually and collectively, they also have causal relationships with food security. Although the focus of this study is on tenure responsive land-use planning and food security, it is important to grasp the meanings of these concepts and the general relationships they share. This is important for understanding the relationship between tenure responsive land-use planning and food security (Figure 1).

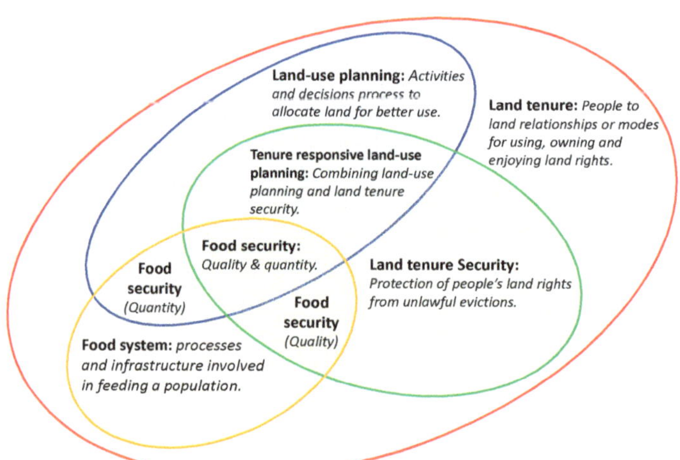

Figure 1. Conceptual delineation of key terms used in the study.

Land tenure is an overarching concept that entails the manner in which people use own and enjoy rights on land (including privileges, obligations, restrictions and responsibilities). It reflects the relationship people, communities, groups and individuals share with land [17,18,23–25,41–54]. Land tenure and land tenure security are two interwoven concepts. On the one hand, land tenure means the "manner in which the rights, restrictions, and responsibilities that people have on the land (and property) are held" [13,55,56]. Land tenure is that which defines and shapes the way people own, hold, and enjoy rights to land. Hence, it embeds local realities (regarding people-to-land relationship) in regard to legal, social and cultural practices relating to land. On another hand, tenure security implies the "rights individuals and groups have to effective protection by the state against forced eviction. Under international law, this entails permanent or temporary removal against the will of persons, families and communities from the homes and land that they occupy, without the provision of, and access to, appropriate forms of legal or other protection" [18]. When land tenure is secure, individuals (and groups) enjoy the ownership, use, and exercise of various rights on land with higher freedom. Tenure security manifests in de facto or de jure ways, and it can be established through legal or social agreement or recognition [10,12,13,17,41–44]. So, it can enable or sustain access to and availability to land (especially when consciously planned to achieve these) for a specific period (or in perpetuity). Moreover, by doing this, it has the potential to enable and sustain access to and availability of food (thereby making food-intake and its stability possible) so far as the individuals involved have the requisite knowledge about nutrition, food governance and issues related to laws and regulations are appropriately conducted [57–60]. In the context of land tenure (considering that food can also be made available via imports and food aid), the availability of food as may be determined by tenure security connotes the quantity of food of acceptable quality possible through food crop production. Access to food implies that individuals and groups can acquire the available food resources for use. Food usage entails that the acquired food is used for balanced nutritional intake. Food sustainability is about having the necessary stability to ensure that access, availability, and use of food are stable enough to ensure a normal state of living. As food production (and its security) is dependent on the access, availability, appropriate use and sustainability of land (regarding access, availability, use, and sustainable management), adequate management of land (through land-use planning) has a direct impact on food security. What makes this relationship possible is tenure security on land. Land tenure exist in any place where there is human to land relationships (so it exists everywhere). However, land-use planning and land tenure security do not exist everywhere except where they have been consciously conditioned to exist.

Land-use planning and tenure responsive land-use planning are two other entangled concepts. Land-use planning, unlike tenure responsive land-use planning, is an old concept in planning literature. Myriad of definitions exist on the concepts of land-use planning from disciplinary and policy angles. However it is defined, land-use planning usually alludes to "activities and decisions concerned with guiding the allocation and use of land in patterns that enable improvements in peoples' way of living" [41]. It is a relevant land intervention measure for development at the local level because "it is one of the most sensitive political issues in any country since it affects people's livelihoods and the essential needs of communities" [12,13,17,42–44]. However, tenure responsive land-use planning is an integrative concept that posits to combine land-use planning and land tenure security (otherwise referred to as tenure security). Just any of the other concept defined in this study, land-use planning is done within an environment of land tenure but not necessarily in an environment that strives to include tenure security. Tenure responsive land-use planning is done inclusive of tenure security. Land tenure security and land-use planning were two separate areas of research until the works of the Global Land Tool Network (of UN-Habitat) which led to the seminal works of Chigbu et al. [10,12,13,17,42–44]. In their works, they [10,12,13,17,42–44] argued for a combination of tenure security and land-use planning concepts (and practice), leading to the coinage of the term, tenure responsive land-use planning. The concept of tenure responsive land-use planning recognizes that "separately, land-use planning and tenure security improvements remain the two strongest forces for influencing spatial transformations and development" and combined they "play crucial roles in achieving development objectives" [13]. Land-use planning assures appropriate use of land while tenure security is the precondition for sustaining livelihoods through sustainable agriculture, among others. This makes its combination with tenure security imperative for food security improvement. Tenure responsive land-use planning is a specific type of land-use planning process that can wholly or incrementally establish land tenure security [12,13]. By incrementally or wholly ensuring land tenure security, it can generate elements of food security when consciously implemented with food security as one of its key objectives.

Food systems and food security are intertwined concepts. A food system includes the various processes (including infrastructure) that involve making food available (in quality or quantity or both) to feeding a population [60–71]. The process within a food system could be land policy and agricultural policy making and implementations for ensuring the eradication of hunger. A food system can interact with other land related processes or conditions (such as land-use planning and land tenure security) to produce food availability (in terms of in quantity or quality. This study views food security as a situation where people have physical and socioeconomic access to sufficient nutritious and safe food that meets their dietary needs (and food preferences) for an active and healthy life [45–47]. This definition of food security captures four critical components of food security, which include food availability, food access, food usage, and stability to continue to enjoy food usage as a healthy way of life [48–53]. These four components can also serve as possible indicators for measuring (quantitatively) or evaluating (qualitatively) food security at national, regional, and at the local levels (household and individual levels). Whenever one of these four aspects are not fulfilled or met, people may suffer from hidden hunger. A food system has the potential to activate and enable land interventions (or development measures) to produce food security outcomes.

2.2. *Tenure Responsive Land-Use Planning as a Generator of Food Security: A Framework*

The major link between the operationalization of tenure responsive land-use planning and the outcome of food security policies is tenure security. A variation in tenure security influences the outcomes of food security policies when tenure security actors interact with food systems [8,12,29,30, 42,43,57,58,60–63,65,67–71]. This link is of primary interest in framing any approaches to improving food security through tenure responsive land-use planning. It is also necessary for analyzing, and designing future (and improving existing) land-use decision-making) processes to ensure that they respond to food security needs of their stakeholders. That is why this study frames food security as a possible outcome of tenure responsive land-use planning (Figure 2).

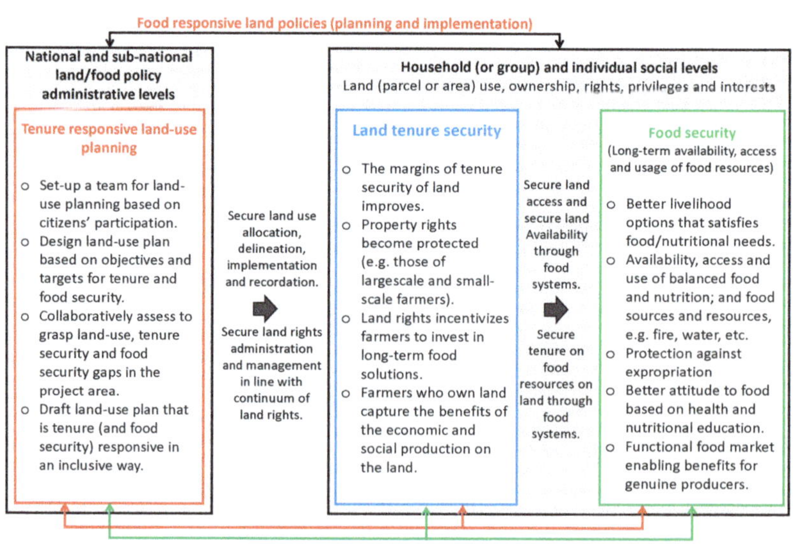

Figure 2. Framework for food security through tenure responsive land-use planning.

Land-use planning, especially in rural areas, usually involve the participation of farmers who own land or the involvement of farmlands owned by people eager to improve their food security situation. In such a situation, the operational framework ensures that such farmers or land users can capture the benefits of expected socioeconomic production on the land to improve their food security. However, this demands a land-use planning process that allows securing tenure and property rights. It calls for a process that allows for, at least three vital systematic aspects: a tenure responsive land-use planning, a land tenure security, and a food security component (Figure 2). The tenure responsive land-use planning component is embedded in policymaking. There are effects in both land tenure security and in food security.

The tenure-responsive component when guided by land (food responsible) policies provides guidance at administrative implementation levels and also enables the sustainability of any outcomes (in this case, food security). Where this is the case (and there is need to do land-use planning), as a starting point will be to set up a team of participants based on the principles of citizens' participation and continuum of tenure (and land rights). Participation will ensure that everyone's rights to engagement are ensured [13]. A continuum of land rights approach implies that participants (and administrators) must recognize that "different types or levels of tenure security may prevail in the everyday lives of people because of the social rules, customary practices and laws within a particular land jurisdiction" [13]. So, land-use planning has "to be implemented with the recognition that land rights are not static, but are manifested in various forms across a continuum of rights–tenure security can be enhanced through the recognition of individual, community and indigenous tenure rights by creating a better understanding of social and legal protections for various means of land tenure" [13]. Taking these steps is not possible unless (in the design of a land-use plan) tenure security objectives and targets are adopted. This implies collaboratively assessing situations to grasp land-use, tenure security, and food security gaps in the project area; and then drafting a new land-use plan that is tenure (and food security) responsive in an inclusive way. Inclusiveness is necessary to ensure access to land to unlock household insecurity by women and other underprivileged groups. This is important to enable land users and farmers who own land to capture the expected socioeconomic benefits from their land.

The second component is influenced by policy (hence, it is an effect susceptible to policy effect). This component can be influenced by secure land use allocation, delineation, implementation and recordation; and secure land rights administration and management in line with a continuum of land rights. It has two parts–the land tenure security and food security–with the former influencing the later through secure availability and access to land; and secure tenure on food resources on land. Where the margins of tenure security of land improve, property rights become protected (e.g., those of large-scale and small-scale farmers), then land rights incentivize farmers to invest in long-term food solutions, and farmers who own land capture the benefits of the economic and social production on the land. This will have a positive effect on food security it will lead to better livelihood options that satisfy food/nutritional needs, protection against expropriation, better attitude to food based on health and nutritional education and a functional food market enabling benefits for genuine producers. This will ultimately result in the availability, access, and use of balanced food and nutrition (and including land related natural resources linked to natural food sources for energy and water access).

The logic behind the framework is that tenure responsive land-use planning ensures protection of property rights to land users (e.g., small-scale farmers) and enables them to take critical steps to invest in long-term solutions that gives them access (and makes land and food available) for use in their quest to fight poverty and hunger. This has been found to be right in Zambia [72], Tanzania [54], Namibia, Laos, Philippines, Brazil, Chile, Ghana and Ethiopia [12,17,73,74]. It, therefore, makes sense to assume that this is possible in a country like Rwanda. This study, therefore, puts focus on the LUC being practiced in Rwanda –with the aim of reaching an improved food security implementation strategy through tenure responsive land-use planning.

3. Materials and Methods

This study follows a mixed methods approach (Qualitative and Quantitative) and a pragmatic world view which fits for a better understanding of complex problematical situations and the consequences of actions. Food security is a very complicated variable which cannot be easily identified by one predetermined method, while the evaluation of consequences of LUC as an action to overcome it similarly needs a multidimensional approach. Creswell [75,76] note that this kind of approach gives the chance to rely on different methodologies and convergent data collection and analysis techniques which gives a better understanding of the problem and enables better solutions. This dictated the use of both qualitative and quantitative methods based on data collected simultaneously from April 2017 to June 2017 as shown in Figure 3.

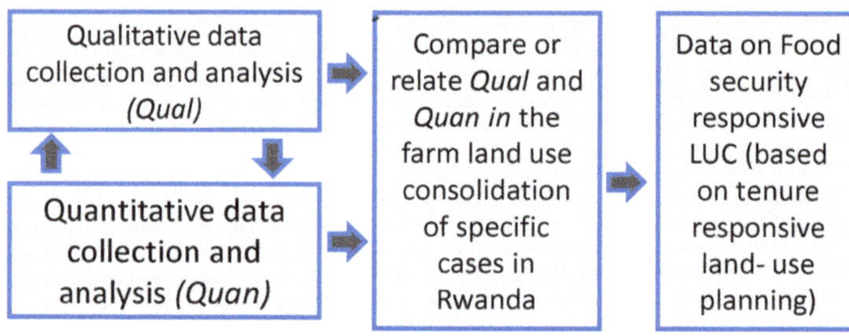

Figure 3. Convergent parallel mixed methods research design.

The data collection relied on a household survey (using semi-structured questionnaires) based on semi-structured face-to-face interview. The survey focused on discerning the values of food security indicators, the perceived driving factors for what influenced food security for individual farmers, their observed impacts of LUC implementation strategies on food security, the type and variations in how actors contributed to the LUC implementation process, and the level of farmer's satisfaction with the LUC program. Field observations of agricultural fields and reviews of food security policy document aided in validating and triangulating the responses of farmers and linking the specific fields to known and documented agro-ecological conditions of the Nyange sector

3.1. Site Selection

The data collection took place in the Nyange Sector of the Musanze District (Northern Province of Rwanda). The choice for this study area was purposively driven by the fact that this Sector is one of the project areas where the LUC program is facing post-implementation resistance from farmers (particularly against the priority crops of the government) in the Northern Province of Rwanda. It is important to mention that the resistance of farmers against LUC in the area cannot be linked with its failure to meet food security, as it is a post-implementation resistance. The resistance is considered as an outcome of the failure of LUC to meet its food security targets, rather than a cause or one of the factors. The farmers in this area have been complaining that the Governments priority crops are unsuitable to their local agro-ecological conditions and less economically valuable than the non-priority crop grown in the area before the introduction of LUC [8,23,24].

Furthermore, the Nyange Sector, as well as the whole Musanze district is a volcanic region. It has experienced changes in weather conditions and rainfall patterns which have led to severe floods as natural risks or disasters often associated with the destruction of food crops and total loss of agricultural production with food insecurity as an obvious consequence over time. Figure 4 shows the location of the case study.

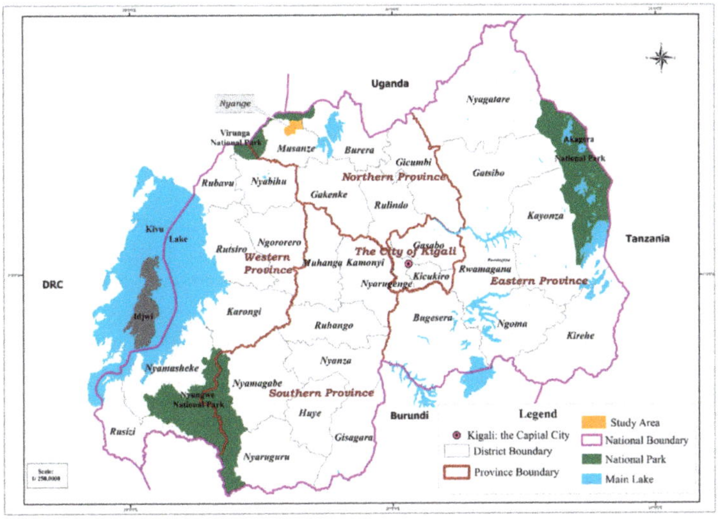

Figure 4. Nyange sector (case study) location in Rwanda.

3.2. *Data Collection and Anylsis Methods*

The data collection was done in five cells (*Kabeza I, Kivugiza I, Ninda, Muhabura, and Kamwumba*) of Nyange sector in Musanze district. The choice of all the five cells was not for the comparison purpose, but for validity and reliability purposes. A combination of stratified random sampling and purposive sampling methods were used for the selection of respondents. In total, 72 respondents representing more than 15% of the total number of households in the sector were randomly selected for the household survey using the formula of Lykken et al. [77] which is appropriate for very homogeneous populations, and six key respondents purposively selected (five cell agronomists, one sector agronomist) respectively were interviewed for primary data collection. Only one household representative member was chosen to respond to a set of questions during the household survey. The choice of the six key informants was motivated by their critical roles in the implementation of LUC and CIP programs at the local level and their likelihood to bear the information about food security in the area as agronomists. The questionnaires were drawn in English language and translated in Kinyarwanda. All the questions were asked in Kinyarwanda as the local language and the answers translated in English by the authors later. This technique of combining households survey (quantitative non-experimental) and other qualitative methods in the assessment of programs and policy impacts is very famous in social sciences and has been widely used by many other researchers in the same context [8,20,24,26,33–35] which makes it a more credible and reliable approach. The Triangulation and Back-checking techniques were used to check the validity and reliability of the collected data, and were chosen for their extensive use in mixed studies like this one.

The information about the failure of LUC to meet food security in the study area was primarily obtained by collecting the perceptions of respondents about the status of household food security before and after the introduction of LUC from the household representatives (Heads) through a household survey and the key informant's interviews. All the households' respondents were asked to choose one answer among four different pre-established indicators of food security in their households after the introduction of LUC. These four indicators include, first, optimal uptake of nourishment which indicates food security in all its aspects (food quantity, food quality, food availability, food accessibility, food utilization or usage, and food sustainability). Second, sufficient balanced food availability which indicates the presence of food security in terms of the availability of diverse foodstuffs (food quality) in sufficient quantity. Third, sufficient access to food which indicates the accessibility of food in terms of quality and quantity for food security. Lastly, insufficient balanced food availability which indicates food insecurity in terms of the unavailability of diverse foodstuffs (food quality) and quantity. On the other hand, the key respondents as more educated people, were asked to compare the status of household food security in all its aspects such as food quantity, food quality, food sustainability/stability, food availability, food accessibility, and food usage/utilization before and after the introduction of LUC in their area. Each key respondent was also required to give a motivation for this comparison. The information about the level of farmer's satisfaction about LUC and the reasons for their satisfaction level has also been gathered through a household survey, in order to check whether food insecurity indicators could be one of the main factors of satisfaction, and which indicators could be linked with the failure of LUC. This helped to answer the questions about whether and how LUC failed to meet food security. Both descriptive statistics along with text description were used to analyze quantitative and qualitative primary data to draw statistical and thematic conclusions about the research questions. A correlation diagram drawn from the primary data, literature review and the authors knowledge about LUC and Food Security nexus was used to show the causal-effects relationships among different factors and variables in different food insecurity scenarios before and after LUC in the study area. This helped to show how LUC failed to meet its food security objectives as a multidimensional concept, rather exacerbated the issue by contributing to other forms of food insecurity scenarios when combined with exogenous factors like climate change and market imperfections among others. In order to verify why and how LUC influences food security targets, the study performed an analysis of the existing LUC policy (objectives and guiding principles) and of its formulation and implementation process in the study area. The main target information was the level of farmer's participation in the process, which could give insights about the independency of land use rights as an indicator of tenure security and factor of food security, as pre-established in the conceptual framework in the literature section. LUC process and main stakeholders' diagram was drawn from the information gathered from the literature and governmental reports, along with the primary data from farmers during household survey and key informant's interviews. The level of farmer's participation was assessed by asking the household members during the household survey whether they participated in 3 main activities of LUC implementation process (provision of parcels and farming activities, choice of the priority crop suitable to their area, and the post harvesting activities). Once the Gap found, it has been correlated/associated with Tenure Responsive Land Use Planning Principles in order to propose a better renewed approach to LUC to make it more food security responsive as solution to the main study problem. A deductive review of the literature about tenure responsive land-use planning principles, LUC guiding principles and Food Security aspects and components was performed to identify the gap in their relationships in the study area, answer the question about the reasons for their failure, and propose new solutions to mitigate food security challenges through tenure responsive land-use planning as secondary data sources to supplement the primary ones.

4. Outcomes and Discussion: Has Land Use Consolidation Improved Food Security? If Not, How and Why?

4.1. Farmers' Perceptions Show That the Outcome of LUC Did Not Support Household Food Security

The study compared the before-and-after situations of the LUC to grasp its food security impacts. The level of food security was determined from the perception of critical informants and household members about it at the household levels (see Figure 5), along with the existing reports about food security status before and after the introduction of LUC.

The majority of household members (49%) report to suffer from insufficient balanced food availability in their daily life after the introduction of LUC, while the majority of the responses from the key informants confirmed the decrease in food quality, food sustainability and food accessibility (at 33%, 33% and 17% of respondents respectively) as indicators of food insecurity at the sector level, despite the increase in, food quantity, food usage/utilization and food availability (at 85%, 67% and

67% respectively). In order to understand why this situation arose, it is necessary to understand the factors that reduced food security in the Nyange Sector.

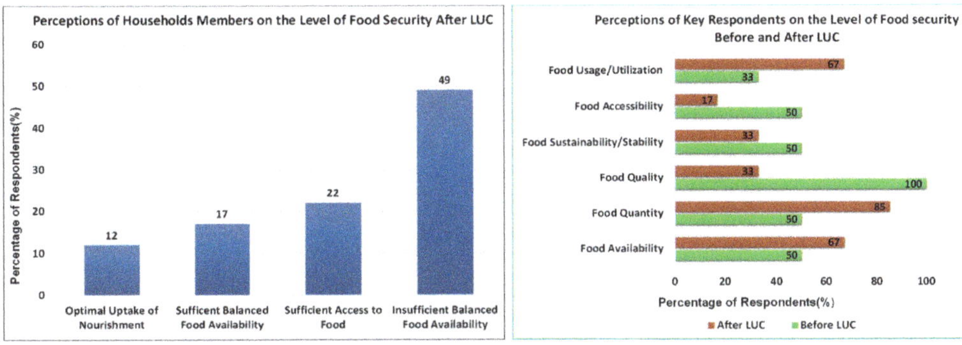

Figure 5. Scenarios of food security in Nyange Sector.

The question that arises is how and why has the LUC not been responsive to food security needs of the rural people? What are the gaps that exist between principles and practices? Answering these questions necessitates a critical examination of food security status and its driving factors before and after its introduction, and its process of implementation. This is necessary to understand the reasons of its unresponsiveness to food security, and possible ways to improve it.

4.2. Specific Scenarios Explaining How and Why LUC Failed to Ensure Food

Understanding how the LUC fail to ensure food security in Nyange Sector demands to outline the specific scenarios and drivers of food insecurity before and after the introduction of LUC. Participation in land decisions is mandatory for achieving food security. The reason for this assertion is that a path to achieving tenure security cum food security through LUC will be to strengthen citizens' involvement by making all activities participatory and inclusive in their operational (and decision making) aspects.

An understanding of the impacts of a specific program on livelihoods needs a careful assessment of the socioeconomic situations of the beneficiaries before and after its introduction. In this study, the views from the key informants and farmers about the level of food security before-and-after LUC alongside a combination of information from the key informants and household survey led to an understanding of the food (in)security status from a before-and-after perspective (Figure 6).

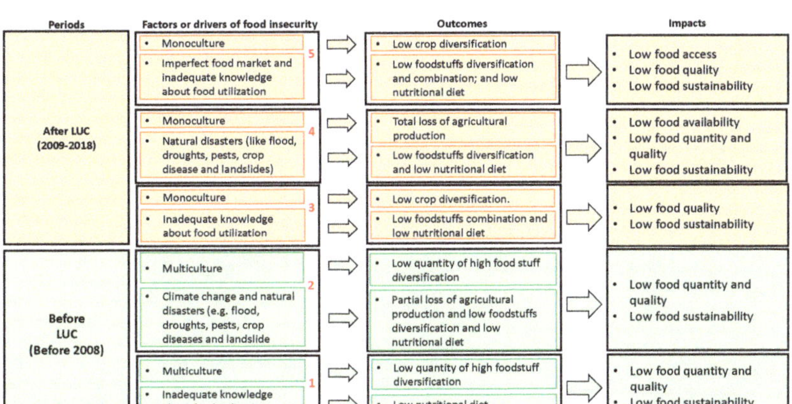

Figure 6. Scenarios showing the failure of LUC to ensure food security in Nyange Sector.

The above diagram depicts scenarios (with a focus on factors, outcomes, and impacts) of food insecurity before and after the introduction of LUC. The scenarios—*Before LUC (1-2)* and *After LUC (3-5)*—a combination of factors led to low levels of food access, food quantity, food quality and the inability of farmers to sustain themselves with what they have (which are indicators of food insecurity at the household level). The two scenarios (*Before LUC*) and the three scenarios (*After LUC*) show how LUC did not improve food security after its introduction respectively below:

- Before LUC 1: In this situation (before the introduction of LUC), farmers were practicing the subsistence agriculture (based on the combination of multiculture agricultural practices), but there was a presence of inadequate knowledge of food utilization by the farmers. This led to low food quality, quantity, and sustainability as indicators of food insecurity at the household level. This was as a result of low foodstuff diversity combination and low balanced nutritional diet intake.
- Before LUC 2: In this situation (also before the introduction of LUC), farmers were practicing the subsistence agriculture (based on the combination of multiculture agricultural practices). However, due to the effects of climate change (change in rainfall patterns and weather conditions) and natural disasters (floods, pest attacks, and droughts), farming led to low food quantity, low food quality, and low food sustainability (food insecurity). This was mainly due to a partial loss of agriculture production of food crops in case of natural hazards occurrence, low foodstuff diversification, and low balanced nutritional diet intake.
- After LUC 3: Under this situation, LUC's dependence on monoculture and the farmers' inadequate knowledge of food utilization lowered food quality intake and sustainability (food insecurity). This was mainly due to low crop diversification (as is always the case with monoculture), low foodstuff combination and low balanced nutritional diet intake.
- After LUC 4: In this situation, LUC's dependence on monoculture, combined with the effects of climate change (change in rainfall patterns and weather conditions) and natural disasters, (floods, pest attacks and droughts) led to low food availability, low food quality, low food quantity and low food sustainability (which are indicators of food insecurity) through the total loss of agriculture production of priority crop in case of natural hazards occurrence. In this case, LUC acts as a bridge of climate change towards food insecurity.
- After LUC 5: In this situation, LUC's dependence on monoculture, combined with the presence of imperfect food market and inadequate knowledge about food usage caused low food quality, accessibility, and sustainability (food insecurity). In this case, LUC induces the reduction in agriculture production of priority crops which increases their prices on the market, makes them unaffordable and inaccessible to the poor farmers for foodstuffs diversification and nutritional balance purpose.

It is important to reiterate that the Government of Rwanda introduced LUC as a way to curb *Before LUC (1-2)* scenarios [3,23,24,28–30]. Hence, *Before LUC (1-2)* scenarios were already known before the introduction of LUC [28–30,33–36,78–80]. A fundamental weakness of *After LUC (3-5)* is that it focuses on food quantity at the national level (as a single aspect of food security) thereby ignoring the aspects of food quality, food availability, food accessibility, and food usage/utilization aspects at the household level. Also, the farmers have the requisite skills for managing climate change situations and do not have the knowledge (and capacities) to deal with natural disasters. These can, in turn, be considered as the main factor of its failure to improve the status of food security after its introduction. It is important to note that the *Before LUC (1-2)* scenarios had food security problems. A major difference between it and the *After LUC (3-5)* scenarios is that multiculture enabled food diversification with low quantity of food production in the *Before LUC (1-2)* whereas monoculture is currently enabling high quantity of food production with low food diversification in *After LUC (3-5)*.

Since the findings show that there were scenarios of food insecurity in both periods (before and after LUC) under two different conditions of Multiculture (Before) and Monoculture (After), the authors argue that LUC cannot be considered as the main driver of food insecurity in the area. Rather, combined with other external factors like climate change, market imperfections, knowledge gaps, it has failed to revert the existing Multiculture-based form of food insecurity and contributed to a new Monoculture based form of food security. Therefore, this raises the question of how to deal with these two extreme scenarios of food insecurity in a responsive way. The authors suggest that the best approach would be finding strategies which could keep both situations (multiculture and monoculture) under different conditions as an optimum solution to this problem. Drawing from the study of Ntihinyurwa et al. [8] about the positive impacts of land fragmentation in Rwanda, we suggest that market oriented and monoculture based land use consolidation be applied in more homogenous areas with less variability in agro-ecological, physical (soil, slope, water, etc.), socio-economic and climatic conditions, and keep the multiculture in more heterogeneous conditions as a risk management strategy, climate change resilient and adaptation strategy to land fragmentation and food insecurity problems in Rwanda, as stipulated by the SDGs (2,13 and 15) in the Agenda 2030.

Furthermore, the explanation for this failure is also shown by the level of farmers' satisfaction with LUC and the reasons for their satisfaction as shown in Figure 7.

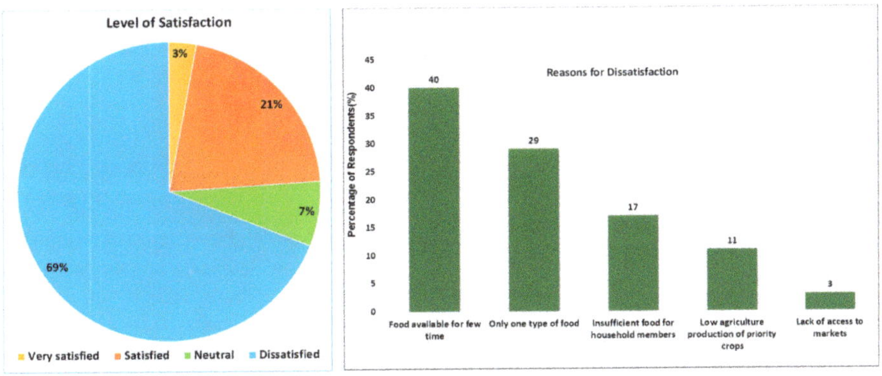

Figure 7. Level of farmers' satisfaction about LUC and reasons behind it.

The level of farmers' satisfaction about LUC is based on pre-established responses. However, the reasons behind their dissatisfaction is not based on pre-established responses, but rather on unguided feedback received from the respondents through follow-up questions. The majority of respondents report being unsatisfied with LUC, accusing it to negatively affect the sustainable/stable availability of balanced food sufficient for their household members (69% and 86% respectively). The majority of the critical informants have also blamed the lack of access to the food markets (market imperfections), leadership problems, low education level of farmers (food usage education) and natural disaster (including drought and climate change) as some other factors that led to LUC failure to meet food security needs of local farmers. However, the key informants note that the introduction of Kitchen Gardens (*Utulima twigikoni*) as new governmental strategies to accompany LUC in its process of meeting food security targets at the household level. In this program, households are taught to make small gardens around their residencies where they can grow small scale vegetables which could help them in their daily life to meet the balanced dietary needs, even though their production is still at the lowest level.

4.3. Gap between LUC Principles and Their Implementation

4.3.1. LUC Objectives and Guiding Principles

The LUC was initiated in Rwanda in 2008 as the central pillar of CIP initiated earlier by the Ministry of Agriculture and Animal Resources (MINAGRI) in September 2007 with a goal to increase agricultural productivity of high-potential food crops and to provide Rwanda with greater food security and self-sufficiency. According to the *Ministerial Order n°14/11.30 of 21/12/2010* of Rwanda [23] which deals with land consolidation models in Rwanda, the ultimate goal/aim of LUC is the rural development and promoting agricultural transformation that increases agricultural production and improves the lives of Rwanda´s people in rural areas. The *Article 5 of the Ministerial order n°14/11.30 of 21/12/2010* [24] defines three models of land consolidation for farm productivity purposes which include: the use of facilitated farming contract; cooperative farming; and farming corporation. Furthermore, Article 14 of the *Ministerial order n°14/11.30 of 21/12/2010 of* Rwanda [26] stipulates that the designing of land use consolidation and its implementation shall respect the following guiding principles:

- Not only the improvement of agricultural production but also the improvement of rural livelihoods;
- To ensure that identified potential land use is market-oriented;
- To determine possibilities of encouraging farmers and private investors to voluntary participate in the project and to support their participation.
- To ensure that women, youth and members of vulnerable groups participate in land consolidation project with the intention of promoting its practical use and to ensure optimum productivity and benefit to them;
- To support any existing off-farm employment opportunities to support the farm laborers that may lose employment due to land consolidation with the intention of promoting its practical use and to ensure optimum productivity;
- To aim at attracting investors who are practically committed;
- To apply democratic principles, use of consultative methods on any issue to be tackled and provide an avenue for members of the community to express their comments on various programs.

The *Official Gazette no 52 of 27/12/2012* of Rwanda [23,24] provides that the selection of an appropriate land (use) consolidation model (with the intention of promoting practical use and ensuring optimum productivity relevant to each particular location) shall be a result of collaboration between the Ministry in charge of agriculture, landowners, land tenants and other stakeholders concerned with the identified local farming areas. The farmer may pay for the inputs at the time of purchase or after harvest, using the proceeds of the sale of the crop.

Based on all of the above information, it is inferable that the LUC objectives, models and principles support food security improvements, but at the national level. Therefore, the inability of LUC to improve food security implies that there are gaps between LUC principles and its implementation.

4.3.2. LUC Process Is Neither Participatory, Tenure Security Responsive, nor Food Security Responsive

A starting point for gaining an insight into the situation of land-use planning (in this aspect, land use consolidation) and food security was to know whether land tenure is secure enough for the rural people of Nyange. Despite that the land in Nyange is formalized, the participants in the LUC believed that their tenure is not "secure enough" to the level that ensures *"participation in economic development"* and *"protection from the loss of full or partial land access or land rights,"* as some of the farmers mentioned. They noted that LUC diffuses their right to farm in their preferred manner which could have helped them to meet their dietary needs and food preferences. One of the farmers advanced that the approach of LUC demands that:

> *"Farmers must forego their traditional intercropping ways of farming and commit to adopting the program's use of single cultivation of priority crops in order. This makes it difficult to produce for household dietary satisfaction, market needs".*

By using the word *"forego"* in expressing their situation, it indicates a lack of a major land-use right which the farmers consider deterrence in their quest to become food secure. Concerning their tenure security situations, another farmer echoed their collective viewpoint, saying:

> *"We do not have the right to cultivate the crop of our choice. We do not have the right to choose the priority crop of the adjacent sites, and we are not allowed to cultivate crops not considered as*

priority crops by the government. If we ignore these rules, we are condemned to pay some penalty fees between 10 to 50 thousand Rwandan Francs based on the size of land cultivated with non-priority and non-chosen priority crops".

The above statement indicates that the farmers do not have tenure security on their land. It also means that they cannot decide on *cultivating what they need most (and best) for their household food needs.* As another farmer put it, *"we cannot afford to decide on what we farm and this situation affects what we eat since not all of us can afford to buy other foodstuffs from the market for us to have diverse food items at home."* The experience expressed by the farmers is indicative of insecurity in the participation of farmers in the program in ways they deem suitable for improving their food security needs. The respondents who participated in LUC gained access to improved agricultural seeds and fertilizers but not necessarily the seeds of their particular needs for household food security improvement. This is because LUC is an "imposed innovation" [81]. As a consequence, food insecurity remains a challenge at the household level. Although different studies reiterate the presence of high security of tenure (land ownership rights) in Rwanda after the national systematic land registration program which left 11.4 million parcels demarcated (more than 90% of the national coverage), and 7.2 million land titles issued [82], these findings from different data sources (key informants, landowners/farmers, and the literature/different reports) question the issue of enjoyment of the use rights over land. This study considers tenure security to mean not only the security of ownership rights, but also the enjoyment of the use rights with freedom—freedom of making decisions about the use of their lands in the way which meets their dietary needs and food preferences. Similar findings were reported by Lengoiboni et al. [83] in their study about the impacts of women land rights registration on household food security in Rwanda. They have mentioned that despite the tenure security of their land rights after the land registration program, women faced tenure insecurity when it comes to the limitations or land use restrictions in the formal land law through the crop intensification program and LUC, by giving them little room for freedom to grow food crops of their choice, thereby negatively impacting their food security at the household level.

The field study and the review of the existing documents have shown the main stakeholders of LUC to include the Ministry of Agriculture and Animal Resources (MINAGRI), Rwanda Agriculture Board (RAB), Ministry of Local Government (MINALOC). Local authorities (at the district, sector, cell and village levels), NGOs, local farmers, private investors and community based associations/organizations are also involved in LUC process as stakeholders as shown in Figure 8.

From the process of LUC represented in Figure 5, it can be seen that local people only implement the instructions of local authorities in a top-down way. The evidence from local authorities reported the problem of performance contracts and targeted figures to meet, as one of the factors inducing them to force local people to grow the government approved priority crops chosen for them at national and district level based on different agro-ecological zones in order to protect their jobs. A significant feature of the process of the LUC is that it lacks citizens' participation in its core areas for instance, in its establishment, crop intensification decisions, crop types to be used, land sizes to be cultivated, land areas to be used and specific activities to be indulged. Despite that land tenure has been regularized in Rwanda, land rights (related to decisions on the *what*-and-*how* aspects of land uses) that would enable them to solve their household specific food security challenges are either unprotected or infringed upon by the LUC process. This is evidence that the farmers involved in the LUC are not secure enough in ways that would secure food for their families. Hence, any efforts towards improving food security must include the widening of the margins of tenure security on land.

These findings are in accordance with the findings of Huggins [34], Kathiresan [35], Konguka [78], and Kathiresan [79] who managed to show the top-down implementation procedure without negotiations with local people and the pressure of local authorities on local people to participate in order to meet their performance contracts. The chosen priority crops in Nyange Sector are Maize, Irish Potatoes, Beans and Wheat respectively.

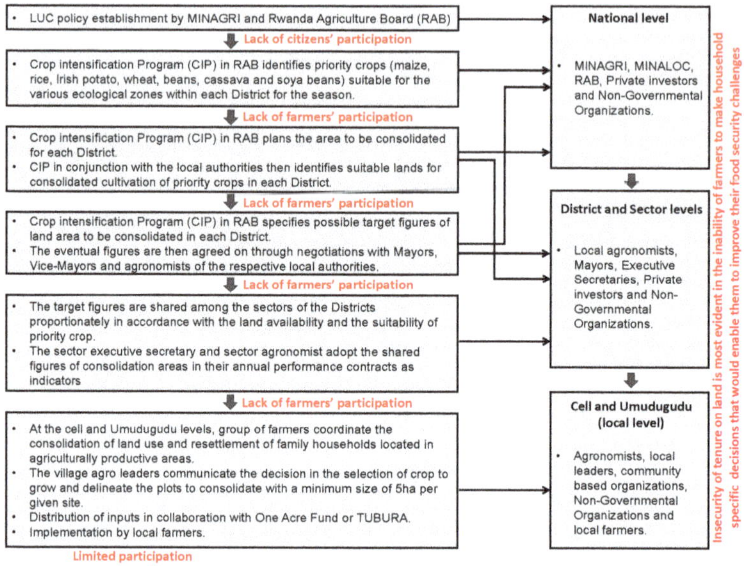

Figure 8. The overall process of the LUC.

Emanating from the process of LUC is the issue of farmers' participation. The information from the household survey (see Figure 9) reveals a limited level of farmer's participation in the decision-making process, particularly in the selection of priority crops.

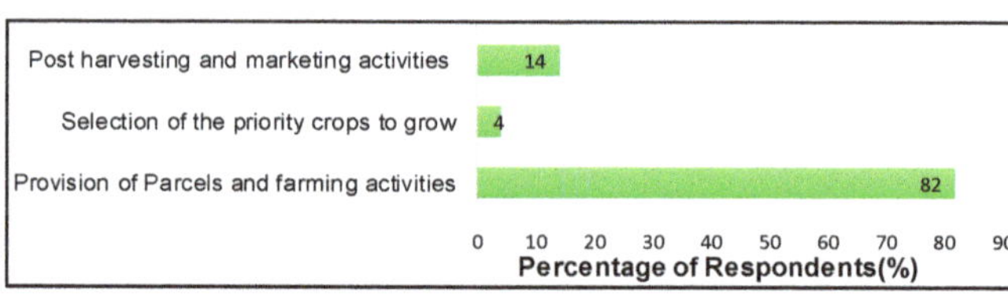

Figure 9. Level of farmers' participating in LUC (based on the household survey).

A majority of respondents declare their participation to be only limited to farming activities of the pre-chosen priority crops. This makes the food security component of the LUC implementation a top-down process (with no farmers' involvement in decision making about the selection of priority crops suitable to their local area). Also, citizens were not involved in the formulation of the project. These imply that some fundamental guiding principles of the LUC are not being followed. These principles include improvement of rural livelihoods, voluntary participation based on democratic and consultative methods as the basis for its bottom-up implementation. These confirm the gap in the LUC implementation, which is why it has not been able to improve food security challenges in Nyange Sector.

5. Making Rwanda's Land Use Consolidation to Generate Food Security: Why Tenure Responsive Land-Use Planning Matters?

The implementation of LUC in Nyange Sector of Rwanda is indicative of why participatory and tenure securing approach to land-use planning is necessary in order to ensure inclusive decision making in land-use and consolidation practices. This is why adopting a tenure responsive land-use planning practice is necessary for making LUC responsive to food security. However, in adopting a tenure responsive land-use planning perspective, it is necessary to ensure that the following critical issues are considered in any operational approach to improving food security:

- Food security focused principles and objectives to be maintained and principles slightly modified: LUC objectives are food security oriented, but the principles seem to focus on improving food quantity at national level, thereby ignoring other aspects like food quality, food accessibility, food sustainability and food utilization as prerequisite of food security at household level. The implication of this is that any approach to improve tenure security must retain existing LUC objectives but improve the principles to make them oriented towards food security at the household level in all its aspects.
- Land ownership: The CIP works through LUC, through the participation of farmers who volunteer to consolidate aspects of their farm operations (while retaining individual ownership of their land) to improve their household food security. This aspect ensures tenure security and has to be retained in any new approach.
- Participation in all food security sensitive aspects of LUC: Although the LUC program is voluntary, participation in the program is quite imposed in many cases. There was a limited role of citizens and farmers because the farmers only had to provide their land parcels and then farm the consolidated plots with no involvement in the aspects of crucial decisions about food security issues. The type of participation observed at the local level has been referred to as "forced and compulsory involvement of local people in the LUC policy done by the local authorities on the pressure to meet their target figures committed in their annual performance contracts with the government" [24]. There is a need to improve participation and inclusiveness in the LUC. Tenure responsive land-use planning provides opportunities for participation and stakeholder engagements in land-related activities.
- Crops selection that is capable of bridging food security gaps: Farmers' commitments to participate in the LUC is on the condition that they abandon their usual intercropping cultivation techniques in favor of cultivating only a single government-approved crop. There is a need to widen the margins of exercise of property rights by farmers to enable them to engage in broader crop selections. Tenure responsive land-use planning enables an adequate understanding of land suitability issues, as well as agricultural decisions related to crops.
- Land tenure decisions that are responsive to food security: Land tenure security is a challenge when farmers commit to LUC conditions because they lose their rights to making household specific food security generating decisions. There is a need to widen the margins of exercising property rights by farmers to enable them to engage in broader decisions based on their specific household food situations and experiences. LUC is based on a generalized Agro-Ecological Zones that do not consider the conditions of local soil (which is usually best known to local farmers). Apart from this current study, many other studies have argued that LUC has had an adverse effect on individual land use rights [8,26,35,36,80,81].
- Household food security focused outcomes: Respondents participating in LUC commonly reported increased crop availability and quantity of the priority crops but confirmed that food insecurity remains a household challenge. This is due to the silos focus of LUC only on boosting food quantity and availability at the national level, ignoring the aspects of quality, accessibility, and sustainability at the household level. Any new approach must embrace the focus on food security improvement in all of its aspects in both principle and practice at the household level.
- External risks caused by natural disaster and market risks: Respondents also reported that their LUC operations face land related risks such as climate change leading to the change in rainfall patterns, floods, drought, and high food prices. Land-use planning is necessary to ensure mitigation against climate change and natural disasters.
- Internal risks caused by knowledge capacity: There is the challenge of poor knowledge capacity on food security issues –such as low awareness of nutrition, food (or foodstuff) combination to

boost balanced nutritional diet. Capacity development is essential for equipping participants with the requisite knowledge on how to deal with market imperfections, nutrition education, and skills necessary for improving and sustaining food security at the household level.

- Market perfections improvements: As LUC has been revealed to be associated with the reduction in agriculture production of priority crops and the increase of their prices on the market, which makes them unaffordable to the poor farmers, and hence their inaccessibility for foodstuffs diversification and nutritional balance purpose, there is a need to create better food market conditions in the new approach. This issue has been raised by many other authors such as Habyarimana and Nkurunziza [84] who have shown how LUC success to meet its food security targets is highly dependent on the level of market perfection as a high market reliant program. Without very well-functioning food market allowing all the participants to afford other foodstuffs besides their production from their priority crops, LUC will never achieve its food security improvement objectives.

Based on the above-listed conditions (which are vital learning points deducible from the case study), this study recommends a renewed LUC process that is framed around land tenure security cum food security improvements. Hence, the need to adopt a tenure responsive land-use planning based approach, which is appropriate for improving the Rwandan situation because it is a process that links land-use, tenure, consolidation issues to food security improvements (Figure 10).

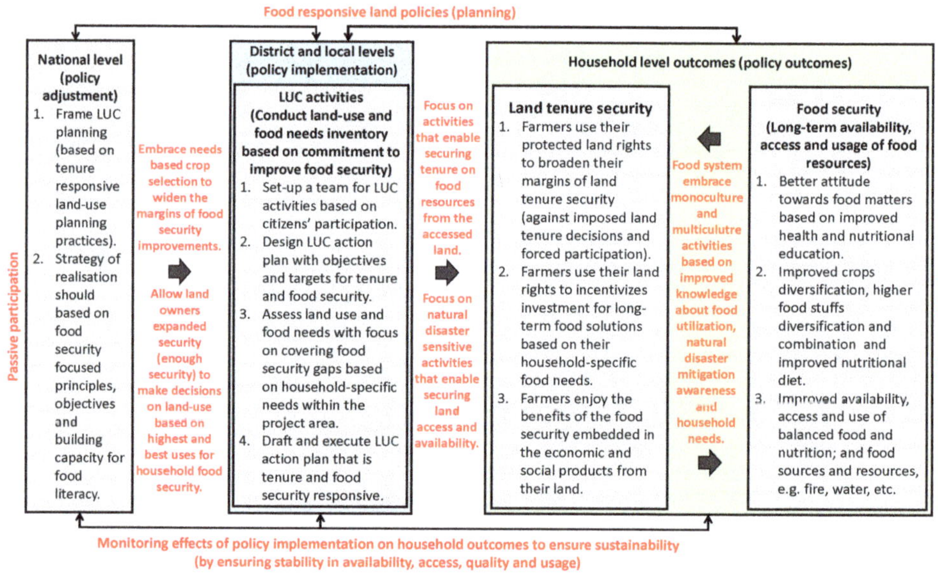

Figure 10. The operational framework for ensuring food security responsive LUC (based on tenure responsive land-use planning).

Any operational approach for LUC to improve food security (through tenure responsive land-use planning), should at the least have four core components of action at national, district and local levels: (1) The aspect of policy (to put in place and enable food responsive land policies that will determine the planning of actions for food security). (2) The aspect of citizens' participation (to ensure that households appropriately undertake food security related decisions). (3) The implementation aspect (to ensure that appropriate activities are carried out with the objective of aligning land-use decisions towards food security objectives). (4) The Monitoring (and evaluation) aspect (to ensure that food security outcomes are sustainable for continuity in improvement). These four aspects are further explained in details.

A starting point for improving food security would be to ensure that the policy aspect is based on land policies that recognize the need to improve food security (food responsive land policies) to ensure that planning induces elements capable of causing food security. Such policies should be made at the national level and translated through the district to the local level to provide the framework for the development of food security in all its aspects of food availability, accessibility, food quantity, and food utilization food quality and food sustainability.

A land policy that is designed to be responsive to land tenure security and food security can provide the implementation pathway for reframing of LUC (based on tenure responsive land-use planning practices). It can provide the strategies and principles, objectives and capacity building measures (for food literacy) for producing food security outcomes. For this to happen, such a policy has to be translated into a district and local level implementation, LUC activities should combine land-use, tenure security, and food security activities by embracing needs-based crop selection necessary for widening the margins of food security improvements at the household level. Policy implementations should also allow landowners to have broader land tenure security (enough security) to make decisions on land-uses (based on highest and best uses) for household food security. With these in place, implementation activities should include conducting land-use, and food needs inventory (with a commitment to improving food security) by fulfilling four actions. First, it will be necessary to set up a team for LUC implementation that is based on citizens' participation. Second, it is vital to design LUC implementation (or action plan) based on objectives and targets for tenure and food security. Third, assess land use and food needs with a focus on covering food security gaps based on household-specific needs within the project area. And, fourth, it is mandatory to draft and execute LUC action plans that are tenure and food security responsive. Focus on activities that enable securing tenure on food resources from the accessed land. It is important to mention the nurturing relationship which land tenure and food security share –improved land tenure security usually results in improved food security, and vice versa. These are possible when the participation of farmers (or landowners) focus on implementing natural disaster sensitive and climate resilient activities that can enable securing land access and availability. The outcome at the household level would be improvements in land tenure security which will enable the farmers to use their land rights to incentivizes investment for long-term food solutions based on their household specific food needs.

Most importantly, it will enable them (farmers) to enjoy the benefits of the food security embedded in the economic and social products from their land. The implication of this is that farmers can practice agricultural land-use activities (whether monoculture and multiculutre), based on their household food priorities, and with improved knowledge about food utilization and natural disaster mitigation. The impact would be long-term availability, access, and usage of food resources (food security). Food security will occur due to a better attitude towards food matters based on improved health and nutritional education; improved crops diversification, higher foodstuffs diversification, and combination; and improved nutritional diet in households.

Food security must be made to have a stable effect on household health and nutrition. This demands making it stable such that it becomes a way of living which can benefit all household members. To ensure this, it is necessary to monitor the effects of policy implementation on household outcomes to ensure sustainability in availability, access, quantity, quality and usage of food.

Operationalizing successful activities for ensuring a food security responsive LUC can only be possible through various kinds of participation ranging from passive to active. While it may not be realistic to expect active participation of rural people in national level strategies, inputs from the local levels should be sought in order to understand their correct food security gaps (passive participation). However, activities at the local level should actively engage farmers so that they are empowered to make critical decisions on how to solve their food security problems and thoroughly enjoy their use rights over their lands. For instance, the government should engage local people in decision making on the choice of crops to grow. More consultative approaches should be applied to make sure the local people are well involved in the procedure after a long capacity building campaign, in order to give them the ability to participate. This requires a bottom-up approach in the implementation process instead of top-down impositions. This means limiting the role of government at the local level to the provision of technical and managerial advice and assistance, with the local people freely using their lands (active participation). Participation is not only one of the critical aspects of the operational procedure for attaining food security, but it is also a precondition for achieving food security. Yet, with more than 90% of all its land resources demarcated and registered in the national cadaster, this can be a very important potential for the success of this new approach to Land Use Consolidation in Rwanda, since a presence of a cadaster is a prerequisite and success factor for land use planning.

6. Conclusions

It is evident that the reason Rwanda implemented the LUC (within its CIP) to ensure food security, but failed to achieve it at the local level in Nyange sector. This study achieved six key objectives. First, identified the general food insecurity scenarios that resulted from the implementation of LUC in Nyange Sector in Rwanda. Second, it identified the gap between LUC principles and their implementation, leading to its poor performance on food security at the household level. Third, it revealed the farmers' perceptions on the issue to show that the outcome of LUC did not support household food security. Fourth, it identified the specific scenarios, and used them to explain how and why LUC failed to ensure food security at the local level in Rwanda. Fifth, it also showed that the LUC process was neither participatory nor tenure security responsive, and so failed to be food security responsive. Sixth (and finally), it provided a renewed perspective for improving the food security situation by adopting a tenure responsive approach to the implementation of LUC at the local level.

It is not uncommon to use the analysis of agriculture policies as a pivotal part of assessing food security. In this study, the LUC (which is the cornerstone of the CIP) is a central part of Rwanda's National Agricultural Policy [85]. The Rwandan National Agriculture Policy recognizes LUC (through CIP) as an instrument for enhancing the "availability, accessibility and optimal use of good quality seeds enhance crop yields and their subsequent contribution to food security, balanced nutrition, value of the product in the market, and economic growth" [85]. This is why this study has focused on the LUC, rather than on the wider agriculture policy of Rwanda. It is also logical to assume that the land policy of Rwanda has a special role in the food security improvement of the country. This study acknowledges that a land policy can set the enabling framework for securing tenure, which has been found to have a direct link to food security. The Rwandan Government [86], in recognition of aspects of its current land policy to ensure full security of tenure on the use of land by its citizens, have initiated a new draft land policy. The draft policy on land is already in place and consultations are underway to come up with a binding document to replace the existing policy which has been in force since 2004 [87]. All of these efforts are indicative that government recognizes that additional efforts are needed to truly make food security a reality.

These policy efforts are necessary as they set up an enabling environment for the operationalization of strategies (or approaches) that can produce food security outcomes. In this regard, this study makes two main contributions to knowledge, each to literature and practice. First, it contributes to planning literature by introducing the concept of tenure responsive land-use planning as a method for food security improvement. By doing this, it demonstrated that tenure responsive land-use planning matters in food security issues because it connects land-use decisions to food security outcomes. Second, it contributes to the how-to aspect (practice) by presenting a framework for operationalizing tenure responsive land-use planning to respond to food security improvements. Furthermore, it confirms previous findings that indicate that Rwanda's LUC is unresponsive to food security [8,26,35,36,80,81]. It also evokes another dimension of land tenure insecurity which has been ignored by other researchers—that is, the capacity for landowners to make critical decisions that affect their food security status. A critical issue emerging from the study is that, the imposition of priority crops on farmers and the denial of farmers to have a say on what crops are planted in their farms or farms adjacent to their farms constitutes tenure insecurity in the context of land-use. Land tenure security can be the legal and social freedom to make decisions on what, where and how to use land (and crops) to achieve household food security objectives. When this kind of freedom is denied, as seen in Rwanda, it affects the production of food crops in accordance to specific household needs. The functionality, efficiency and sustainability of farmland environments, provides farmers with the challenge to innovate and secure food in quantitative and qualitative manner, as well as to sustain production. This is because appropriate land-use and tenure security are capable of inducing food security [1,7,8,10–26,30–37,39–54,78–87].

Finally, the data presented in this study are not peculiar to Rwanda alone. Most countries of the Global South (especially in sub-Saharan Africa) share some of the experiences identified in this study –a situation where food insecurity is exacerbated due to lack of tenure security on agricultural land-use decisions. Apart from the operational framework for improving food security (through tenure responsive land-use planning), there is one central lesson that can be drawn by countries where local communities are food insecure. That lesson is that land tenure security, in the context of food security, goes beyond access and use of land. It entails making household specific decisions on land-uses with the purpose of improving food security situations.

Author Contributions: U.E.C. contributed to this research in the following aspects: conceptualization, methodology, validation, formal analysis, investigation, resources, writing—original draft preparation, writing—review and editing, visualization and supervision. P.D.N. contributed to this research in the following aspects: methodology, validation, formal analysis, investigation, resources, writing—original draft preparation, writing—review and editing, visualization, data curation. E.I.N. contributed to this research in the following aspects: validation, formal analysis, investigation, data curation. W.T.d.V. contributed to this research in the following aspects: writing—review and editing, validation, resources, and supervision.

Funding: This research received no external funding.

Acknowledgments: The authors would like to acknowledge the Editor of this Special Issue, and the anonymous reviewers who contributed to shaping the quality of this article.

Conflicts of Interest: The authors declare no conflict of interest.

References

1. Chigbu, U.E.; Izugbara, C.O.; de Vries, W.T. Land, Culture, Culture Loss and Community: Rural Insights from Sub-Saharan Africa. In *The Routledge Handbook of Community Development: Perspectives from Around the Globe*; Kenny, S., McGrath, B., Phillips, R., Eds.; Routledge: London, UK, 2018; pp. 98–114, ISBN 978-1-31-567410-0.
2. Fensholt, R.; Rasmussen, K.; Kaspersen, P.; Huber, S.; Horion, S.; Swinnen, E. Assessing land degradation/recovery in the African Sahel from long-term earth observation based primary productivity and precipitation relationships. *Remote Sens.* **2013**, *5*, 664–686. [CrossRef]
3. Naidoo, S.; Olaniran, A. Treated wastewater effluent as a source of microbial pollution of surface water resources. *Int. J. Environ. Res. Public Health* **2014**, *11*, 249–270. [CrossRef]
4. Walsh, C.M.; Van Rooyen, F.C. Household food security and hunger in rural and urban communities in the Free State Province, South Africa. *Ecol. Food Nutr.* **2015**, *54*, 118–137. [CrossRef] [PubMed]
5. Orcherton, D.; Mitchell, D.; McEvoy, D. Perceptions of climate vulnerability, tenure security and resettlement priorities: Insights from Lami Town, Fiji Islands. *Aust. Geogr.* **2016**, *48*, 235–254. [CrossRef]
6. Bayisenge, J. Women's Experiences of Land Conflicts in the Context of the Land Tenure Reform Program in Rwanda. *Int. J. Gend. Women's Stud.* **2015**, *3*, 118–133. [CrossRef]
7. Zevenbergen, J.; de Vries, W.; Bennett, R.M. *Advances in Responsible Land Administration*; CRC Press: New York, NY, USA, 2016; ISBN 9781498719599.
8. Ntihinyurwa, P.D.; de Vries, W.T.; Chigbu, U.E.; Dukwiyimpuhwe, P.A. The positive impacts of farm land fragmentation in Rwanda. *Land Use Policy* **2019**, *81*, 565–581. [CrossRef]
9. Tusting, L.S.; Bottomley, C.; Gibson, H.; Kleinschmidt, I.; Tatem, A.J.; Lindsay, S.W.; Gething, P.W. Housing improvements and malaria risk in sub-Saharan Africa: A multi-country analysis of survey data. *PLoS Med.* **2017**, *14*, e1002234. [CrossRef] [PubMed]

10. Chigbu, U.E. Concept and Approach to Land Management Interventions for Rural Development in Africa. In *Geospatial Technologies for Effective Land Governance*; El-Ayachi, M., El Mansouri, L., Eds.; IGI Global: Hershey, PA, USA, 2019; pp. 1–14.
11. Chigbu, U.E. Masculinity, men and patriarchal issues aside: How do women's actions impede women's access to land? Matters arising from a peri-rural community in Nigeria. *Land Use Policy* **2019**, *81*, 39–48. [CrossRef]
12. Chigbu, U.E.; Haub, O.; Mabikke, S.; Antonio, D.; Espinoza, J. *Tenure Responsive Land Use Planning: A Guide for Country Level Implementation*; UN-Habitat: Nairobi, Kenya, 2016. Available online: https://www.researchgate.net/profile/Uchendu_Chigbu/publication/315483297_Tenure_Responsive_Land_Use_Planning_A_Guide_for_Country_Level_Implementation/links/5ba5fd60a6fdccd3cb69efee/Tenure-Responsive-Land-Use-Planning-A-Guide-for-Country-Level-Implementation.pdf?origin=publication_detail (accessed on 15 April 2019).
13. Chigbu, U.E.; Schopf, A.; de Vries, W.T.; Masum, F.; Mabikke, S.; Antonio, D.; Espinoza, J. Combining land-use planning and tenure security: A tenure responsive land-use planning approach for developing countries. *J. Environ. Plan. Manag.* **2017**, *60*, 1622–1639. [CrossRef]
14. UN-Habitat. *Global Campaign for Secure Tenure: A Tool for Advocating the Provision of Adequate Shelter for the Urban Poor*; UN-Habitat: Nairobi, Kenya, 2004.
15. Mitchell, D.; Enemark, S.; van der Molen, P. Climate resilient urban development: Why responsible land governance is important. *Land Use Policy* **2015**, *48*, 190–198. [CrossRef]
16. Sait, M.A.; Chigbu, U.E.; Hamiduddin, I.; de Vries, W.T. Renewable Energy as an Underutilised Resource in Cities: Germany's Energiewende' and Lessons for Post-Brexit Cities in the United Kingdom. *Resources* **2019**, *8*, 7. [CrossRef]
17. Chigbu, U.E.; Alemayehu, Z.; Dachaga, W. Uncovering land tenure insecurities: Tips for tenure responsive land-use planning in Ethiopia. *Dev. Pract.* **2019**, *29*, 371–383. [CrossRef]
18. UN-Habitat. *Secure Land Rights for All*; GLTN/UN-Habitat: Nairobi, Kenya, 2008.
19. Zevenbergen, J.A.; Augustinus, C.; Antonio, D.; Bennett, R.M. Pro-Poor Land Administration: Principles for Recording the Land Rights of the Underrepresented. *Land Use Policy* **2013**, *31*, 595–604. [CrossRef]
20. Uwayezu, E.; de Vries, W.T. Indicators for Measuring Spatial Justice and Land Tenure Security for Poor and Low Income Urban Dwellers. *Land* **2018**, *7*, 84. [CrossRef]
21. Bizimana, C.; Nieuwoudt, W.; Ferrer, S. Farm Size, Land Fragmentation and Economic Efficiency in Southern Rwanda. *Agrekon* **2004**, *43*, 244–262. [CrossRef]
22. Ansoms, A.; Verdoodt, A.; Van Ranst, E. *The Inverse Relationship between Farm Size and Productivity in Rural Rwanda*; University of Antwerp: Antwerp, Belgium, 2008.
23. Brown, M.; Hughes, A.K. Island tenure "secure enough" in rural Rwanda? In Proceedings of the World Bank Conference on Land and Poverty, Washington, DC, USA, 20–24 March 2017. Available online: https://www.chemonics.com/wp-content/uploads/2017/03/Secure-Enough-Land-Tenure_2017.pdf (accessed on 15 April 2019).
24. Ntihinyurwa, P.D. An Evaluation of the Role of Public Participation in Land Use Consolidation (LUC) Practices in Rwanda and Its Improvement. Master's Thesis, Technical University of Munich, München, Germany, 2016.
25. National Institute of Statistics of Rwanda. *Rwanda Poverty Profile Report. 4th Population and Housing Census*; National Institute of Statistics of Rwanda: Kigali, Rwanda, 2015. Available online: www.statistics.gov.rw (accessed on 15 April 2019).
26. Ntihinyurwa, P.D.; Masum, F. Participatory Land Use Consolidation in Rwanda: From Principles to Practice. In Proceedings of the FIG Working Week, Helsinki, Finland, 29 May–2 June 2017. Available online: https://www.fig.net/resources/proceedings/fig_proceedings/fig2017/papers/ts04h/TS04H_ntihinyurwa_masum_9008_abs.pdf (accessed on 15 April 2019).
27. Government of Rwanda. *Urbanization and Rural Settlement Sector Strategic Plan 2013–2018*; Government of Rwanda: Kigali, Rwanda, 2013. Available online: www.rha.gov.rw/fileadmin/user_upload/Documents/Final_Urbanization_Sector_Strategic_Plan_131210_.pdf (accessed on 15 April 2019).
28. Marara, J. *Constraints and Handicaps of Rural Development in Rwanda*; National University of Rwanda: Kigali, Rwanda, 2015. Available online: http://www.ide.go.jp/library/English/Publish/Download/Workshop/pdf/02_04.pdf (accessed on 15 April 2019).

29. Pritchard, M.F. Land, power and peace: Tenure formalization, agricultural reform, and livelihood insecurity in rural Rwanda. *Land Use Policy* **2013**, *30*, 186–196. [CrossRef]
30. Ngenzi, E.I. *The Impacts of Land Use Consolidation on Food Security in Rwanda: A Case Study of Nyange Sector in Musanze District*; Institut d'enseignement Supérieur de Ruhengeri: Ruhengeri, Rwanda, 2017.
31. Government of Rwanda. *Rwanda National Land Use Planning Guidelines*; Government of Rwanda: Kigali, Rwanda, 2017. Available online: http://www.rlma.rw/uploads/media/LUP_Guidelines_Final_Published.pdf (accessed on 15 April 2019).
32. Ngoga, T.H. *Rwanda's Land Tenure Reform: Non-Existent to Best Practice*; CABI: Wallingford, UK, 2017. Available online: https://www.cabi.org/Uploads/CABI/OpenResources/41037/Rwanda%20Land%20Tenure%20Reform%20Ngoga.pdf (accessed on 15 April 2019).
33. Musahara, H.; Birasa, N.; Bizimana, C.; Niyonzima, T. Land use consolidation and poverty reduction in Rwanda. In Proceedings of the World Bank Conference on Land and Poverty, Washington, DC, USA, 24–27 March 2014.
34. Huggins, C. Consolidating land, consolidating control: What future for smallholder farming in Rwanda's 'Green Revolution'? In Proceedings of the International Conference on Global Land Grabbing II, Ithaca, NY, USA, 17–19 October 2012.
35. Kathiresan, A. *Farm Land Use Consolidation in Rwanda: Assessment from the Perspectives of Agriculture Sector*; Ministry of Agriculture and Animal Resources: Kigali, Rwanda, 2012.
36. Musahara, H.; Huggins, C. Land reform, land scarcity and post-conflict reconstruction: A case study of Rwanda. *Eco-Confl. Policy Brief* **2004**, *3*, 1–4.
37. Government of Rwanda. *Rwanda National Food and Nutrition Strategic Plan 2013–2018*; Ministry of Health: Kigali, Rwanda, 2014. Available online: http://www.moh.gov.rw/ (accessed on 15 April 2019).
38. PR-Index. *Prindex Comparative Report*; Prindex/Overseas Development Institute: London, UK, 2018. Available online: https://www.prindex.net/reports/ (accessed on 15 April 2019).
39. Food and Agricultural Organization of United Nations. *Guidelines for Land-Use Planning*; FAO: Rome, Italy, 1993.
40. World Bank. *Doing Business 2010*; World Bank: Washington, DC, USA, 2009.
41. Chigbu, U.E.; Kalashyan, V. Land-Use Planning and Public Administration in Bavaria, Germany: Towards a Public Administration Approach to Land-Use Planning. *Geomat. Land Manag. Landsc.* **2015**, *1*, 7–18. [CrossRef]
42. Chigbu, U.E.; Leitmeier, A.; Masum, F.; Baume, M.; Antonio, D.; Mabikke, S.; Espinoza, E.; Hernig, A. Land Use Planning for Tenure Security: An E-Learning Tool for Developing Countries. In Proceedings of the World Bank Conference on Land and Poverty, Washington, DC, USA, 23–27 March 2015.
43. Chigbu, U.E.; Masum, F.; Leitmeier, A.; Antonio, D.; Mabikke, S.; Espinoza, E.; Hernig, A. Securing tenure through Land Use Planning: Conceptual framework, evidence, and experiences from selected countries in Africa, Asia, and Latin America. In Proceedings of the World Bank Conference on Land and Poverty, Washington, DC, USA., 23–27 March 2015.
44. Chigbu, U.E.; Masum, F.; Schopf, A.; Mabikke, S.; Antonio, D.; Espinoza, J.; Graefen, C. Tenure responsive land use planning: Critical steps for actions in tackling urban poverty in developing countries. In Proceedings of the World Bank Conference on Land and Poverty, Washington, DC, USA, 14–18 March 2016.
45. Food and Agricultural Organization of United Nations. *Trade Reforms and Food Security: Conceptualizing the Linkages*; FAO: Rome, Italy, 2003. Available online: http://www.fao.org/docrep/005/y4671e/y4671e00.htm#Contents (accessed on 15 April 2019).
46. World Food Summit. *Rome Declaration on World Food Security*; FAO: Rome, Italy, 1996.
47. UN Committee on World Food Security. *Assessment of the World Food Security Situation*; FAO: Rome, Italy, 2005.
48. Maxwell, S.; Smith, M. Household food security: A conceptual review. In *Household Food Security: Concepts, Indicators, Measurements: A Technical Review*; Maxwell, S., Frankenberger, T., Eds.; UNICEF: New York, NY, USA; IFAD: Rome, Italy, 1993; pp. 1–72.
49. Ellis, F. The determinants of rural livelihood diversification in developing countries. *J. Agric. Econ.* **2000**, *51*, 289–302. [CrossRef]
50. Pinstrup-Andersen, P. Food security: Definition and measurement. *Food Secur.* **2009**, *1*, 5–7. [CrossRef]
51. Van der Molen, P. Food security, land use and land surveyors. *Surv. Rev* **2017**, *49*, 147–152. [CrossRef]
52. Manjunatha, A.V.; Anik, A.R.; Speelman, S.; Nuppenau, E.A. Impact of land fragmentation, farm size, land ownership and crop diversity on profit and efficiency of irrigated farms in India. *Land Use Policy* **2013**, *31*, 397–405. [CrossRef]

53. Handayani, W.; Rudianto, I.; Setyono, J.S.; Chigbu, U.E.; Sukmawati, A.N. Vulnerability assessment: A comparison of three different city sizes in the coastal area of Central Java, Indonesia. *Adv. Clim. Chang. Res.* **2017**, *8*, 286–296. [CrossRef]
54. Gwaleba, M.J.; Masum, F. Participation of Informal Settlers in Participatory Land Use Planning Project in Pursuit of Tenure Security. *Urban Forum* **2018**, *29*, 169–184. [CrossRef]
55. International Federation of Surveyors. *The Bathurst Declaration on Land Administration for Sustainable Development*; FIG: København, Denmark, 1999.
56. Chigbu, U.E.; Paradza, G.; Dachaga, W. Differentiations in Women's Land Tenure Experiences: Implications for Women's Land Access and Tenure Security in Sub-Saharan Africa. *Land* **2019**, *8*, 22. [CrossRef]
57. Holden, S.T.; Shiferaw, B. Land degradation, drought, and food security in a less-favoured area in the Ethiopian highlands: A bio-economic model with market imperfections. *AgEcon* **2004**, *30*, 31–49. [CrossRef]
58. Alden, W.L. 'The law is to blame': The vulnerable status of common property rights in Sub-Saharan Africa. *Dev. Chang.* **2011**, *42*, 733–757. [CrossRef]
59. Ali, D.A.; Deininger, K.; Goldstein, M. Environmental and gender impacts of land tenure regularization in Africa: Pilot evidence from Rwanda. *J. Dev. Econ.* **2014**, *110*, 262–275. [CrossRef]
60. Holden, S.T.; Ghebru, H. Land tenure reforms, tenure security and food security in poor agrarian economies: Causal linkages and research gaps. *Glob. Food Secur.* **2016**, *10*, 21–28. [CrossRef]
61. Breene, K. Food Security and Why It Matters. Formative Content on World Economic Forum Website. 2016. Available online: https://www.weforum.org/agenda/2016/01/food-security-and-why-it-matters/ (accessed on 15 April 2019).
62. United Nations Organisation. *Transforming Our World: The 2030 Agenda for Sustainable Development*; United Nations: New York, NY, USA, 2015.
63. Prosterman, R. Land tenure, food security and rural development in China. *Development* **2001**, *44*, 79–84. [CrossRef]
64. Field, E.; Torero, M. *Do Property Titles Increase Credit Access among the Urban Poor? Evidence from Peru*; Department of Economics, Harvard University: Boston, MA, USA, 2004.
65. Otsuka, K. Efficiency and equity effects of land markets. In *Handbook of Agricultural Economics 4*; Evenson, R.E., Pingali, P., Eds.; Elsevier: Amsterdam, The Netherlands, 2010; pp. 2671–2703.
66. Wiig, H. Joint titling in rural Peru: Impact on women's participation in household decision-making. *World Dev.* **2013**, *52*, 104–119. [CrossRef]
67. Deininger, K.; Jin, S. Tenure security and land-related investment: Evidence from Ethiopia. *Eur. Econ. Rev.* **2006**, *50*, 1245–1277. [CrossRef]
68. Godfray, H.C.J.; Beddington, J.R.; Crute, I.R.; Haddad, L.; Lawrence, D.; Muir, J.F.; Pretty, J.; Robinson, S.; Thomas, S.M.; Toulmin, C. Food security: The challenge of feeding 9 billion people. *Science* **2010**, *327*, 812–818. [CrossRef] [PubMed]
69. Lambin, E.F.; Meyfroidt, P. Global land use change, economic globalization, and the looming land scarcity. *Proc. Natl. Acad. Sci. USA* **2011**, *108*, 3465–3472. [CrossRef]
70. Holden, S.T.; Otsuka, K. The roles of land tenure reforms and land markets in the context of population growth and land use intensification in Africa. *Food Pol.* **2014**, *48*, 88–97. [CrossRef]
71. Ngoga, T.H. *The Potential for Tenure—Responsive Land Use Planning in Kampala*; International Growth Center Policy Brief: London, UK, 2018. Available online: https://www.theigc.org/wp-content/uploads/2018/02/Ngoga-2018-policy-brief.pdf (accessed on 15 April 2019).
72. Mulenga, C. Securing Tenure through Land Use Planning: Evidence from Lupande Game Management Area, Zambia. Master's Thesis, Technical University of Munich, Munich, Germany, 2015.
73. Gebrie, T.N. Assessment of Land Use Planning to Improve Tenure Security in Squatter Settlement Neighborhoods of Addis Ababa. Master's Thesis, Technical University of Munich, Munich, Germany, 2015.
74. Alemayehu, Z.M. Pro-Poor Land Use Planning: Mapping Stakeholders' Opinions as the Base for Participation in Ethiopia. Master's Thesis, Technical University of Munich, Munich, Germany, 2016.
75. Creswell, J.W. *Research Design: Qualitative, Quantitative, and Mixed Approaches*, 2nd ed.; Sage: Thousand Oaks, CA, USA, 2003.
76. Creswell, J.W. *Research Design: Qualitative, Quantitative, and Mixed Methods Approach*; Sage Publications, Inc.: Thousand Oaks, CA, USA, 2014.

77. Lykken, D.T.; Iacono, W.G.; Haroian, K.; McGue, M.; Bouchard, T.J. Habituation of the skin conductance response to strong stimuli: A twin study. *Psychophysiology* **1988**, *25*, 4–15. [CrossRef] [PubMed]
78. Konguka, G.O. Land Consolidation in Rwanda. In Proceedings of the World Bank Conference on Land and Poverty, Washington, DC, USA, 8–11 April 2013.
79. Kathiresan, A. *Enabling Self Sufficiency and Competitiveness of Rwanda Rice: Issues and Policy Options*; Ministry of Agriculture and Animal Resources: Kigali, Rwanda, 2010.
80. University of Rwanda. *Assessment of the Economic, Social, and Environmental Impacts of the Land Use Consolidation Component of the Crop Intensification Program in Rwanda*; USAID Rwanda LAND Project: Kigali, Rwanda, 2014.
81. Dawson, N.; Martin, A.; Sikor, T. Green Revolution in Sub-Saharan Africa: Implications of Imposed Innovation for the Wellbeing of Rural Smallholders. *World Dev.* **2015**, 204–218. [CrossRef]
82. Government of Rwanda. *Land Tenure Regularisation Programme-Closure Workshop*; Ministry of Environment (MoE), Kigali Convention Centre: Kigali, Rwanda, 2019. Available online: http://www.environment.gov.rw/index.php?id=7&tx_news_pi1%5Bnews%5D=25&tx_news_pi1%5Bcontroller%5D=News&tx_news_pi1%5Baction%5D=detail&cHash=23eb2322b7a771a6eb90886fa02c0733 (accessed on 15 April 2019).
83. Lengoiboni, M.N.; Groenendijk, E.M.; Mukahigiro, A. Land tenure regularization in Rwanda: Registration of land rights for women and its impacts on food security. *J. Land Adm. East. Afr.* **2015**, *3*, 399–414.
84. Habyarimana, J.B.; Nkunzimana, T. Policy Reforms and Rural Livelihoods Sustainability: Challenges and Opportunities—Empirical Evidence from the Adoption of the Land Use Consolidation (LUC) Policy in Rwanda. *Afr. Dev. Rev.* **2017**, *29*, 96–108. [CrossRef]
85. Government of Rwanda. *National Agricultural Policy*; Ministry of Agriculture and Animal Resources: Kigali, Rwanda, 2017. Available online: http://extwprlegs1.fao.org/docs/pdf/rwa174291.pdf (accessed on 15 April 2019).
86. Government of Rwanda. *National Land Policy*; Ministry of Lands, Environment, Forests, Water and Mines: Kigali, Rwanda, 2004. Available online: https://rema.gov.rw/rema_doc/Policies/National_land_policy_english_version_.pdf (accessed on 15 April 2019).
87. Nsabimana, E. New land policy to address outstanding issues in sector. *The New Times* 2018. Available online: https://www.newtimes.co.rw/news/new-land-policy-address-outstanding-issues-sector (accessed on 15 April 2019).

International Journal of
Environmental Research and Public Health

MDPI

Article

The Environmental Impact and Formation of Meals from the Pilot Year of a Las Vegas Convention Food Rescue Program

Samantha To [1,2,*], Courtney Coughenour [2] and Jennifer Pharr [2]

1 Coconino County Public Health Services District, Flagstaff, AZ 86001, USA
2 School of Public Health, University of Nevada Las Vegas School of Public Health, Las Vegas, NV 89154, USA; Courtney.Coughenour@unlv.edu (C.C.); Jennifer.Pharr@unlv.edu (J.P.)
* Correspondence: sto@coconino.az.gov; Tel.: +1-928-679-7290

Received: 26 March 2019; Accepted: 12 May 2019; Published: 16 May 2019

check for updates

Abstract: Annually, millions of tonnes of leftover edible foods are sent to landfill. Not only does this harm the environment by increasing the release of greenhouse gases which contribute to climate change, but it poses a question of ethics given that nearly 16 million households are food insecure in the US, and hundreds of millions of people around the globe. The purpose of this study was to document the amount of food diverted from landfill in the pilot year of a convention food rescue program and to determine the amount of greenhouse gas (GHG) emissions avoided by the diversion of such food. In the pilot year of the convention food rescue program 24,703 kg of food were diverted. It is estimated that 108 metric tonnes of GHG emmisions were avoided as a result, while 45,383 meals for food insecure individuals were produced. These findings have significant implications for public and environmental health, as GHG emissions have a destructive effect on the earth's atmosphere and rescued food can be redistributed to food insecure individuals.

Keywords: food rescue; convention center; greenhouse gas emissions; food security; landfill diversion

---

1. Introduction

'Food waste' is defined as any edible item that goes uneaten at the retail or consumer level [1]. An estimated 31% of all food sold and prepared at these levels is ultimately wasted [1]. Reasons for this waste include over-purchasing by both retailers and consumers, and over-sized portions [2]. This often results in the over preparation of food for meals, which results in food waste [3,4].

Wasted food is one of the largest contributors to landfill in United States (US) and has major environmental and economic impacts. A recent report by the US Environmental Protection Agency (EPA) stated that of the municipal solid waste sent to landfill in 2014, 29 million tonnes, or 21.6% of all waste, was discarded food. This is a concern, as by-products have the potential to contaminate land, water, and air [5]. Greenhouse gases (GHG) released by food decay contribute to climate change, increasing global surface temperatures, and have the potential to effect health and wellbeing around the world [6]. GHG include carbon dioxide (CO_2), methane (CH_4), and nitrous oxide (N_2O), which combined are calculated as CO_2 equivalents (CO_2E) [6]. The continual release of GHG into the atmosphere increases solar energy absorption, and the global temperature cyclically climbs [7], effecting animal and plant biodiversity, reducing freshwater resources, worsening acidification of oceans, and depleting the ozone layer [8]. Impacts on human health include more natural disasters, famine-induced malnutrition, and severe mortality rates during extreme weather events [8]. Positive associations have been reported between climate change and a rising risk of infectious diseases [8], especially vector-borne and water-borne illnesses [9]. Mosquito-borne diseases, cholera, and salmonellosis outbreaks are expected to amplify with increased atmospheric temperatures; even seasonal allergies will surge with extended pollen seasons [9]. With such severe implications to environmental and human health, approaches to prevent GHG emissions and the resultant damage to the atmosphere should be paramount and incorporated into all industries.

The EPA reports that in the US, the decaying of solid waste in landfill accounts for 34% of human-related methane production [10], 16% of which is from uneaten food alone [7]. Tonini and colleagues estimate the global warming impact of avoidable food waste in the United Kingdom (UK) to be between 2000 and 3600 kilograms (kg) CO_2E per tonne of food waste [11], while Salemdeeb estimated that preventing avoidable food waste in the UK could lead to a reduction in GHG by a magnitude of 706 to 896 kg CO_2E per tonne of food waste [12]. It is also estimated that, in the US, 1.4 kg of CO_2E per person per day are produced by food wasted at the consumer and retail levels, excluding food that is nonedible. This amount of CO_2E produced from food waste is equal to the annual emissions of 33 million average passenger vehicles [13].

Additionally, it is estimated that the economic impact of food waste globally is 750 billion (US) dollars per year [14,15], while in the US it is nearly 200 billion dollars [10,13]. The economic cost of food waste varies based on the size and state of development of the country, but some of the estimates are 7.7 billion (US) dollars per year in South Africa [16], 18.3 billion in the UK [17], 21.1 billion in Canada [18], and 5.8 billion in Australia [19]. Preventing the production of CO_2E and the wasting of economic resources by diverting food waste is a controllable way to reduce the impact of global warming and food insecurity [7,20].

Food insecurity is defined as the "limited or uncertain availability of nutritionally adequate and safe foods, or [the] limited or uncertain ability to acquire acceptable foods in socially acceptable ways" [21]. In 2015, it was recorded that 15.8 million households (12.7%) in the US were food insecure [22], a portion of the over 700 million undernourished people globally [15]. At any point of a given year, about one in ten Americans will experience food insecurity, while approximately half of all children will, at some point during their childhood, live in a household that requires food assistance [21,23].

Being food insecure is not always consistent. In some households, the food intake and eating patterns of one or more members is reduced or disrupted at various times of the year due to a lack of money or resources [10]. An investigation into this subject reveals that households are not lacking food on all days of the year, but experience insecurity for around seven months at a time [24].

Annual income and lack of financial management skills can play an important role in the level of food insecurity a household experiences, but other risk factors that have been associated include being at risk of homelessness, a household head who is American Indian, not receiving child support, a noncustodial father that does not regularly visit, seasonality, residing in a state with unemployment rates higher than average, and having at least one cigarette smoker in the home [25]. A report by Feeding America noted that food-insecure families reported altering eating habits in order to afford non-food items, raising concerns about the potential health risks to families with children [26]. Surveys revealed that families placed less importance on paying for food and medication and prioritized paying for rent, water and utility bills, and transportation costs [26]. In one US study of the households surveyed who had cut back on food, 24% did so to afford basic household goods, such as toothpaste, laundry detergent, diapers, or shampoo [26].

A lack of adequate food and poor nutrition have several negative impacts on the physical and mental health of children, adults, and seniors. Food insecure mothers-to-be are more likely to suffer from oral health and mental health problems, as well as increased risk of birth defects. Negative physical health outcomes including exacerbated chronic illnesses, oral health problems, increased risk of asthma, frequent stomachaches, headaches, and iron deficiencies, as well as adverse mental health impacts including behavioral troubles, more visits to a psychologist, and impaired intellectual proficiencies occur in children who experience food insecurity [27]. Adults in food-insecure households are more likely than their food-secure counterparts to experience long-term physical health problems, higher levels of depression, and to have more advanced levels of chronic disease [28]. In senior adults, food-insecurity is associated with higher rates of diabetes, hypertension, heart disease, asthma, depression, functional impairment, limitations to activities of daily living, and a lack of social support [24,29]. These negative health outcomes influence health care costs and hospitalizations, leading to job instability, and a lower quality of life [30,31]. Increasing access to wholesome, nutritious foods using foods that would have been discarded could be an opportunity to positively affect the health and well-being of food-insecure individuals.

Allowing edible food to decompose in landfills should be reconsidered when so many people suffer from hunger and food insecurity every year. An estimated 31% of all the food sold and prepared at retail and consumer levels is wasted annually in the US [1]. Food surplus is routinely generated by food retailers due to the seasonality, variability, and unpredictability of customer demand, however a notable reason for such high rates of restaurant waste is due to large portion sizes. Restaurants entice customers with larger portion sizes to convey that the payer is receiving a bargain; however, this trend creates more waste when the customer cannot finish the meal [3]. Trends of large meals encourage a culture of overconsumption and waste [3]. The EPA food recovery hierarchy prioritizes collecting unspoiled, healthy food to redistribute to hungry people, second only to source reduction [2]. While strides have been made in the US to reduce food waste, it is improbable that this will prevent it entirely [10]. The most frequently promoted solution to such waste focuses on the management of the existing surplus through efficient recovery [32], creating an appropriate setting to incorporate food recovery practices.

Las Vegas, Nevada is a hub for convention gatherings, accommodating over six million visitors per year [33]. Aria, a major resort and convention center owned by MGM Resorts International, realized the potential of its untouched convention food to improve both community wellbeing and environmental health. Hotels and convention centers such as this one take pride in and attract endless clients by providing lavish banquets resources, including quality food. As a result, meals prepared for convention attendees are overestimated to ensure that sufficient portions can be offered to each guest. This overabundance of high-quality food inspired a pilot food rescue program between Aria and a local food bank.

Aria's convention center offers corporations a selection of event spaces that can accommodate over five thousand attendees at a time [34]. To ensure the efficacy and safety of the program, a partnership between the MGM Resorts International, the local health district, and Three Square Food Bank (a member of Feeding America) developed policies and protocols to ensure that when an abundance of food had been prepared for conventioneers, it was suitable for donation. Events that have more than 500 attendees are expected to generate enough surplus meals to initiate the food rescue program process [35]. The surplus food that has not left the kitchen and has maintained a safe temperature in the hot box, in its original tray, never having been served to a guest gets donated [36]. A team from the Three Square Food Bank arrives to the convention center with a refrigerated truck, tests the contents to confirm that the temperature and quality of ingredients are safe, and the leftover foods are then transferred to trays provided by the food bank. The quantity is catalogued, and then transported to a Three Square Food Bank warehouse in hotboxes to maintain appropriate temperature [35]. At the warehouse, the trays are labeled and placed into a blast chiller and stored in freezers [35]. Details of the recovered products (what types of food, how many palettes, date available) are entered into an inventory system, from which partners of the food bank can browse, select, and order food items to be used at their sites in congregate meals. Meals are distributed to clients by a network of charitable, nonprofit organizations [37].

This program is unique because, to the best of our knowledge, food rescue efforts of edible, prepared convention food has not been executed on such a large scale before. The purpose of this study was to document the amount of food in pounds diverted from landfill in the pilot year of the convention food rescue program, estimate the number of meals created from the diverted food, and determine the amount of GHG emissions avoided by the diversion of such food.

2. Materials and Methods

2.1. Rescued Food

This study quantified the amount of food recovered in pounds from Aria's convention center during the pilot year (August 2016 to July 2017). The amount was determined by weighing each tray of food prior to being transported from the Aria and chilled at Three Square Food Bank's warehouse, and then donated to food pantries. Using the United States Department of Agriculture's (USDA) estimate that each meal consists of about one half (0.544) kg of food, the number of meals provided to food-insecure individuals in Las Vegas, NV by agency partners from this rescued food was estimated [38].

2.2. GHG Emissions Avoided

The estimated amount of GHG emissions avoided due to the redirection of food was determined using the EPA's Waste Reduction Model (WARM) [39], a tool provided by the EPA to calculate the change in amount of GHG saved by utilizing alternate scenarios. Prevented GHG emissions are calculated by comparing the emissions associated with the baseline material management (i.e., current practices) with an alternative scenario, as opposed to simply multiplying the quantity of materials managed by an emission factor [39]. The EPA has conducted extensive research to determine "the life-cycle GHS and energy factors for materials across several categories (e.g., plastics, metals, woods [... food waste])". This life cycle varies based on many factors related to the local landfill operations. The WARM program incorporates the conditions under which the specific landfill operates, which are outlined below.

The total weight of rescued food was used as the baseline amount of waste generated, and entered as food waste, tonnes landfilled in step 1 of the WARM program. The alternative management scenario for the current study was source reduction, as the waste would have been diverted from landfill; the total weight of rescued food was again entered as food waste, tonnes source reduced in step 2. Parts 3 through 9b are related to the conditions under which the specific landfill operates. The landfill that the food would have been transported to is Apex Landfill. This landfill is managed by Republic Services, occupies over 2200 acres and is located in Lincoln County, NV [40]. The specific conditions under which this landfill operates are as follows: this region is considered a mountainous region in the WARM program. The food waste is not a product of recycling and is therefore considered virgin. Landfill gas is recovered at this site with typical efficiency and is used for energy [41]. The decay rate of this landfill is most accurately described as moderate and dry [41]. Food waste digestion is not done separately from the landfill cell and therefore is not cured. Lastly, WARM incorporates emissions released during transportation to the waste management facility in an attempt to offset the CO_2E created by the garbage haul process; it is 47.3 km between the Aria convention center and the local landfill. From the calculated CO_2E, proportions of CO_2, CH_4, and N_2O were determined based on percentages published in previous research [6].

3. Results

The amount of convention food rescued was collected by Three Square Food Bank by weighing the amount of food brought to their warehouse to be chilled. From August 2016 through July 2017 the program had yielded 24,703 kg of donated food [38]. Obtaining the exact number of meals created from the rescued food was not possible since agency partners are not required to weigh the distributed portions served at congregate meals. However, using the USDA's estimate that each meal consists of about 0.544 kg of food [42], it is estimated that 45,383 meals were produced from donated convention food in the 12-month time frame.

The pilot year of the convention food rescue program yielded 24,703 kg, or 24.7 metric tonnes, of donated food. This was used as the baseline amount of waste generated and source reduction was entered as the alternative management scenario, as the waste that was diverted from landfill.

The WARM program estimated that by diverting 24,703 kg of food waste from the landfill, the resulting change in GHG emissions were 108 metric tonnes (MT) of CO_2E avoided annually (see Table 1). A GHG equivalencies calculator determined that 108 MT of CO_2E is equivalent to GHG emissions from 22.9 passenger vehicles driven for one year, CO_2 emissions from 46,004 liters gasoline consumed, 16.2 homes' annual electricity use, 250 barrels of oil consumed, or 53,597 kg of coal burned [43] (See Figure 1).

Table 1. WARM outputs and greenhouse gas emission proportions for the pilot year of Aria's convention food rescue program [39].

GHG Emission	Proportions of CO_2E (%)	Amount of GHG Emission ($MTCO_2E$)
Carbon Dioxide	76.7	82.8
Methane	14.37	15.5
Nitrous Oxide	7.9	8.5
Total change in GHG Emissions ($MTCO_2E$):		108

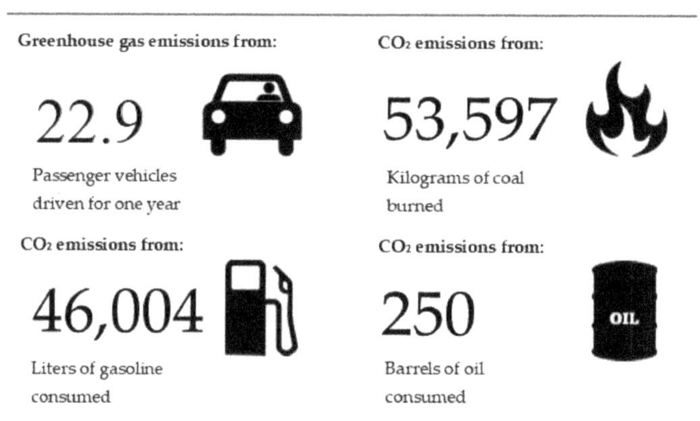

Figure 1. Examples of human related behaviors and actions that would have resulted in Greenhouse Gas Emissions of 108 MT CO_2E [43].

The proportions of CO_2, CH_4, and N_2O are reported to be 76.7%, 14.4%, and 7.9% of CO_2E, respectively [6]. The ratios of carbon dioxide, methane, and nitrous oxide prevented by diverting convention food rescue program food waste from Apex Landfill can be viewed in Table 1.

4. Discussion

Addressing food waste by incorporating a rescue program into a food supply chain is a responsible method of promoting a sustainable food system. Building collaborations between organizations that reduce food waste while increasing access to adequate meals is an important initiative [44]. Our study shows that a convention center food rescue program's pilot year resulted in avoidance of GHG emissions and provided many meals for food insecure individuals.

Food waste is a public health concern due to environmental implications, and various methods to offset waste are considered and executed. Apex Landfill has made state-of-the-art strides towards changing the environmental impact it has on the community. The institution presently collects methane from the landfill and sends the gas to a power plant where it is used to produce enough electricity to power over 9000 homes in the Las Vegas Valley [41]. Composting is not a current option offered by the landfill: however, some food waste that is transported from the Las Vegas Strip is consolidated separately at the landfill and is used to supply feed for hogs at a neighboring pig farm. Matured pigs are available for sale to neighboring cities, and future intentions at the site include the incorporation of hog bedding into composting mounds, from which nutrient-rich soil will feed on-site plant nurseries where fruit and vegetables will be grown and offered to casino restaurants [41]. While the intentions are reasonable, livestock manure is the fourth largest source of methane production, just behind landfills, with an increasing trend of emissions from swine and dairy cow [45]. These efforts by Apex and other regional landfills to reduce the environmental impact are genuine, methane will continue to be produced as a byproduct from the animals. Therefore, while diverting food scraps to animal farms is an option in the EPA's food recovery hierarchy more superior to landfilling and composting, source recovery and feeding hungry people is still most preferred, especially with surplus edible food products. Researchers have found that food rescue was "more economically costly than landfill or composting, [though it] is a lower cost method of obtaining food for the food insecure than direct purchasing" [46]. Additionally, Phillips et al. suggest that the "costs [of food rescue] can be reduced (and supply increased) simply by recruiting additional donors to participate [47]." This is noteworthy for the current study, as many additional convention centers are interested in becoming involved in food rescue efforts.

Redirecting nutritious foods to populations in need may result in positive changes in their health and well-being. Because food costs influence everyday spending, a program that has the potential to impact a household's food security status, such as congregate meal sites, has the potential to impact health and wellbeing [27]. Participating in food assistance programs has been shown to liberate funds that might have been spent on food, reallocating resources to be spent on other costs such as health care, strengthening a positive impact on inhabitants' health, and increasing independence among struggling families [27]. Similarly, the stress caused by the lack of monetary ability to provide food to their family increases maternal stress and may impact mental health status. Researchers explain that causality of food insecurity and poor health conditions may go in both directions; for example, limited nutrition intake in the food-insecure could lead to exacerbation of chronic conditions such as diabetes or HIV, while having diabetes or HIV and the resulting medical costs may trigger someone to be food insecure [48]. Access to no-cost, nutritious food through food assistance programs can relieve some of these stressors. Additionally, improving nutritional intake could break the cycle, eliminating or minimizing the negative health outcomes and alleviating costs of long-term medical care.

Increasing access to nutritious meals have beneficial effects on the health and well-being of individuals of all ages. Living in a food secure household can positively impact on a child's health [27], as well as make a difference in the child's academic performance and social skills [49]. Also, if single mothers are more food secure, they can be more financially secure and independent from damaging, non-custodial partners. The liberation from a negative influence removes numerous other factors that are detrimental to the growth of their children [50].

The majority of food waste in high income countries occurs at the consumption stage, therefore the implementation and expansion of programs similar to Aria's convention food rescue program are necessary and appropriate [51]. Research, such as that done by Raji et al. has shown that limited freeze-thaw cycles have limited nutritional loss [52]. Conversations with convention and food bank stakeholders and compliance to best practices ensure the safety and nutritional quality of the meals that have been chilled, frozen, and reheated once, and should be encouraged elsewhere.

The convention food rescue program is just beginning and not yet performing at full capacity. Current efforts are underway to expand the convention food rescue program and landfill diversion practices to other convention centers at various casino properties. Considering that Las Vegas is one of the leading convention destinations in the US, expansion of convention food rescue in Las Vegas, alone, has the ability to contribute significantly to reductions in GHG. It is predicted that the convention food rescue program will set precedence for other large-scale food establishments and convention facilities, both within and outside of the Las Vegas Valley. Increasing the amount of recovered food would prevent future food waste degradation from releasing tonnes of harmful GHG into the atmosphere. Climate change is a critical, all-encompassing public health issue and approaches to reduce the emissions that harm the atmosphere should be explored.

Convention food rescue programs can be used in conjunction with other food waste prevention strategies and policies to reduce food waste's contribution to GHG emissions. These strategies include (1) food redistribution policies to reattribute edible retail and commercial food to food banks, (2) education of the public and commercial/retail food employees about food waste, portion size, food purchasing and planning, and food preparation, (3) logistical improvements, such as serving plated meals rather than buffet-style [14]. Potential food waste prevention policies could be tax incentives for food donation, limited liability regulations for donors, logistical support for the collection and transportation of edible food, tax incentives for the prevention of food waste, and support for research and development of new innovative initiatives to reduce food waste [14].

Study limitations should take into account that calculations of CO_2E are approximate, and there is likely to be variation in the exact amount of GHG diverted as a result of this program. For example, while emissions resultant from transporting waste between the facility and the landfill are incorporated, potential GHG emissions from transporting surplus food to warehouses, food pantries, or households has not been factored in, nor does the model take into account any GHG emissions that may result from convention center attendees traveling to Las Vegas. Also, the volume of GHG that is generated from a section of landfill will dissipate over time or cease if the site is permanently capped off and waste is no longer added; alternatively, emissions may escalate as the landfill continues to grow [8]. Additionally, the number of meals created from the rescued food is an estimate. All meals served during the pilot project were congregate style, and servings were not measured for exact weights. Although the demand for surplus meals at the food pantries is very high, catering for 500–700 patrons per day, it cannot be guaranteed that each patron entirely consumed their meal [41]. Another consideration is that source reduction is a preferred method of food waste reduction over redistribution. After a scrutinizing examination of the amount of recoverable foods from convention centers, it is predicted that the amount of excess food will be decreased. While this may negatively impact the efforts to feed hungry and food insecure individuals, it is likely that the positive impact on climate change will be enhanced.

This program is a downstream solution to food insecurity. In order for real progress to be made against food insecurity, individuals need education, employment with livable wages, and affordable health care. An investigation by Martin, Colantonio, Picho & Boyle (2016) explains that food-insecure families experience these challenges due to lack of confidence and inability to be self-sufficient [53]. Large and small organizations alike could do more to tackle food insecurity by helping secure at-risk populations in long-term employment, offering financial literacy programs to the public, or being involved in policy change that will encourage a decrease in food insecurity. Employers should also ensure that their employees are paid sufficient wages and provided access to healthcare.

Studies investigating the impact of the provision of prepared meals from recovered convention foods are limited. Aria's convention food rescue program is unique in that it is the first of its kind at such a large scale and has the intentions to grow even larger. Alternatives like this program that better utilize recoverable food and divert food waste from landfills are important to human and environmental health.

5. Conclusions

Study findings indicated that roughly 45,383 meals were provided using food collected by this convention food rescue program between August 2016 and July 2017. According to a Hunger in America Client Survey, about 428,900 visits to the Three Square Food Bank's meal program occur annually [51]. This means that meals produced from the convention food rescue program during this time period was enough to supply 10.6% of the meals provided by the congregate meal kitchens in Southern Nevada. This program is still in its beginning stages, and food rescue is only taking place at one convention center. As the program develops and becomes more efficient and more convention centers join the rescue efforts, the number of meals assembled and mouths fed could increase substantially given the number of conventions held annually in Las Vegas. The successful model for this rescue program should be mimicked in other similarly positioned cities around the world.

There is an obligation to prevent the disposal of usable, edible food when many people are suffering from hunger and food insecurity. An increase in access to low- or no-cost meals for populations in need can decrease the harmful effects that being food-insecure can inflict. More meal availability can be predicted to decrease behavioral problems and increase school performance in children, decrease severity of chronic diseases in the elderly and people with disabilities, and reduce rates of stress and depression in adults. Also, as mentioned before, not only are mental and physical health outcomes effected by limited food, but financial, housing, hygiene, and health care needs are more likely to be met when food limitations are alleviated, all the more intensifying the benefits of diminishing food insecurity.

Diverting food from landfills prevents the decomposing food waste from releasing GHG into the environment. Our findings revealed that 108 MT of CO_2E were avoided by diverting recoverable food from the convention food rescue program in the pilot year away from the landfill. Findings are significant, as this food rescue program only included food from one convention center. As the convention food rescue program expands to include multiple convention properties, the prevention of GHG emissions will be amplified, and so too will the potential positive consequences on public and environmental health.

Author Contributions: Conceptualization, C.C. and S.T.; methodology, S.T. and C.C.; formal analysis, S.T.; investigation, S.T.; writing—original draft preparation, S.T.; writing—review and editing, S.T., C.C., and J.P.; visualization, S.T.; project administration, S.T. and C.C.

Funding: This research received no external funding.

Acknowledgments: The publication fees for this article were supported by the UNLV University Libraries Open Article Fund.

Conflicts of Interest: The authors declare no conflict of interest.

References

1. Blondin, S.A.; Carmichael Djang, H.; Metayer, N.; Anzman-Frasca, S.; Economos, C.D. 'It's just so much waste.' A qualitative investigation of food waste in a universal free school breakfast program. *Public Health Nutr.* **2014**, *18*, 1565–1577. [CrossRef] [PubMed]
2. Vogliano, C.; Brown, K. The state of America's wasted food and opportunities to make a difference. *J. Acad. Nutr. Diet.* **2016**, *116*, 1199–1207. [CrossRef] [PubMed]
3. Lipinski, B.; Hanson, C.; Lomax, J.; Kitinoja., L.; Waite, R.; Searchinger, T. *Food Waste Footprint: Impacts on Natural Resources*; Installment 2 of Creating a Sustainable Food Future; Food and Agriculture Organization of the United Nations: Rome, Italy, 2013.
4. Oakdene, H. Responsible Hospitality Partnership, and the Waste & Resources Action Programme. In *Overview of Waste in the UK Hospitality and Food Service Sector*; WRAP (Waste & Resources Action Programme): Banbury, UK, 2013.
5. Environmental Protection Agency. *Advancing Sustainable Materials Management: 2014 Fact Sheet*; Environmental Protection Agency: Washington, DC, USA, 2016.
6. Zuberi, M.J.; Ali, S. Greenhouse effect reduction by recovering energy from waste landfills in Pakistan. *Renew. Sustain. Energy Rev.* **2015**, *44*, 117–131. [CrossRef]
7. Lee, S.; Kim, J.; Chong, W. The causes of the municipal solid waste and the greenhouse gas emission from the waste sector in the United States. *Waste Manag.* **2016**, *56*, 593–599. [CrossRef]
8. Rossati, A. Global warming and its health impact. *Int. J. Occup. Environ. Med.* **2017**, *8*, 7–20. [CrossRef]
9. McMichael, A.; Woodfurr, R.; Hales, S. Climate change and human health: Present and future risks. *Lancet* **2006**, *367*, 11–17. [CrossRef]

10. Buzby, J.C.; Wells, H.F.; Hyman, J. *The Estimated Amount, Value, and Calories of Postharvest Food Losses at the Retail and Consumer Levels in the United States*; EIB-121; US Department of Agriculture, Economic Research Service: Washington, DC, USA, 2014.
11. Tonini, D.; Albizzati, P.F.; Astrup, T.F. Environmental impacts of food waste: Learnings and challenges from a case study on UK. *Waste Manag.* **2018**, *76*, 744–766. [CrossRef]
12. Salemdeeb, R.; Vivanco, D.F.; Al-Tabbaa, A.; Ermgassen, E.K. A holistic approach to the environmental evaluation of food waste prevention. *Waste Manag.* **2017**, *59*, 442–450. [CrossRef]
13. Venkat, K. The climate change and economic impacts of food waste in the United States. *Int. J. Food Syst. Dyn.* **2011**, *2*, 431–446.
14. Thyberg, K.L.; Tonjes, D.J. Drivers of food waste and their implications for sustainable policy development. *Resour. Conserv. Recycl.* **2016**, *106*, 110–123. [CrossRef]
15. Food and Agriculture Organizations of the United Nations. *The State of Food Insecurity in the World*; FAO: Rome, Italy, 2015; Volume 2, p. 15.
16. Nahman, A.; de Lange, W. Costs of food waste along the value chain: Evidence from South Africa. *Waste Manag.* **2013**, *33*, 2493–2500. [CrossRef]
17. Quested, T.; Ingle, R.; Parry, A. *Executive Summary: Household Food and Drink Waste in the United Kingdom*; Waste and Resources Action Programme: Banbury, UK, 2007.
18. Gooch, M.; Felfel, A.; Marenick, N. *Food Waste in Canada*; Value Chain Management Centre: Oakville, ON, Canada; George Morris Centre: Guelph, ON, Canada, November 2010.
19. Facts on Food Waste. Food Wise. 2015. Available online: http://www.foodwise.com.au/foodwaste/food-waste-fast-facts/ (accessed on 23 October 2015).
20. Block, D. Reducing greenhouse gases at landfills. *BioCycle* **2000**, *41*, 40–46.
21. Berraca, M.; Callahan, K. A fulfilling resolution. *Casino J.* **2006**, *19*, 35.
22. Coleman-Jensen, A.; Rabbitt, M.; Gregory, C.; Singh, A. *Household Food Security in the United States in 2015*; U.S. Department of Agriculture, Economic Research Service: Washington, DC, USA, 2016.
23. Rank, M.; Hirschl, T. Estimating the risk of food stamp use and impoverishment during childhood. *Arch. Pediatr. Adolesc. Med.* **2009**, *163*, 994–999. [CrossRef]
24. Gundersen, C.; Ziliak, J. Food insecurity and health outcomes. *Health Aff.* **2015**, *34*, 1830–1839. [CrossRef] [PubMed]
25. Gundersen, C.; Engelhard, E.; Waxman, E. Map the meal gap: Exploring food insecurity at the local level. *Appl. Econ. Perspect. Policy* **2014**, *36*, 373–386. [CrossRef]
26. Waxman, E.; Santos, R.; Daley, K.; Flese, B.; Koester, B.; Knowles, E. *In Short Supply: American Families Struggle to Secure Everyday Essential*; Executive Summary; Feeding America: Chicago, IL, USA, 2014.
27. Gundersen, C.; Kreider, C. Bounding the effects of food insecurity on children's health outcomes. *J. Health Econ.* **2009**, *28*, 971–983. [CrossRef]
28. Gundersen, C.; Kreider, B.; Pepper, J. The economics of food insecurity in the United States. *Appl. Econ. Perspect. Policy* **2011**, *33*, 281–303. [CrossRef]
29. Afulani, P.; Herman, D.; Coleman-Jensen, A.; Harrison, G. Food insecurity and health outcomes among older adults: The role of cost-related medication underuse. *J. Nutr. Gerontol. Geriatr.* **2015**, *34*, 319–342. [CrossRef]
30. Laraia, B.; Siega-Riz, A.; Gundersen, C.; Dole, N. Psychosocial factors and socioeconomic indicators are associated with household food insecurity among pregnant women. *J. Nutr.* **2006**, *136*, 177–182. [CrossRef]
31. Brucker, D. The association of food insecurity with health outcomes for adults with disabilities. *Disabil. Health J.* **2017**, *10*, 286–293. [CrossRef]
32. Mourad, M. Recycling, recovering and preventing "food waste": Competing solutions for food systems sustainability in the United States and France. *J. Clean. Prod.* **2015**, *126*, 461–477. [CrossRef]
33. 2016 Las Vegas Year-To-Date Executive Summary. 2016. Available online: http://epubs.nsla.nv.gov/statepubs/epubs/651944-2016.pdf (accessed on 4 September 2017).
34. Meeting Spaces. Aria. Available online: https://www.aria.com/en/meetings-groups/meeting-space.html (accessed on 5 August 2017).
35. Jesnik, A.; (Three Square Food Bank, Las Vegas, NV, USA). Personal communication, 2017.
36. Johnson, M.F.; (Three Square Food Bank, Las Vegas, NV, USA). Personal communication, 2017.
37. Schmit, M.; (Vice President of Food Service at Catholic Charities, Las Vegas, Nevada, USA). Personal communication, 2017.

38. U.S. Department of Agriculture, Agricultural Research Service. *Nutrient Intakes from Food and Beverages: Mean Amounts Consumed per Individual, by Gender and Age, What We Eat in America, NHANES 2011–2012*; U.S. Department of Agriculture, Agricultural Research Service: Washington, DC, USA, 2014.
39. Waste Reduction Model (WARM): Environmental Protection Agency. Available online: https://www.epa.gov/warm (accessed on 13 October 2017).
40. Carter, G. Over the hills to the promised landfills. *Vegas Seven* **2015**, *271*, 20–21.
41. Clinker, M.; (Republic Services, Las Vegas, Nevada, USA). Personal communication, 2017.
42. Whaley, R.; (Three Square Food Bank, Las Vegas, NV, USA). Personal communication, 2017.
43. Greenhouse Gas Equivalencies Calculator. Energy and the Environment. Environmental Protection Agency. Available online: https://www.epa.gov/energy/greenhouse-gas-equivalencies-calculator (accessed on 20 October 2017).
44. Food and Agriculture Organizations of the United Nations. Global Initiative on Food Loss and Waste Reduction. Available online: http://www.fao.org/3/a-i4068e.pdf (accessed on 22 March 2018).
45. Environmental Protection Agency. Inventory of U.S. Greenhouse Gas Emissions and Sinks: 1990–2015 (Full Report). 2017. Available online: http://www3.epa.gov/climatechange/emissions/usinventoryreport.html (accessed on 2 September 2017).
46. Reynolds, C.J.; Piantadosi, J.; Boland, J. Rescuing Food from the Organics Waste Stream to Feed the Food Insecure: An Economic and Environmental Assessment of Australian Food Rescue Operations Using Environmentally Extended Waste Input-Output Analysis. *Sustainability* **2015**, *7*, 4707–4726. [CrossRef]
47. Phillips, C.; Hoenigman, R.; Higbeem, B.; Reed, T. Understanding the Sustainability of Retail Food Recovery. *PLoS ONE* **2013**, *8*, e75530. [CrossRef]
48. Whittle, H.; Palar, K.; Seligman, H.; Napoles, T.; Frongillo, E.; Weiser, S. How food insecurity contributes to poor HIV health outcomes: Qualitative evidence from the San Francisco Bay Area. *Soc. Sci. Med.* **2016**, *170*, 228–236. [CrossRef]
49. Jyoti, D.; Frongillo, E.; Jones, S. Food Insecurity Affects School Children's Academic Performance, Weight Gain, and Social Skills. *J. Nutr.* **2005**, *135*, 2831–2839. [CrossRef]
50. Hernandez, D. The impact of cumulative family risks on various levels of food insecurity. *Soc. Sci. Res.* **2015**, *50*, 292–302. [CrossRef]
51. Mills, G.; Weinfield, N.; Borger, C.; Gearing, M.; Macaluso, T.; Mendonca, J.; Zedlewski, S. *Hunger in America 2014: Food Bank Report Prepared for Feeding America*; Feeding America: Chicago, IL, USA, 2014.
52. Raji, A.O.; Akinoso, R.; Raji, M.O. Effect of Freeze-thaw cycles on the nutritional quality of some selected Nigerian Soups. *Food Sci. Nutr.* **2015**, *4*, 163–180. [CrossRef]
53. Martin, K.; Colantonio, A.; Picho, K.; Boyle, K. Self-efficacy is associated with increased food security in novel food pantry program. *Popul. Health* **2016**, *2*, 62–67. [CrossRef]

International Journal of
Environmental Research and Public Health

MDPI

Article

Impact of Organic Manure on Growth, Nutrient Content and Yield of Chilli Pepper under Various Temperature Environments

Botir Khaitov [1], Hye Jin Yun [1], Yejin Lee [2], Farrukh Ruziev [1], Thi Hien Le [1], Mirjalol Umurzokov [1], Aung Bo Bo [1], Kwang Min Cho [1] and Kee Woong Park [1,*]

[1] Crop Science Department, Chungnam National University, Daejeon 34134, Korea
[2] Soil Management Division, NIAST, Rural Development Agency, Wanju 55365, Korea
* Correspondence: parkkw@cnu.ac.kr; Tel.: +82-42-821-7823

Received: 18 June 2019; Accepted: 17 August 2019; Published: 21 August 2019

Abstract: Expected climatic changes likely elicit serious challenges for crop production. Therefore, it is indispensable to investigate the response of crop growth parameters and yield under temperature variability environments. The current experiment on chilli pepper growth was conducted in a field, rain-shelter plastic house, and plastic greenhouse, with accumulated temperatures of 2832 °C, 2967 °C, and 3105 °C in 2017; and 2944 °C, 3091 °C, and 3168 °C in 2018 growing seasons. Based on soil analysis, 132.7 kg ha^{-1} (1× of livestock manure compost as an optimum and 265.4 kg ha^{-1} (2×) as a double amount of organic matter were applied to each simulated temperature condition. The results showed that organic manure application favorably affects the growth attributes and nutrient uptake of chilli pepper with the highest values found in the plastic greenhouse, followed by the rain-shelter house, over the open field cultivation condition. The highest growth of chilli pepper was at the 2× rate of organic manure application, whereas the highest yield was found at the 1× rate of organic manure application. The application of organic manure at the 1× rate in the greenhouse increased root, shoot, and fruit dry weights of chilli pepper by 21.4%, 52.4%, and 79.7%, respectively, compared to the control values. These results indicate that the rational use of organic amendments might be the best solution for chilli pepper production under variable climate conditions.

Keywords: chilli pepper; organic manure; application rate; elevated temperature; rain-shelter plastic house; greenhouse; field

1. Introduction

Global warming poses a great threat to crop production all around the world. The increase in atmospheric temperature in response to global climate change might have serious consequences for crop production and be dangerous for food security [1]. Furthermore, continuously increasing populations in the world put pressure on agriculture to produce more crops. The consequences of climate change on crop production, cropping system, and soil fertility are expected to be huge but uncertain [2]. According to scientific predictions, climate change will be responsible for increase in temperature and precipitation up to 1.0–5.0 °C and 9–11%, respectively, depending on regions by the end of this century [3]. This will very likely bring a negative impact to crop production [4,5]. While, simultaneous rise in atmospheric CO_2 and temperature are already showing some impact on nature, in the meantime, it is prudent to study the performance of crop species under simulated high temperature conditions.

The effectiveness of organic amendments is also theorized to change along with climate change [6]. Application of organic manure increases organic elements' availability in soil, thereby improving the nutrient use efficiency (NUE) of crops and alleviating the harmful impact of climate change on crop production [7]. However, the application rate of chemical fertilizers has soared extensively during the last decades to enhance crop yield in order to meet the increasing demand of populations. The consequences are already visible on serious ecological disturbances, especially in pollution of soil and water resources. In recent years, the scientific community has been informing the advantages of implementing organic farming in agricultural production to ensure sustainable nutrient management for crops, food safety, and soil health [8]. This strategy improves NUE, while maintaining high productivity of crops, soil quality, and agricultural sustainability.

Soil is one of the most important environmental factors with indispensable significance in plant physiology [9]. Thus, maintaining soil quality is of great importance for crop growth and enhancing productivity. Organic manure is a natural source of nitrogen (N) formation process in the soil. Livestock manure returns essential macroelements including N (2.42%), P (1.51%), and K (0.41%), as well as micronutrients such as magnesium, calcium, sulfur, and manganese to the soil while maintaining its fertility. Several researchers have reported that nutrients available for plant uptake increase rapidly in the experimental soil after organic manure application, although bacterial populations were different in various soils [10,11]. Microorganisms living in the soil are important for decomposing, mineralizing, and recycling organic matters [12]. Microbial populations intensively induce the production of phytohormones such as gibberellin and auxin in plant roots grown in fertile soil with rich organic manures which stimulate plant growth [13].

Chilli pepper (*Capsicum annuum*), the most known species of the plant genus Capsicum (peppers), originated in Mexico, the southern part of North America. Nowadays, this crop is commonly cultivated in all continents of the world and mainly used for food and pharmaceutical uses. Superior levels of bioactive and antioxidant compounds such as carotenoids, phenolic compounds, and ascorbic acid are contained in fresh chilli pepper fruits [14]. However, the level of these components in the fruit depends on genetic and environmental factors. If chilli pepper is grown for a food ingredient, its quality and market value are determined by the level of fruit pungency [15]. Some essential components are enhanced in organically cultivated crops, for instance, quercetin (flavonoid) in spinach, Chinese cabbage, and Welsh onion [16]; phenol and flavonoid in *Labisia pumila* [17]; protein and carotene in *Capsicum chinense* [18]; and capsaicin content in pepper [11].

In Korea, the cultivation area of chili peppers is ~30 thousand hectares, and the fruit production reached ~72 thousand tones in 2018, increasing by 16 thousand tones (28.4%) from 56 thousand tones in 2017 [19]. It is well-known that climate change causes modifications in crop physiology, NUE efficacy, yield, and other parameters [20]. Therefore, it is important to elucidate the impact of climate change on crop productivity for sustainable agricultural production.

Our goal in this study was to evaluate the effect of 1× rate (132.7 kg ha^{-1}) and 2× rate (265.4 kg ha^{-1}) livestock manure treatments compared to control (without manure) on the growth and yield parameters of chilli peppers grown under three different environmental conditions such as a plastic greenhouse, rain-shelter plastic house, and open field conditions. The objective of this research was to study how chilli peppers react to the different rates of organic manure application under expected climate change scenarios. Furthermore, we were curious to understand how plant physiology parameters such as chlorophyll content, plant height, shoot, and root dry weights change in response to the applied organic manure under various temperature environments.

2. Material and Methods

2.1. Experimental Conditions and Plant Material

The climate of South Korea with four seasons allows the cultivation of many crop species. The average annual air temperature ranges between −5 °C to −2.5 °C in January and 22.5 °C to 25 °C in July. The annual precipitation is ~1380–1400 mm, the main part of it falls between November and May, with an average humidity of 66%.

A two-year experiment, during 2017 and 2018 growing seasons, was conducted in the field, rain-shelter plastic house, and plastic greenhouse conditions simulating three different temperature environments. Climate parameters in the plastic greenhouse and rain-shelter plastic house were recorded with an Em50® series data logger (METER Group, Inc., Pullman, WA, USA). During the growing seasons, the highest air temperature was observed in the greenhouse, followed by rain-shelter plastic house compared to the field. However, the highest amount of rainfall precipitated in the field experiment. The accumulated temperatures for 130 days were 2832 °C, 2967 °C, and 3105 °C in 2017; and 2944 °C, 3091 °C, and 3168 °C in 2018 in plastic greenhouse, rain-shelter plastic house, and open field conditions, respectively. The detail climatic records for both growing season are presented in Table 1.

Table 1. Weather data on air temperature, rainfall, and relative humidity of the study area, Chungnam province (2017 to 2018 growing seasons' data).

Year	Month of the Year					
	May	**Jun.**	**Jul.**	**Aug.**	**Sep.**	**Oct.**
			Air Temperature (°C)			
2017 Field	22.2	26.4	24.6	24.9	19.5	13.3
Rain-shelter house	22.4	23.7	27.8	26.7	21.8	14.1
Greenhouse	23.3	24.6	28.3	27.1	22.4	14.8
2018 Field	18.8	20.8	25.1	25.1	18.5	12
Rain-shelter house	19.3	21.6	26.6	26.4	19.5	13.1
Greenhouse	20.1	22.4	27.3	26.7	19.7	13.9
			Rainfall (mm)			
2017	35.8	51.4	326.7	358.4	97.1	51.9
2018	90.4	225.5	100	437.1	91.3	90.2
			Relative Humidity (%)			
2017 Field	93.4	94.5	94.9	95.1	95.6	95.7
Rain-shelter house	92.9	93.6	93.8	93.9	94	94.2
Greenhouse	94	94.1	94.6	94.7	95.1	95.2
2018 Field	95.8	95.7	95.8	95.7	96.1	96.5
Rain-shelter house	94.7	94.5	94.6	94.5	94.9	95.3
Greenhouse	95.1	94.9	95.1	95	95.4	95.8
			Soil Moisture			
2017 Field	0.28	0.242	0.221	0.234	0.231	0.224
Rain-shelter house	0.231	0.179	0.166	0.181	0.174	0.166
Greenhouse	0.187	0.193	0.185	0.171	0.185	0.201
2018 Field	0.26	0.264	0.28	0.278	0.279	0.285
Rain-shelter house	0.201	0.174	0.18	0.201	0.196	0.193
Greenhouse	0.249	0.229	0.218	0.233	0.221	0.217

Source: Meteorological Station of Chungnam province, Korea.

Seeds of chilli pepper (*Capsicum annuum* L.) variety—Keunsarang (Nongwoo Bio, Sunwon, Korea)—were obtained from Crop Physiology department, Chungnam National University, Korea. Prior to starting the experiment, the seeds were sorted: broken seeds were discarded and good quality seeds were chosen for further use in the experiment. The seeds' surface was sterilized with a 0.2% Hg_2Cl solution for 3 min, rinsed in distilled water thoroughly, and left to dry at room temperature (25 °C). Then, the seeds were sown directly into 2 cm depth soil in pots (30 cm in diameter). Each pot was filled with 4 kg of soil (55% clay, 20% silt, and 25% sand) and the experiment was commenced in the greenhouse of Crop Physiology Department, Chungnam National University. The seeds germinate under favorable environmental conditions in the 2nd week of April each year (day length 15–16 h, temperature 25–30 °C, and humidity 65–70%). When the germination was complete after 7 days, the percentage of seed germination was calculated and the seedling density was thinned to four seedlings per pot.

The seedlings of chilli pepper were transplanted into a plastic greenhouse, rain-shelter plastic house, and open field at the experimental station located in Unbong-eup, Namwon-si, Jeollabuk-do, Republic of Korea. The experiment was set in randomized complete block design with three replicated plots (15 m × 4 m) for each treatment.

The chemical analyses of organic manure were conducted to estimate N, P_2O_5, and K_2O contents before application. The organic manure consisted of 67.06% organic matter, 2.42% N, 1.51% P_2O_5, and 0.41% K_2O. Two different levels of organic matter were applied to each temperature treatments as per experimental layout. Based on a soil analysis, 132.7 kg ha^{-1} (1×) of livestock manure compost as an optimum and 265.4 kg ha^{-1} (2×) was applied as the double rate of organic manure.

Ammonium nitrate (34%), fused phosphate (17%), and muriate of potash (52%) were applied as inorganic sources of N, P, and K. Recommended doses of the chemical fertilizers N, P (P_2O_5), and K (K_2O) were 80 : 50 : 40 kg ha^{-1}, and this amount was divided into three equal portions. The first part was applied as a basal dose and the two remaining portions were given during the vegetation period. All other agronomic operations such as plant protection, weeding, and irrigation measures were conducted similarly in all plots.

2.2. Soil Analyses

The soil in the experimental area belongs to Podzolic (ash-gray forest soil), slightly acidic with pH 5.9–6.6 and EC 1.3–2.4 dS/m. It consists of clay 540–585 g kg^{-1}, silt 260–291 g kg^{-1}, organic matter 18.7–25.7 g kg^{-1}, and sand 150–172 g kg^{-1}. Soil chemical analyses indicated that NO_3–N, 52.9 mg/kg; NH_4^+–N, 13.2 mg/kg; P_2O_5, 671.4 mg/kg; K, 1.2 cmol/kg; Ca, 8.5 cmol/kg; Mg, 2.6 cmol/kg; and Na, 0.3 cmol/kg.

Soil samples were randomly collected at 0–30 cm depth in sealable plastic bags from each replicated plot every month during the experiment. Air-dried soil samples at room temperature were ground and sieved through a 2-mm mesh before chemical analysis.

Soil and plant samples were analyzed according to the methods developed by the National Institute of Agricultural Science and Technology (NIAST, 2000). Soil pH and EC parameters were measured with pH and EC meters using 1 : 5 ratio of soil and distilled water. The organic matter was extracted using the Tyurin method, available phosphate was extracted using the Lancaster method [21], and substitutional cation was extracted using 1M NH_4OAC (pH 7.0) and analyzed with a coupled plasma spectrophotometer (Integra XL, GBC, Toronto, Canada).

Plant materials were dried and pulverized, and then 1 mL of concentrated sulfuric acid and 10 mL of 50% perchloric acid were added into 0.5 g of the sample, followed by decomposition by heating on a hot plate. Total N, P_2O_5, and K_2O were analyzed with Kjeldahl distillation, Vanadate method, and inductively coupled plasma spectrophotometer, respectively [22]. Furthermore, the ECH2O 5TE Sensor (Decagon devices, Pullman, WA, USA) was deployed for temperature and moisture analyses. Ambient temperature, humidity, and radiation were determined by ATMOS14 (Decagon devices, USA).

2.3. Plant Sampling, Fruit Yield, and Yield Attributing Features of Chilli Peppers

Investigations of plant growth parameters were conducted at 30, 50, 70, 100, and 130 days after transplanting (DAT). Harvesting chilli fruits began two weeks after the beginning of the mature stage, around 95 DAT.

The parameters of plant height, dry weight of shoot and root, length of internode, chlorophyll contents of leaf, stem diameter, and number of fruit branches were measured in three replicates. The chilli pepper fruit was harvested every two weeks from each treatment. The number of fruits, fruit weight, and color were recorded until the final harvest. Fruit samples were stored in a −4 °C refrigerator until used for further chemical analyses. Fruit quality attributing parameters, including crude protein and total N, were determined by the modified Kjeldahl method [23]. The yield was calculated by estimating the number of fruits harvested for 130 days.

2.4. Statistical Analysis

The effects of organic manure application on growth attributes, nutrient content, and chilli pepper yield were determined under three different temperature conditions in two growing seasons (2017 and 2018), and were subjected to statistical significance using the ANOVA CROPSTAT program. No significant differences in any traits were observed between the two experimental years, and data was pooled before statistical analysis. Student's t-test was used for comparative analyses between treatments. The mean comparisons were conducted using a least significant difference (LSD) test ($p \leq 0.05$).

3. Results

3.1. Effects of Organic Manure and Temperature Increase on the Growth of Chilli Peppers

The application of organic manure had a significant positive effect on plant growth parameters such as plant height, length of internode, number of fruit branches, and stem diameter, and the effect was more pronounced under elevated temperature conditions (Tables 2 and 3). Organic manure treated at the 2× rate yielded the highest plant growth (plant height, length of internode, number of fruit branches, and stem diameter), was and was substantially higher than those of the 1× rate treatment at all temperature environments. Although a significant difference was not detected, the chlorophyll content of leaf and stem diameter parameters increased with application rate increase. However, compared to the field condition plant growth parameters such as plant height, stem diameter, dry weight of shoot and root, and number of fruit branches tended to be significantly higher under the greenhouse followed by rain-shelter house conditions. The application of organic manure at the 1× rate increased plant height and the number of fruit branch parameters by 28.9% and 37.8% in the greenhouse and by 23.3% and 26.1% in the rain-shelter house, respectively, compared to the field. Likewise, when organic manure was applied at the 1× rate, the length of internode increased by 44.7% and 34.3% in the greenhouse and rain-shelter house, respectively, compared to the field values (Table 2).

Table 2. Plant growth attributes at 70 DAT.

Main Plot	Treatments	Height (cm)	Chlorophyll	Length of Internode (cm)	Stem Diameter (mm)	Fruit Branch (No.)
Field	control	86.7 g	62.6 a	6.0 d	9.0 a	22.3 cd
	X1	89.0 g	65.1 a	6.7 cd	9.6 a	23.0 cd
	X2	92.0 fg	68.7 a	8.0 cd	11.1 a	21.0 d
Rain-shelter house	control	106.0 de	66.1 a	8.3 cd	9.1 a	26.3 cd
	X1	109.7 cd	64.6 a	9.0 bc	9.9 a	29.0 bc
	X2	116.7 ab	67.5 a	9.8 ab	10.3 a	24.7 cd
Green house	control	111.0 bc	62.3 a	7.7 cd	10.2 a	31.7 ab
	X1	114.7 bc	62.3 a	9.7 ab	11.0 a	31.7 ab
	X2	122.7 a	65.0 a	10.7 a	11.6 a	38.0 a
LSD 0.05		14.4	6.99	2.16	3.93	7.72
CV (%)		8.2	6.2	8.1	11.6	6.4

Means separated by same lower case letter in each column are not significantly different at $p < 0.05$ among treatments. LSD = least significant difference.

Table 3. Dry weights of root, shoot, leaf, and fruits at 70 DAT.

Main Plot	Treatments	Root (g)	Shoot (g)	Leaf (g)	Fruit (g)
Field	control	3.7 g	21.6 g	21.6 e	67.4 g
	X1	4.5 cd	27.6 ef	25.2 e	83.8 de
	X2	4.3 de	23.9 g	22.2 e	77.9 ef

Table 3. *Cont.*

Main Plot	Treatments	Root (g)	Shoot (g)	Leaf (g)	Fruit (g)
Rain-shelter house	control	4.0 f	31.0 ef	26.4 de	72.3 fg
	X1	4.6 bc	37.8 bc	33.0 bc	95.7 b
	X2	4.7 ab	37.5 cd	34.7 ab	87.2 cd
Green house	control	4.2 ef	34.6 de	23.8 de	85.7 de
	X1	4.7 ab	37.9 ab	31.0 cd	114.6 a
	X2	4.9 a	41.8 a	35.1 a	92.6 bc
	LSD 0.05	2.83	6.45	5.46	10.47
	CV (%)	11.0	13.5	13.1	8.3

Means separated by same lower case letter in each column are not significantly different at $p < 0.05$ among treatments. LSD = least significant difference.

Similarly, the observation conducted in 130 DAT presented the effectiveness of organic manure on plant growth parameters of chilli pepper (Figures 1–3). However, the application of organic manure at the 2× rate significantly increased plant growth attributes, with a slight decrease observed at the 1× rate in all environment conditions.

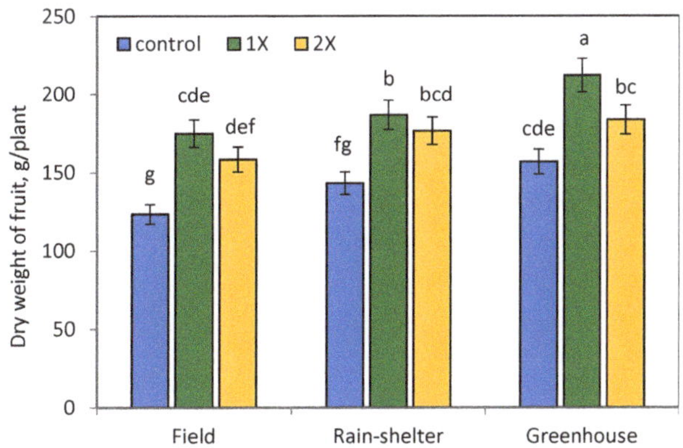

Figure 1. Dry weight of fruit per plant (g) at 130 DAT. Means separated by same lower case letter in each column are not significantly different at $p < 0.05$ among treatments.

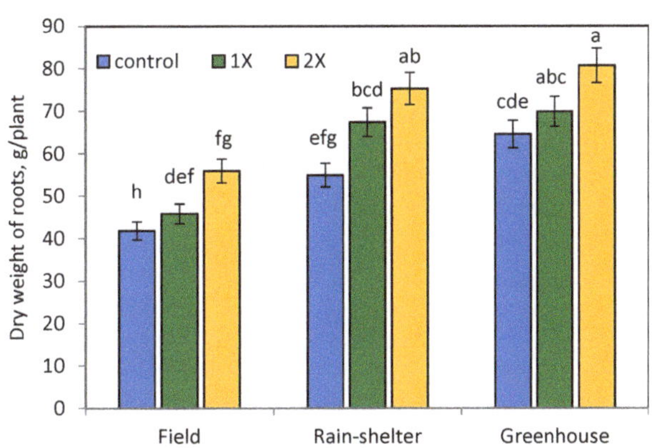

Figure 2. Dry weight of root weight per plant (g) at 130 DAT. Means separated by same lower case letter in each column are not significantly different at $p < 0.05$ among treatments. LSD = least significant difference.

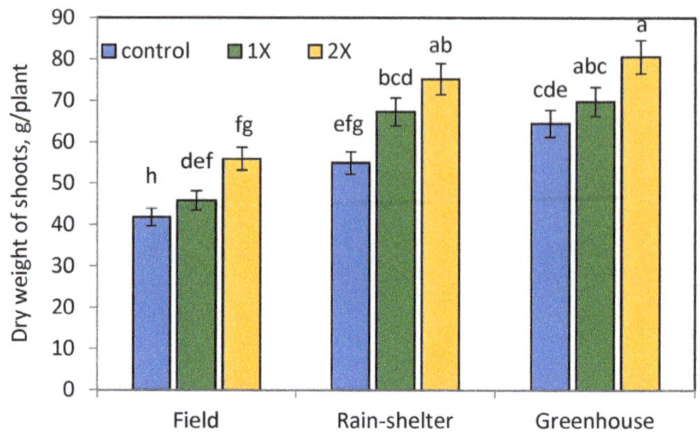

Figure 3. Dry weight of shoot weight per plant (g) at 130 DAT. Means separated by same lower case letter in each column are not significantly different at $p < 0.05$ among treatments.

The enhanced plant development in response to organic manure application is well-known [24], exhibiting a significant steady increase in plant growth due to the improved nutrient conditions in the root rhizosphere. Although organic manure application did not bring any significant difference at the beginning of growing season, the growth parameters were significantly higher at both rates of organic manure treatment at 130 DAT (Table 4). The application of organic manure at the 1× and 2× rates significantly increased chilli pepper growth and N uptake, but the 1× rate treatment stabilized the biomass ratio x plant yield interrelations. However, the organic manure based nutrient management scheme efficiently increased plant growth in all combinations of elevated temperature.

Table 4. Plant growth attributes at 130 DAT.

Main Plot	Treatments	Height (cm)	Chlorophyll	Length of Internode (cm)	Stem Diameter (mm)	Branch (No.)
Field	control	88.3 f	67.8 a	7.3 b	13.3 f	26.7 d
	X1	95.0 f	73.1 a	8.7 ab	14.9 ef	28.0 d
	X2	98.7 f	77.3 a	9.0 ab	15.7 de	33.7 cd
Rain-shelter house	control	113.0 de	69.6 a	9.7 ab	14.9 ef	29.3 cd
	X1	116.3 cd	71.7 a	10.0 ab	16.7 bc	35.7 bc
	X2	117.7 bc	71.2 a	10.3 ab	16.1 cd	35.7 bc
Green house	control	110.7 de	69.3 a	9.7 ab	14.7 ef	31.0
	X1	121.3 ab	73.9 a	10.3 ab	17.2 ab	39.0 ab
	X2	126.0 a	74.6 a	11.7 a	17.6 a	42.7 a
LSD 0.05		11.6	6.47	2.92	1.85	8.62
CV (%)		6.0	5.2	17.7	6.9	14.9

Means separated by same lower case letter in each column are not significantly different at $p < 0.05$ among treatments. LSD = least significant difference.

3.2. Effects of Organic Manure and Elevated Temperature on Dry Weight and Yield of Chilli Pepper

The highest average of dry weight chilli pepper fruit yield was found at the 1× rate of organic manure application, especially under the greenhouse condition (Figures 1–3). Our results show that the dried fruit of chilli pepper at the 1× rate organic manure application was significantly higher by 78.9% in the greenhouse and 20.5% in the rain-shelter house compared to the field condition. Chilli peppers grown in the rain-shelter house at the 2× rate of organic manure treatment showed the highest increase by 24.5%, 34.5%, and 13.5% of root, shoot, and fruit weight attributes, respectively, when compared to the appropriate control in the field. Likewise, root, shoot, and leaf dry weights at the 2× organic manure application in the greenhouse increased by 43.5%, 44.2%, and 36.0%, respectively compared to those of the control values.

However, the highest chilli peppers' yield was observed at the 1× rate of organic manure application under the greenhouse and followed by rain-shelter house conditions (Figure 3), the nutrient content in the soil probably was enough to meet the requirement of stabilizing crop growth. N sources positively influence crop productivity. However, a higher N input is accompanied by adverse effects for crop yield, soil health, and greenhouse gas emissions [25]. The quantity and quality of organic manure have tremendous effect on N balance of the soil, while other abiotic and biotic factors need to be considered to maintain N efficiency [26]. These results suggest that organic manure application at the 1× rate was sufficient to achieve the maximum chilli pepper yield in the greenhouse followed by the rain-shelter house, compared in the field (Figure 3). The increased temperature in the greenhouse affected the growth of chilli peppers and resulted in the highest yield at the 1× rate of organic manure application, suggesting that there is no need for extra organic manure application. The 2× rate organic manure application compared to the 1× rate caused a 31.3% and 5.8% reduction of fruit yield per plant in the greenhouse and rain-shelter house conditions, respectively. Significantly higher chilli pepper fruit yield harvested in the greenhouse gave a cue that the increased temperature synergistically optimized growth conditions and nutrient availability to produce a maximum chilli pepper yield at the 1× rate of organic manure treatment. The oversupply of organic manure to the soil led to an adverse effect for chilli pepper productivity and decreased its efficiency.

According to recent findings by Gu et al. [27], manure amendment can efficiently reduce ammonia volatilization in different cropping systems. On the other hand, a steady and significant increase in crop yield over the years with the application of organic manures was observed in many previous studies [10]. This might be explained by the fact that the organic manure application at the 2× rate resulted in higher N content in crop vegetative and generative organs than the 1× rate (Tables 5–8). Furthermore, manure application at the 1× rate sustained a steady and smooth supply of N that might create favorable condition for plant growth and lead to higher N uptake from the soil. In addition, it is reasonable to predict that manure amendments reduce N loss, which might enhance N uptake.

Table 5. Root nutrient uptake at 130 DAT.

Main Plot	Treatments	Ca	K	Mg	Na	P	N
				(mg g^{-1})			
Field	control	11.3 h	80.3 c	9.4 cd	3.6 b	3.8 d	46.6 g
	X1	12.9 g	114.9 ab	10.4 cd	2.6 f	3.8 d	51.4 f
	X2	16.2 c	119.8 ab	9.4 cd	2.7 e	4.2 b	52.8 d
Rain-shelter house	control	15.2 e	67.2 c	22.9 a	3.7 b	2.9	36.9 i
	X1	15.5 d	106.8 b	15.7 b	4.1 a	3.8 d	44.5 h
	X2	17.3 a	118.2 ab	11.4 c	3.7 b	4.3 a	56.2 c
Green house	control	14.4 f	75.9 c	20.9 ab	3.1 d	3.3 e	34.2 j
	X1	15.4 d	96.6 bc	15.8 b	3.3 c	4.1c	61.2 b
	X2	16.9 b	126.7 a	15.8 b	2.6 f	4.1c	79.6 a
LSD 0.05		2.3	10.4	4.9	1.3	0.5	0.6
CV (%)		10.4	13.7	13.2	8.1	8.5	8.1

Means separated by same lower case letter in each column are not significantly different at $p < 0.05$ among treatments.

Table 6. Shoot nutrient uptake at 130 DAT.

Main Plot	Treatments	Ca	K	Mg	Na	P	N
				(mg g^{-1})			
Field	control	11.1 d	129.9 e	6.1 cd	0.60 bc	3.68 d	44.5 d
	X1	14.4 c	134.6 cd	7.4 c	0.57 bc	3.78 cd	49.9 cd
	X2	25.2 ab	137.9 cd	8.7 ab	0.66 b	4.01 bc	50.1 c

Table 6. *Cont.*

Main Plot	Treatments	Ca	K	Mg	Na	P	N
		(mg g^{-1})					
Rain-shelter house	control	18.5 bc	166.3 bc	9.5 ab	0.65 b	3.93 c	44.0 e
	X1	25.8 ab	176.3 b	6.7 bc	0.68 ab	4.12 b	49.7 cd
	X2	27.3 a	166.3 bc	6.9 bc	0.78 a	4.04 bc	50.2 b
Green house	control	20.8 b	155.3 c	7.7 b	0.68 ab	4.04 bc	40.3 f
	X1	24.3 ab	174.1 b	8.1 ab	0.70 ab	4.42 ab	51.7 bc
	X2	27.7 a	219.2 a	10.2 a	0.81 a	4.49 a	54.0 a
LSD 0.05		3.8	29.8	2.48	0.16	0.29	0.57
CV (%)		11.9	12.6	9.6	6.3	4.9	8.1

Means separated by same lower case letter in each column are not significantly different at $p < 0.05$ among treatments.

Table 7. Fruit nutrient uptake at 130 DAT.

Main Plot	Treatments	Ca	K	Mg	Na	P	N
		(mg g^{-1})					
Field	control	2.0 f	26.5 bc	1.7 e	1.5 b	3.8 g	22.0 f
	X1	2.7 d	31.8 bc	2.3 c	1.8 a	4.2 d	27.0 c
	X2	2.9 c	34.6 ab	2.5 b	1.8 a	4.3 c	27.9 b
Rain-shelter house	control	2.0 f	26.9 bc	1.7 e	1.2 d	3.9 f	23.5 e
	X1	2.7 d	33.5 ab	2.7 a	1.5 b	3.7 h	27.9 b
	X2	3.2 a	36.9 ab	2.3 c	1.5 b	4.4 b	26.0 d
Green house	control	2.1 e	27.1 bc	1.6 f	0.9 e	4.0 e	21.2 g
	X1	3.1 b	32.2 b	2.1 d	1.4 c	4.2 d	27.1 c
	X2	3.2 a	38.8 a	2.7 a	1.4 c	4.5 a	30.4 a
LSD 0.05		0.43	6.1	0.4	0.4	0.8	0.5
CV (%)		11.2	13.1	12.9	10.7	13.1	13.3

Means separated by same lower case letter in each column are not significantly different at $p < 0.05$ among treatments.

Table 8. Leaf nutrient uptake at 130 DAT.

Main Plot	Treatments	Ca	K	Mg	Na	P	N
		(mg g^{-1})					
Field	control	28.9 d	120.1 d	12.2 d	0.66 b	5.1 d	58.2 d
	X1	49.3 bc	182.5 b	15.3 bc	0.59 c	6.7 b	78.6 c
	X2	50.8 b	221.4 a	15.9 b	0.57 c	7.9 a	84.0 a
Rain-shelter house	control	39.3 c	166.2 bc	14.7 c	0.65 b	5.2 d	55.5 e
	X1	59.8 ab	240.5 a	17.2 a	0.77 a	6.7 b	81.6 b
	X2	67.5 a	243.2 a	17.1 a	0.75 a	7.5 ab	79.0 c
Green house	control	49.5 bc	142.6 c	15.0 c	0.73 a	5.3 d	53.2 f
	X1	66.2 a	183.6 b	17.6 a	0.77 a	6.1 c	78.4 c
	X2	67.5 a	243.2 a	17.4 a	0.75 a	7.5 ab	79.0 c
LSD 0.05		11.8	41.1	1.1	0.12	0.5	0.66
CV (%)		15.4	15.0	4.9	15.3	5.2	6.2

Means separated by same lower case letter in each column are not significantly different at $p < 0.05$ among treatments.

3.3. *Effects of Organic Manure and Temperature on the Nutrition of Chilli Peppers*

The application of organic manure increased plant Ca, K, Mg, Na, total P, and N contents at all temperature environments, especially in the greenhouse condition with significant differences (Tables 5–8). In the roots, the highest nutrient uptake was observed at the 2× rate of organic manure application in all three temperature conditions. However, the nutrient content in the plant roots was more pronounced in the greenhouse condition (Table 5). Ca, K, and Mg contents increased by 49.8%, 57.8%, and 68.6%, respectively, in the greenhouse condition compared to the respective control in the field condition, while Na and P concentrations were slightly higher in the chilli roots grown under the rain-shelter house condition.

The organic manure application increased nutrient uptake into the roots, thereby nutrient content of the shoots increased (Table 6). Ca, K, and Mg contents of chilli pepper grown at the 2× rate organic manure treatment in the greenhouse condition were higher by 10%, 58.9%, and 17.6% compared to the respective control in the field condition. Similarly, Na, total P, and N contents at the 2× rate of organic manure application were higher by 23.8%, 11.8%, and 7.8%, respectively, compared to their respective control.

A similar trend of increasing nutrient contents by the 2× rate application of organic manure was also detected in the fruit content (Table 7). The increased temperature effect comparing the two rates of organic manure application was significant, suggesting that higher temperature and organic manure application generate greater nutrient content in chilli pepper fruits. Averaged across, organic manure application as a subplot significantly increased total Ca, K, and Mg contents by 11.2%, 12.1%, and 8.7%, respectively, compared to the respective control. A similar trend was observed on total P and N contents in chilli pepper fruits. However, Na content was slightly higher at the 1× rate organic manure application in the greenhouse condition.

Analyses of leaf chemical content of chilli pepper exhibited a similar trend to the increase of nutrient levels depending on the organic manure application (Table 8). The applied organic manure and elevated air temperature interactions promoted nutrient uptake by chilli peppers, which resulted in increased plant biomass and yield parameters. As expected, organic manure application initially improved soil properties and promoted plant growth, which resulted in the increase of plant nutrient uptake, growth, and yield [27]. The application of organic manure and elevated temperature interaction resulted in the abundance of nutrients in the soil, probably associated with the impact of beneficial microbes and favorable conditions for plant growth. These results indicate that organic manure application can alter mineral nutrient profiles in all plant parts, and improve their uptake and mobility within plants.

3.4. Effects of Different Rates of Organic Manure Application on the Soil's Chemical Contents

The application of organic manure significantly increased soil organic matter (OM) and nutrients content in all experimental plots (Tables 9 and 10). Averaged across soil depths, organic manure application, especially at the 2× rate, significantly increased soil OM content. At the end of the growing season, the highest OM content was observed in the field condition (38.5 mg kg^{-1}) followed by greenhouse (37.6 mg kg^{-1}), and rain-shelter house (31.5 mg kg^{-1}) conditions at the 2× rate of organic manure treatments. Soil pH levels tended to decrease in the field and rain shelter house conditions at the end of the growing season, but slightly increased in the greenhouse condition in all organic manure treatments. The soil's EC parameter increased in the field condition, but the decrease was observed under the rain-shelter house and greenhouse conditions. Although no significant difference was observed in soil K and Na contents among treatments, some increase was noticed depending on the organic manure application rate, especially in the greenhouse condition. Furthermore, Ca and Mg contents of the soil increased with the increase in organic manure application rate, but it was statistically not significant in most cases.

In the control, OM and N contents slightly decreased at the end of growing season in all temperature conditions. In the case of no organic manure input to the soil, plants uptake available nutrients from the soil and decrease the content of these essential elements. The highest N content in the soil was determined with the highest application rate of organic manure. However, the increasing rates of N in the soil did not exert any significant effects on the concentrations of K and Na, whereas consistently increased Ca and Mg contents. The increase of pepper plant nutrients content and yield was in response to the highly availability of organic nutrients in the soil after organic manure application. Temperature is also known to have a positive effect on the mineralization of soil OM and residues that may have affected nutrient availability in the elevated temperature under greenhouse condition.

It is well documented that higher N fertilization is usually accompanied with increased demand for other macro and micronutrients uptake by plants to maintain a balance. This phenomenon accelerates plant growth and yield formation processes in crops.

Table 9. Effects of different rates of organic manure application on soil chemical contents at the end of the experiment.

Main Plot	Treatments	pH (1 : 5)	E.C (dS/m)	OM (g/kg)	NO_3-N (ppm)	NH_4^+-N (ppm)
Field	control	5.8 c	1.2 b	32.4 b	10.7 f	3.1 de
	X1	6.1 b	1.1 b	37.7 a	21.0 d	7.7 c
	X2	6.5 ab	0.7 c	38.5 a	21.5 d	1.6 e
Rain-shelter house	control	6.3 b	1.4 ab	28.6 c	69.0 a	22.7 b
	X1	6.4 ab	1.5 a	31.9 b	51.1 b	23.6 b
	X2	6.8 a	1.4 ab	31.5 b	22.4 d	40.7 a
Green-house	control	5.9 bc	0.8 c	25.6 c	35.8 c	5.7 cd
	X1	6.3 b	1.1 b	31.6 b	19.7 de	4.0 cd
	X2	6.5 ab	1.5 a	37.6 a	17.7 de	7.0 c
LSD 0.05		4.5	0.3	4.6	8.4	4.6
CV (%)		5.1	4.9	5.8	4.8	5.7

Means separated by same lower case letter in each column are not significantly different at $p < 0.05$ among treatments.

Table 10. Effects of different rates of organic manure application on soil chemical contents at the end of the experiment.

Main Plot	Treatments	P_2O_5, (mg/kg)	K, (cmol/kg)	Ca, (cmol/kg)	Mg, (cmol/kg)	Na, (cmol/kg)
Field	control	374.7 i	1.2 c	7.2 f	1.9 d	0.1 c
	X1	480.6 g	1.2 c	7.8 d	2.5 b	0.1 c
	X2	542.1 f	1.3 bc	7.8 d	2.5 b	0.1 c
Rain-shelter house	control	412.3 h	1.3 bc	8.3 c	2.3 bc	0.2 b
	X1	569.9 e	1.3 bc	8.7 b	2.3 bc	0.2 b
	X2	588.0 d	1.2 c	8.8 a	2.4 bc	0.3 a
Green-house	control	727.6 c	1.3 bc	7.6 e	2.5 b	0.1 c
	X1	772.9 b	1.4 b	8.7 b	2.6 ab	0.2 b
	X2	862.8 a	1.9 a	8.8 a	3.0 a	0.2 b
LSD 0.05		13.4	1.5	0.95	0.5	0.09
CV (%)		6.5	6.3	7.4	9.7	5.6

Means separated by same lower case letter in each column are not significantly different at $p < 0.05$ among treatments.

4. Discussion

Theoretically, incorporating organic manure in soil should improve the growth attributes of plants. However, in this study we demonstrate that an optimum amount of organic manure is advantageous for higher yields. Furthermore, elevated temperatures substantially improved chilli pepper fruit yield in the greenhouse and rain-shelter house conditions. Likewise, plant growth parameters such as the dry weight of leaf, shoot, and root tended to be higher in the greenhouse followed by rain-shelter plastic house compared to the field conditions. Favorable elevated temperature conditions were the reason for the intensive plant growth in the greenhouse, although several authors have declared the negative impact of elevated temperatures on crop productivity [6,28]. However, it is essential to consider the crop's biological properties and soil-climatic conditions. Chilli peppers originated in Mexico, therefore, this plant prefers high temperatures by its biological properties, and the elevated temperature in the greenhouse may have promoted metabolic processes in the plant cells. Therefore, higher accumulated temperatures of 3091 °C and 3168 °C in the rain-shelter plastic house and plastic greenhouse, respectively, were more effective compared to the lower temperature (2944 °C) in the field (Table 1). Similarly, air and soil temperatures were higher under greenhouse followed by rain-shelter plastic conditions than those in the field condition (Table 1). The accumulated heat from sunlight maintained daily 2–3 °C higher temperature in the greenhouse and rain shelter house during the entire period of the experiment. The elevated air temperature in the greenhouse triggers an increase in soil temperature. This implies that mineralization processes of the applied organic manures in the greenhouse may have been faster than at other ambient conditions.

In the field, the fruit yield did not differ substantially at the 1× and 2× rates of organic manure treatments with values of 179.4 and 158.3 g plant^{-1} at 130 DAT, respectively, with a significant difference compared to the control treatment (123.5 g plant^{-1}). On the other hand, these values in the greenhouse condition were substantially higher, recorded 212.2 and 183.6 g plant^{-1} at the 1× and 2× rates of organic manure treatments, respectively. These results show that organic manure along with elevated temperature integration were the main factors that might have stimulated plant growth and chilli pepper yield. The present study is in agreement with results by Guo et al. who reported that elevated temperatures have a positive effect on crop yield in the North China Plain [27].

The organically amended soil is generally reflected in the enhanced chemical content of the plant vegetative parts. The results indicate that the application of organic manure significantly enhances Ca, K, Mg, Na, total P, and N contents in all plant vegetative parts, which is reflected on the growth and yield attributes of chilli peppers. As a subplot effect in this study, an increment in plant growth features was in response to the improved soil quality by the input of organic manure as a fertility amendment. Recently, Surendran et al. reported that increased temperature and moisture cause rapid mineralization of soil organic carbon in tropical regions [29,30]. The results of this study also show that the highest soil moisture was observed in the field followed by greenhouse and rain shelter conditions.

Organic manure application improves soil physical–chemical properties by actively facilitating bacterial growth as well [31,32]. As reported by Gomiero et al., besides enhancing soil quality, organic farming can also improve water use efficiency and this can lead to an increase in yield by 70–90% [2]. In order to have a clear view of organic manure application, a long-term experiment must be conducted. For instance, the experiment conducted in Northeast of China showed that 18 years of manure improved maize yield by 218% when compared to the control value [33]. Furthermore, Diacono and Montemurro found that long-term application of organic amendments increased crop yield by up to 250% and improved crop yield quality [24]. Rational implementation of organic manure is accompanied by the highest productivity of agricultural crops and without environmental damage even under adverse climatic conditions [34,35].

5. Conclusions

The beneficial effect of organic manure associated with elevated temperatures was found to have high efficiency in improving chilli peppers' nutrient content and crop productivity. The application of 2× rate (265.4 kg ha^{-1}) organic manure substantially increased the growth parameters of chilli peppers at all temperature conditions, whereas the 1× rate (132.7 kg ha^{-1}) exhibited a higher yield increase, especially under elevated temperatures in the greenhouse conditions. Therefore, applying the 1× rate of organic manure is considered the optimum for obtaining a maximum chilli pepper fruit yield. The excessive use of (2× rate) organic manure may result in the accumulation of more vegetative biomass than the increase in chilli pepper yield value. Thus, the rational use of organic amendments for chilli pepper cultivation is highly recommended depending on soil quality and environmental conditions.

Further studies are required to understand the long-term effect of organic manure application on chilli pepper production under variable temperature conditions. Furthermore, it is important to develop a proper simulation model to elucidate the long-term effect of an elevated temperature environment and organic amendments on growth and yield of chilli peppers. Despite forecast uncertainties, the project model allows improving the efficiency of agricultural management practices in various surroundings.

Author Contributions: K.W.P., was the Principal Investigator of the project, initiated and supervised the experiment, and contributed intellectually to the development of the final manuscript; Y.L., conceived the original idea for the analysis and participated in the interpretation of the results; B.K., H.J.Y., and K.M.C., undertook the literature search and drafted the manuscript; F.R., T.H.L., M.U., and A.B.B., helped with experiments and participated in the interpretation of the results. All authors have read and approved the final paper.

Funding: This study was conducted by support from the "Cooperative Research Program for Agriculture Science & Technology Development (Project No. PJ01283702)" Rural Development Administration, Republic of Korea.

Conflicts of Interest: The authors declare no conflict of interest.

References

1. Lobell, D.B.; Schlenker, W.; Costa-Roberts, J. Climate trends and global crop production since 1980. *Science* **2011**, *333*, 616–620. [CrossRef] [PubMed]
2. Gomiero, T.; Pimentel, D.; Paoletti, M.G. Environmental Impact of Different Agricultural Management Practices: Conventional vs. *Org. Agric. Crit. Rev. Plant Sci.* **2011**, *30*, 95–124. [CrossRef]
3. Meehl, G.A.; Stocker, T.F.; Collins, W.D.; Friedlingstein, P.; Gaye, T.; Gregory, J.M.; Kitoh, A.; Knutti, R.; Murphy, J.M.; Noda, A.; et al. Global climate projections. In *Climate Change 2007, The Physicalscience Basis*; Solomon, S., Qin, D., Manning, M., Chen, Z., Marquis, M., Averyt, K.B., Tignor, M., Miller, H.L., Eds.; Cambridge University Press: Cambridge, UK, 2007; pp. 283–296.
4. Challinor, A.J.; Watson, J.; Lobell, D.B.; Howden, S.M.; Smith, D.R.; Chhetri, N. A-meta-analysis of crop yield under climate change and adaptation. *Nat. Clim. Chang.* **2014**, *4*, 287–291. [CrossRef]
5. Graß, R.; Thies, B.; Kersebaum, K.C.; Wachendorf, M. Simulating dry matter yield of two cropping systems with the simulation model HERMES to evaluate impact of future climate change. *Eur. J. Agron.* **2015**, *70*, 1–10. [CrossRef]
6. Lotter, D.W.; Seidel, R.; Liebhardt, W. The performance of organic and conventional cropping systems in an extreme climate year. *Am. J. Altern. Agric.* **2003**, *18*, 146–154. [CrossRef]
7. Liang, S.; Li, Y.; Zhang, X.; Sun, Z.; Sun, N.; Duan, Y.; Xu, M.; Wu, L. Response of crop yield and nitrogen use efficiency for wheat-maize cropping system to future climate change in northern China. *Agric. For. Meteorol.* **2018**, *262*, 310–321. [CrossRef]
8. Dungan, R.S.; Gan, J.; Yates, S.R. Effect of temperature, organic amendment rate and moisture content on the degradation of 1, 3-dichloropropene in soil. *Pest Manag. Sci.* **2001**, *57*, 1107–1113. [CrossRef] [PubMed]
9. Chludil, H.; Corbino, G.B.; Leicach, S.R. Soil quality effects on Chenopodium album flavonoid content and antioxidant potential. *J. Agric. Food Chem.* **2008**, *56*, 5050–5056. [CrossRef] [PubMed]
10. Parham, J.A.S.P.; Deng, S.; Raun, W.; Johnson, G. Long-term cattle manure application in soil. *Biol. Fertil. Soils* **2002**, *35*, 328–337.
11. Wang, W.J.; Zhu Dan, Y.; Wang, H.S. Effect of different amount of organic fertilizer in the yield of organic vegetables. *North Hortic.* **2010**, *17*, 29–30.
12. Nguyen, N.L.; Kim, Y.J.; Hoang, V.A.; Subramaniyam, S.; Kang, J.P.; Kang, C.H.; Yang, D.C. Bacterial Diversity and Community Structure in Korean Ginseng Field Soil Are Shifted by Cultivation Time. *PLoS ONE* **2016**, *11*, e0155055. [CrossRef] [PubMed]
13. Albiach, R.; Canet, R.; Pomares, F.; Ingelmo, F. Microbial biomass content andenzymatic activities after the application of organic amendments to a horticultural soil. *Bioresour. Technol.* **2000**, *75*, 43–48. [CrossRef]
14. Marin, A.; Ferreres, F.; Tomas-Barberan, F.A.; Gill, M.I. Characterization and quantification of antioxidant constituents of Sweet pepper (*Capsicum annuum* L.). *J. Agric. Food Chem.* **2004**, *53*, 3861–3869. [CrossRef] [PubMed]
15. Gangadhar, B.H.; Mishra, R.K.; Pandian, G.; Park, S.W. Comparative study of color, pungency, and biochemical composition in chili pepper (*Capsicumannuum*) under different light emitting diode treatments. *Hortic. Sci.* **2012**, *47*, 1729–1735.
16. Ren, H.; Endo, H.; Hayashi, T. The superiority of organically cultivated vegetables to ones regarding mutagenic activities. *Mutat. Res.* **2001**, *496*, 83–88. [CrossRef]

17. Ibrahim, H.; Jaafar, M.; Karimi, H.Z.E.; Ghasemzadeh, E.A. Impact of organic and inorganic fertilizers application on the phytochemical and antioxidantactivity of Kacipfatimah (*Labisia pumila* Benth). *Molecules* **2013**, *18*, 10973–10988. [CrossRef]
18. DAT, S.; Teja, K.C.; Duary, B.; Agrawal, P.K.; Bhattacharya, S.S. Impact of nutrient management, soil type and location on the accumulation of capsaicin in *Capsicum chinense* (Jacq.): One of the hottest chili in the world. *Sci. Hortic.* **2016**, *213*, 354–366.
19. Statistics Korea. 2018. Available online: http://kostat.go.kr (accessed on 20 August 2019).
20. Dahal, K.; Knowles, V.L.; Plaxton, W.C.; Hüner, N.P.A. Enhancement of photosynthetic performance, water use efficiency and grain yield during long-term growth under increased CO_2 in wheat and rye is growth temperature and cultivar dependent. *Environ. Exp. Bot.* **2014**, *106*, 207–220. [CrossRef]
21. NIAST. *Method of Soil and Plant Analysis*; National Institute of Agricultural Science and Technology: Suwon, Korea, 2000.
22. Chapman, H.D.; Pratt, F.P. *Determination of Minerals by Titration Method: Methods of Analysis for Soils, Plants and Water*; California University, Agriculture division: Oakland, CA, USA, 1982.
23. Cottenie, A.; Verloo, M.; Kiekens, L.; Velghe, G.; Camerlynck, R. *Chemical Analysis of Plant and Soils*; Laboratory of Analytical and Agrochemistry, State University of Ghent: Ghent, Belgium, 1982.
24. Diacono, M.; Montemurro, F. Long-term effects of organic amendments on soil fertility. In *Sustainable Agriculture*; Springer: Dordrecht, The Netherlands, 2011; Volume 2, pp. 761–786.
25. Fageria, N.K.; Baligar, V.C. Enhancing nitrogen use efficiency in crop plants. *Adv. Agron.* **2005**, *88*, 97–185.
26. Hati, K.M.; Swarup, A.; Mishra, B.; Manna, M.C.; Waniari, R.H.; Mandal, K.G.; Misra, A.K. Impact of long-term application of fertilizer, manure and lime under intensive cropping on physical properties and organic carbon content of an Alfisol. *Geoderma* **2008**, *148*, 173–179. [CrossRef]
27. Gu, L.; Liu, T.; Wang, J.; Liu, P.; Dong, S.; Zhao, B.; So, H.B.; Zhang, J.; Zhao, B.; Li, J. Lysimeter study of nitrogen losses and nitrogen use efficiency of Northern Chinese wheat. *Field Crop. Res.* **2016**, *188*, 82–95. [CrossRef]
28. Manna, M.C.; Swarup, A.; Wanjari, R.H.; Ravankar, H.N.; Mishra, B.; Saha, M.N.; Singh, Y.V.; Sahi, D.K.; Sarap, P.A. Long-term effect of fertilizer and manure application on soil organic carbon storage, soil quality and yield sustainability under sub-humid and semi-arid tropical India. *Field Crop. Res.* **2005**, *93*, 264–280. [CrossRef]
29. Surendran, U.; Vijayan, A.K.; Bujair, V.; Joseph, E.J. Influence of open and polyhouse conditions on soil carbon dioxide emission from Amaranthus plots with different nutrient management practices under changing climate scenario. *Curr. Sci.* **2018**, *114*, 1311–1317. [CrossRef]
30. Guo, R.; Lin, Z.; Mo, X.; Yang, C. Responses of crop yield and water use efficiency to climate change in the North China Plain. *Agric. Water Manag.* **2010**, *97*, 1185–1194. [CrossRef]
31. Wang, X.; Jia, Z.; Liang, L.; Yang, B.; Ding, R.; Nie, J.; Wang, J. Impacts of manure application on soil environment, rainfall use efficiency and crop biomass under dryland farming. *Sci. Rep.* **2016**, *6*, 20994. [CrossRef] [PubMed]
32. Weitza, A.; Linderb, E.; Frolkingc, S.; Crillc, P.; Kellerc, M. N2O emissions from humid tropical agricultural soils: Effects of soil moisture, texture and nitrogen availability. *Soil Biol. Biochem.* **2001**, *33*, 1077–1093. [CrossRef]
33. Zhang, X.Z.; Gao, H.J.; Peng, C.; Li, Q.; Zhu, P. Effects of combined application of organic manure and chemical fertilizer on maize yield and nitrogen utilization under equal nitrogen rates. *J. Maize Sci.* **2012**, *20*, 123–127.
34. Bai, Z.; Caspari, T.; Gonzalez, M.R.; Batjes, N.H.; Mäder, P.; Bünemann, E.K.; de Goede, R.; Brussaard, L.; Xu, M.; Ferreira, C.S.S.; et al. Effects of agricultural management practices on soil quality: A review of long-term experiments for Europe and China. *Agric. Ecosyst. Environ.* **2018**, *265*, 1–7. [CrossRef]
35. Bir, M.S.H.; Eom, M.Y.; Uddin, M.R.; Park, T.S.; Kang, H.W.; Park, K.W. Weed Population Dynamics under Climatic Change. *Weed Turfgrass Sci.* **2014**, *3*, 174–182. [CrossRef]

International Journal of
Environmental Research and Public Health

MDPI

Article

Handmade Comal Tortillas in Michoacán: Traditional Practices along the Rural-Urban Gradient

Esperanza Arnés * and Marta Astier

Centre for Research in Environmental Geography (CIGA), National Autonomous University of Mexico (UNAM), Morelia Campus, Michoacán 58190, Mexico

* Correspondence: esperanza.arnes@gmail.com

Received: 26 July 2019; Accepted: 28 August 2019; Published: 3 September 2019

Abstract: Certain components of global food security continue to be threatened. Globalization has impacted food patterns, leading to greater homogenization of diets and the standardization of processes of food transformation, both in the countryside and in the cities. In Mexico, this has led to a drop in the use of native corn landraces and in the value associated with traditional practices around their growing and the processing and consumption of tortillas. The aim of this work was to analyze the main characteristics of the handmade comal tortilla system along the rural-urban gradient taking into account: (1) The type of seed and production, (2) manufacturing processes, (3) marketing channels and purpose of sales, and (4) perceptions regarding the quality of the product. Research was conducted on 41 handmade tortilla workshops located in rural areas in the Lake Pátzcuaro Basin and in urban and peri-urban areas of a medium-sized city in Michoacán (Mexico). Results showed that the origin of the grain follows a gradient-like pattern: In rural areas, tortillas are made with local and native corn predominate, while in urban contexts most tortillas come from hybrid corn produced in Sinaloa or Jalisco. There is a generalized preference for white tortillas, but blue tortillas are used for personal consumption in rural areas and as a gourmet product in the city. 100% of the rural workshops make their own nixtamal, while almost 50% of the peri-urban and urban businesses buy pre-made nixtamal dough. Surprisingly, 50% of the rural handmade tortilla workshops admit that they add nixtamalized corn flour and/or wheat flour to their tortilla mix. We conclude that not all handmade comal tortillas are produced equally and, although in rural areas traditions are better preserved, these also have contradictions. We also conclude that it is important to promote the revaluation of agrobiodiversity, traditional gastronomy, and food security without sacrificing quality, nutrition, and flavor.

Keywords: maize; food security; agrobiodiversity; traditional food systems; local knowledge; cultural practices; sustainable agriculture

1. Introduction

1.1. Food Security and Food Patterns

The current global food model is clearly inefficient and harmful, as attested by the world's 821 million hungry and 672 million obese [1]. These numbers point to a problem that goes beyond the lack of availability or access to food, but rather to its misutilization. Utilization, along with availability, access, and stability is one of the four components that make up the concept of food and nutrition security (food and nutrition security is defined, in a broad sense, as *"when all people at all times have physical, social, and economic access to food, which is safe and consumed in sufficient quantity and quality to meet their dietary needs and food preferences (...)* [2,3].

These maladjustments are the consequence of a change in global food patterns, characterized by an increase in the consumption of meat, dairy, and processed products (high in fats and sugar), and a decrease in cereals, legumes, and vegetables [4,5]. The tendencies of this food transition began to be documented in the 1990s and, were initially linked mainly to urban environments [6]. However, similar patterns can be currently observed in rural areas, and it is increasingly becoming directly related to the socioeconomic level of the population [7,8]. This change in food patterns in the last decades, driven by the demands of globalization, responds to a wide number of complex dynamics that are evident in specific local context [9]. Some authors call it glocalization and it may be related to ignorance about the composition and origin of foods, not valuing traditional cooking techniques, easy access (physical and economically) to processed foods of highly publicized brands and difficult access to healthy foods in low-income areas (the current dichotomy of food-deserts/food-swamps) [10,11].

Mexico is not an exception to this state of affairs, and it lives with the paradox of malnutrition (obesity and hunger) both in rural and urban areas [5,8]. Mexico has the second-highest obesity rate in the world (32.4% of people over 15 years old) and the first rate of child obesity [12]. According to studies, this is due to an increase in the consumption of saturated fats and it coincides in timing with the incorporation of Mexico to the North American Free Trade Agreement (NAFTA) in 1994 [13]. Moreover, globalization has also contributed to the homogenization of the diet [14] and the standardization of processes of food transformation. This has been due to the need to adapt to international export standards and lower costs of food provision and access [15].

Although there are many studies that explain the link between globalization and the change in both diet and traditional food practices in the cities [6,16], the dynamics around agro-alimentary systems in the rural-urban gradient remain unexplored [8].

1.2. Diets and Traditions. The Cultural Value of the Traditional Mexican Diet

Food, apart from its nutritional value, is a key element of human identity. Food is culture when it is produced, through the farmers' knowledge of the management of the field, crops and their varieties (agriculture), when it is transformed, using culinary techniques (cooking), and when it is consumed, bringing together economic, nutritional, and/or symbolic aspects [17].

In 2010, the Intergovernmental Committee for the Safeguarding of Intangible Cultural Heritage (ICH) of UNESCO, approved the candidacy of Mexico under the title: "Traditional Mexican cuisine—ancestral, ongoing community culture, the Michoacán paradigm" and it was included in the Representative List of the Intangible Cultural Heritage of Humanity [18]. This award is an acknowledgment of the agro-alimentary traditions of the country, the maximum expression of which can be seen in the Michoacán region. It highlights the value of the preservation of species and varieties that have their center of origin in Mexico (maize, beans, pumpkin, and chili), traditional growing methods, such as the *milpa*, and the culinary processes that transform the crop into food, such as nixtamalization, employing specialized tools, such as *metates*, stone mortars, and *comal* (prehispanic thin, disc-shaped device made from unglazed clay or metal used to cook maize tortillas) [19].

However, other studies have shown that traditional knowledge in general, and, specifically, culinary knowledge, which has been preserved mostly in farming systems in rural contexts since pre-Hispanic times, is eroding and transforming at an accelerated rate [20,21]. Currently, in Mexico, several institutions are working to strengthen the rural sector for the preservation of traditional agro-alimentary systems, from production to consumption [5]. However, a deeper analysis of the state of these systems is needed in order to respond to the challenges that exist both in the countryside and the cities and understand the relationship between them.

1.3. Maize Tortillas

Corn tortillas have been and continue to be the basic food of the Mexican diet and its main source of energy and protein. The center of origin and domestication of the *Zea mays* is Meso-America [22], specifically Mexico, where, currently, 59 native races and 300 varieties of maize are still preserved [23].

Despite the above-mentioned changes in food patterns of the population, this crop continues to be the basic pillar of the Mexican culture and food. With about 7.5 million hectares, it is responsible for most of the sown area in the country, which sets Mexico as the seventh largest producer and the fifth largest consumer of maize in the world [24]. Eighty-seven percent of the maize that is produced is white corn for human consumption, and 70.5% is cultivated under rainfed conditions [25]. The states with the largest volume of production are Sinaloa (19.5%), Jalisco (13.2%), and Michoacán (7.4%) [25].

It is estimated that in rural areas the consumption of maize is of about 274 kg/person/year [26] and that of tortilla is of up to 0.455 kg/person/day [27]. Although these numbers have not diminished, the way in which these are consumed has changed. Today, maize is present in a wide range of processed foods that do not require exhaustive labeling to indicate the type of maize that was used, nor its provenance. This has a direct effect on the loss of native races and the values associated with the traditional growing, transformation, and consumption practices. But what happens with tortillas?

Tortillas are a type of food made from nixtamalized maize which occupies a central place in the culture and daily diet in Mexico and Central America. Although the term "tortilla" was coined after the conquest (as a reference to the round-shaped Spanish omelet), the process of nixtamalization dates back to pre-Hispanic times—between 1200 and 1500 BC, according to the earliest evidence, found in South Guatemala [28].

From the náhuatl *nixtli* (ashes) and *tamalli* (dough), nixtamalized tortilla is made from just three ingredients: Maize, water, and limestone. The process begins by mixing three parts of water with 1% lime ($Ca(OH)_2$) and one part of maize. The preparation is cooked for 30–90 min (the exact time depends on the type of maize, the moisture of the grain and the "strength" of the lime) [29]. The maize is then left to soak in the cooking water for 8–12 h, during which it increases from 12% to 100% of its final weight. After soaking, the cooking water, called *nejayote* is discarded and the maize is rinsed two or three times, without removing the pericarp or the germ. Cooking and rinsing should not be excessive, or the dough will lose its *correa* (colloquial term that refers to the elasticity of the dough, which prevents the tortillas from breaking) and become rubbery. On the other hand, if it is not rinsed enough, it can become coarse and develop a slightly bitter taste [29]. The result is a soft but cohesive dough that can have up to 45% of moisture [30,31].

Although maize is widely consumed worldwide, it is only nixtamalized in Meso-America. This process transforms the food physically, chemically and nutritionally through its alkaline action [32]. Some of its benefits include the increase in bioavailability of amino acids and greater content of phosphorus and calcium. It is also especially important in that it increases niacin availability [33].

However, not all maize tortillas are equal. Not even handmade comal tortillas, which, according to the collective imagery, are synonym with traditional tortillas [19]. The quality and nutritional content of tortillas depend on the decisions taken in the field and on the processes of food transformation. Michoacán is one of the states with the greatest diversity of native maize and the third state in maize production in the country [25]. However, it is important to remark that, although some authors claim that the richness of native maize has not diminished, the cultivated surface has, as it has been substituted with hybrid improved maize, introduced by large seed companies [34]. This diminishment has led to the lack of availability of native grain in some areas of the state (especially in urban centers) and in specific times of the year, as well as making it difficult for people to consume tortillas made with these types of maize. This is something that impacts the revalorization of both the agrobiodiversity and the traditional gastronomic culture and traditional diet.

The question that emerges from the points above would be: Is there, in the state of Michoacán, a gradient in the loss of traditions and quality in all aspects of the maize–tortilla system as we get away from rural environments and approach the city? In what ways has a medium-sized city such as Morelia, which is close to rural areas where native maize still exists and is cultivated and traditional transformation practices are maintained, been a victim to the influx of globalization (considering globalization as a phenomenon, that has influenced the homogenization of diets)?

Based on this, the objective of the present research is to analyze the main characteristics of the handmade comal tortilla system (from sowing to sale and/or consumption) in urban (Morelia), peri-urban and rural contexts in Michoacán, considering: (1) The type of seed and production, (2) the process of transformation, (3) channels of commercialization and purpose of sale, and (4) perceptions around the quality of the product.

2. Study Area and Methodological Approach

Three study areas were defined in the State of Michoacán, Mexico. One was an urban area located within the metropolitan limits of the city of Morelia. Another was a peri-urban area located southwest of the city of Morelia. The third study area was a rural area encompassing several communities located in the Lake Pátzcuaro Basin (LPB) (Figure 1). Throughout the study, results will be shown comparing the average of some areas with others. In some cases, the particularities observed in a specific area will be explained in detail.

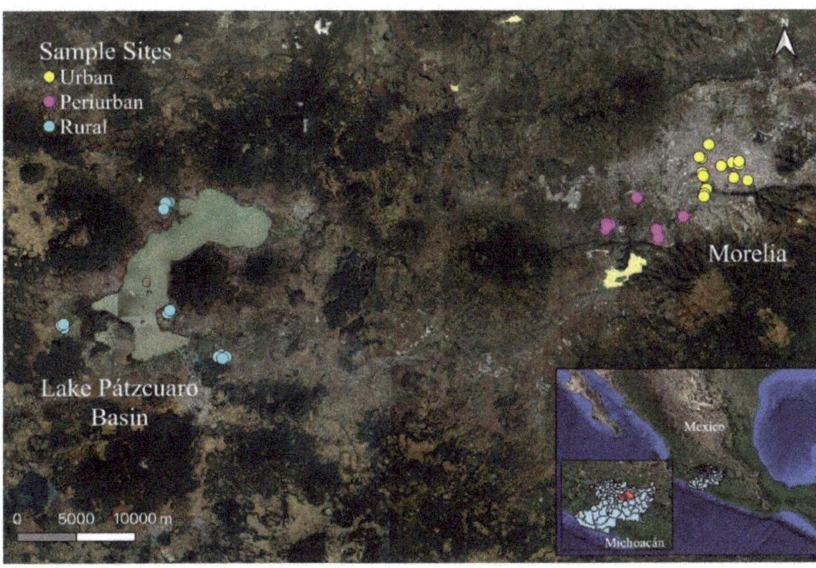

Figure 1. Map of the urban, peri-urban and rural areas of Morelia and the Lake Pátzcuaro Basin (LPB).

Morelia, a medium-sized city, is the capital of the State of Michoacán. Around 600,000 people live in its metropolitan area [35]. Since the 1980s, it has shown an accelerated and heterogeneous population expansion [36]. The areas adjacent to the city, defined as diffuse urbanization, expanded periphery or peri-urban environments (in this study we will use the latter term) have particular characteristics, forming an imprecise gradient between rural and urban regions [37]. In the selected peri-urban area, crop fields coexist with urban settlements, and the inhabitants lead a mixed way of life. The selected rural area is located approximately 60 km from Morelia, on the shores of Lake Patzcuaro. Four rural centers were included, each with a population of less than 2500 inhabitants and belonging to different municipal entities: San Francisco Uricho (Mun. Erongarícuaro), Tzurumútaro (Mun. Pátzcuaro), San Pedro Cucuchucho (Mun. Tzintzuntzan), and San Andrés Tziróndaro (Mun. Quiroga).

The selection of the study areas was made based on the degree of similarity or difference in the production, processing, and sale of handmade comal tortillas (from now on HCT). The rural communities of the LPB have a long history of cultivating native varieties of maize using traditional methods and making HCT [38,39]. The peri-urban area is located along the main road that connects Morelia (urban area) with the LPB. There is thus a continuum in the access to and exchange of products and knowledge between urban and rural areas. It is important to note that other peri-urban areas in the city of Morelia (those harboring housing complexes for the upper classes, or those bordering the municipality of Tarímbaro to the north, the state's main hybrid maize production area) are subject to different socioeconomic dynamics, depending on their geographical location or economic function.

Our object of study was the little establishments selling handmade tortillas, made and cooked by women on a comal. From now on, we will refer to these establishments as handmade comal tortilla establishments (HCTE). These establishments range from informal workshops (established by women who make tortillas in their homes, alone or helped by a family member, and who intermittently go out to sell them in their own community or in other communities) to more established companies that keep accounting and billing records and make use of hired labor (Figure 2). The criterion used to select the establishments was the use of the comal as a traditional tool for making tortillas, as opposed to the industrial tortilla shops that rely on mechanized processes.

Figure 2. Handmade comal tortilla establishment in urban, peri-urban, and rural areas. Source: Authors.

We used a non-probability sampling method to determine the number of HMTE to be sampled [40]. A total of 41 semi-structured interviews were conducted in the three study areas. Fifteen interviews were conducted in HCTE in the urban area, 10 interviews were conducted in the peri-urban area, and 16 interviews were conducted in establishments or with people selling HCT in the four rural communities mentioned above. The semi-structured interviews were conducted during December 2018 and January 2019. They were divided into four main parts. The first part was linked to maize cultivation, the second to the techniques of maize transformation and tortilla processing. The third part was related to the tortillas' sale strategies and their consumption and the fourth part were qualitative questions about the quality of their product and their life.

Peralta de Legarreta [21] proposes an approach for analyzing gastronomic culture based on four elements: (1) The link between man and land and natural resources, (2) the selection of the edible and the non-edible, (3) the culinary transformation of food, (4) consumption. Using this approach, the author analyzes the path or chain followed by food "from the land to the table". The same approach is part of other analytic methods, such as the food value chains used by the FAO [41]. In the present study, we also differentiated a series of elements associated with the various activities carried out along the food chain. The first and second elements are related to the production and transformation of food, the third, unlike the work of Peralta de Legarreta [21], is related to the sale of the food product and, to a lesser extent, to its consumption, since the purpose of this study is to analyze establishments selling HCT.

3. Results

3.1. Type of Seeds and Tortilla Production Systems

ORIGIN and VARIETY: The results of our inquiries regarding the type of seeds used by HCTE show a gradient in terms of origin and variety of grains. In rural areas, tortillas made with local and native maize predominate, while in urban and peri-urban areas, most HCT are made with hybrid maize seeds produced in Sinaloa or Jalisco (Table 1). As has been noted by other authors, much of the production of native maize is not used, or only in a small percentage, by the Mexican agri-food industry. It is mainly produced for self-consumption and, to a lesser extent, for local trade [24]. This trend explains that 50% of the HCT sold in rural environments (and between 93% and 80% of those sold in urban and peri-urban environments, respectively) are made with hybrid maize from distant origins, especially during certain periods of the year. These data, however, come from four rural communities, Orozco-Ramírez et al. [42] determined that maize production in two of them was decreasing. Another regional study reported that 56% of the homemade and traditional tortillas consumed between 51% and 70% of the local native maize [43]. These data were confirmed 11 years later by Astier et al. (2019) in a study based on surveys applied in 19 rural communities. The strategy usually used by local HCTE in rural areas is to use local native maize until it runs out, and then buy maize from stores supplied from the Morelia Supply Center.

Table 1. Main characteristics of maize for handmade comal tortillas (HCT) in rural, peri-urban, and urban context.

Characteristic		Rural		Peri-Urban		Urban	
		% HCTE	N° of HCTE	% HCTE	N° of HCTE	% HCTE	N° of HCTE
Origin	Local	50	8	30	3	13.5 *	2
	Regional	0	0	15	1.5	6.5 *	1
	Sinaloa/Jalisco	50	8	55	5.5	80	12
	Total	100	16	100	10	100	15
Variety	Native	50	8	20	2	7	1
	Hibrid	50	8	80	8	93	14
	Total	100	16	100	10	100	15
Color	White	90	14.5 *	100	10	97	13.5 *
	Blue	10	1.5 *	0	0	3	1.5 *
	Total	100	16	100	10	100	15
Production system	Organic	6	1	0	0	7	1
	Conventional	94	15	100	10	93	14
	Total	100	16	100	10	100	15

* In those establishments where several types of maize were used, the value was divided by the number of variables of that characteristic. Source: Semi-structured interviews.

The Morelia Supply Center (MSC) is located in the northeastern periphery of the city. The MSC contains several maize-dispensing depots, most of them selling white hybrid maize from different origins. Sinaloa maize is highly demanded and appreciated for its homogeneity, whiteness, and consistency. This maize is produced under irrigation during the autumn-winter cycle, which makes it possible to supply the market when rainfed maize is scarce. Up to 90% of the maize used in Mexico is Sinaloa maize [44]. Its price remained unchanged between January and April 2019 at 5.4 MXN/kg in the MSC. Barca maize (Jalisco) is smaller than Sinaloa maize, but is valued for its homogeneity and for its lower cost of 5 MXN/kg (also unchanged between January and April 2019). Barca maize is rainfed maize, but due to the edaphoclimatic conditions of the area, it usually gives very satisfactory yields of approximately 7.11 t/ha [45]. Finally, Region maize (the term "Region" is used both in the MSC and by HCTE to refer to maize that comes from the northeastern area of the State of Michoacán and southeast of the State of Guanajuato), rainfed grown temporarily and characterized by its lower cost (4.7–4.9 MXN/kg) is not as preferred due to its low homogeneity and the presence of residues, compared to the previous two varieties. Astier et al. [38] made an interesting spatial analysis of the origin of the maize used to make handmade tortillas in the LPB.

COLOR: Tortillas made with white maize are generally preferred by consumers compared to tortillas made with blue or red maize, to the point of being impossible to find blue tortillas for sale in the peri-urban area. Blue tortillas (a distinctive sign of native maize), is only sold in three Morelia HCTE (in two of them, only upon request and subject to grain availability). Consumers of blue tortillas pay higher prices for what is considered an artisanal product. It is worth mentioning that in urban establishments selling blue tortillas (Table 2), the amount of blue tortillas sold does not exceed 6% of the total sales of tortillas.

Table 2. Characteristics of blue comal tortillas in the urban area of Morelia.

Urban Establishment (UE)	Purchase Price of Blue Corn (MNX/kg)	Sale Price of Blue Tortillas (MXN/kg)	Sales of Blue Tortillas (kg/week)	Blue Tortillas as Percentage of the Total Volume of Sales
UE No. 10	10	30	2.5	2
UE No. 11	13	40	25	0.35
UE No. 12	10	18	30	6

In rural areas, blue tortillas are sold in small and informal establishments at the same price as white tortillas. These tortillas are mainly used for self-consumption, since selling them in rural areas does not generate an extra profit, as occurs in urban areas. This has been reported by other studies conducted throughout the country [46,47].

PRODUCTION SYSTEM: While the use of agrochemicals in maize production is not as widespread and intense as in the production of berries or vegetables, it is difficult to find farmers who grow 100% organic maize in Michoacán. Organic maize is produced on demand and is mostly destined for foreign markets in Canada and the US [48]. Some enterprises in the Purépecha Plateau, such as Marku Achekoren and Coyote Rojo, have been producing organic maize for decades. However, the present study identified an increasing (although still irregular) demand for products made with organic maize (especially tortillas) in urban areas of Michoacán, but supply is still scarce. In Mexico, organic production processes are regulated by the Organic Products Law, and export products must have the corresponding certification [49]. Participatory certification mechanisms exist at the local level, with greater flexibility in terms of time, lower costs and closer ties with producers, transformers, and consumers. Red Tsiri is an initiative working in the LPB that tries to improve the living conditions of producers of organic native maize and of the people who transform this maize into artisanal food products by selling directly to consumers or through short marketing channels in Morelia [50].

3.2. *Transformation and Elaboration of Tortillas*

The elaboration of HCT consists of the following main steps: (1) Nixtamalization, (2) grinding and use of metate, (3) pressing or *palmear* ("palmear" involves shaping the dough by hand until it forms a tortilla), (4) cooking on the comal. The type of maize grain used, the use or not of different flours, the type of fuel used for cooking and the elaboration of the tortillas can vary widely depending on the environment, urban or rural, and the natural resources (maize and firewood) available (Table 3).

NIXTAMAL: All rural HCTE prepare their own nixtamal using firewood and maize grains produced by themselves or bought from someone else. Nixtamalized dough is highly perishable and must be prepared daily for it to be fresh. Between 50% and 60% of urban and peri-urban HCTE buy the nixtamalized dough from third-parties to make the tortillas they sell. None of these HCTE were able to identify the origin or production process of the maize used to produce the dough they acquired.

This ignorance is also shared by a large proportion of the consumers of the final product (tortillas) in Mexico, in contrast with the trends and the legislation regarding traceability found in the European Union (traceability is the ability to track the movements of a food product, from its production to its consumption, through all the links in the food chain, thanks to an identification and control system), although not all food products face the same requirements [49,51].

Table 3. Main features in HCT elaboration process.

Characteristic			Rural		Peri-Urban		Urban	
			% HCTE	N° of HCTE	% HCTE	N° of HCTE	% HCTE	N° of HCTE
Nixtamal		Corn	100	16	40	4	46.7	7
		Dough	0	0	60	6	53.3	8
		Total	100	16	100	10	100	15
Technologies	Heat source	Firewood	75	12	60	6	0	0
		Gas	25	4	40	4	100	15
	Total		100	16	100	10	100	15
	Grinding	Metate	-	7 *	0	0	0	0
		External mill	87.5	14	100	10	60	9
		Own mill	12.5	2	0	0	40	6
	Total		100	16	100	10	100	15
	Pressing	Hand	43	7	0	0	0	0
		Press	57	9	100	10	100	15
	Total		100	16	100	10	100	15
Use of flours		Yes	50	8	20	2	13.3	2
		No	50	8	80	8	86.7	13
		Total	100	16	100	10	100	15

(*) N° of HCTE that use *metate* as another source of grinding (not considered in the total sum). Source: Semi-structured interviews.

TECHNOLOGIES: As mentioned earlier, the traditional process of making tortillas involves manual and unsophisticated operations. Firewood (for cooking the nixtamal and heating the comal) is used in 75% of rural HCTE, the remaining 25% use gas. As we approached urban environments, the use of gas became more widespread: 40% of peri-urban and 100% of urban HCTE. We measured the investment cost of using firewood and gas to elaborate one dozen of HCT in three peri-urban and rural HCTE. The average cost in MXN was 2.37 and 3.83, respectively. It is essential to underline that in many rural households and establishments, firewood is collected, therefore it doesn't represent a monetary cost. It is worth noting that 33% of the wood-burning stoves used in rural areas were Patsari-type stoves (the *Patsari* stove is the product of a participatory process of technological innovation. The design of this stove's combustion chamber and tunnels is optimized, and its parts, including a metal chimney support and metal comales (pans to place the pots) are custom-designed for durability. The stove is constructed using a metallic mold to ensure that critical dimensions are maintained. The exterior structure is made of brick, and the internal body is made of a mixture of mud, sand, and cement. All of these materials are locally available, and the custom-made stove parts are also manufactured by small local industries), which improve the efficiency of firewood and pose lower health risks for the people who use them [52].

The use of a press, as opposed to the elaboration of tortillas by hand (*palmear*), is found in most HCTE, up to 57% of those located in rural areas. In urban and peri-urban areas, all HCT sold are made with a press (except for those sold in the markets by individual rural women and which are not the object of study of this work). Using a press saves time and effort, increasing the speed of production six-fold. The results of the present study indicate that consumers in all environments have a clear preference for pressed tortillas (thinner and more uniform). Novelo and García [53] recorded social differences in the consumption of tortillas at the beginning of the 20th century, with thin and white tortillas being consumed by the upper classes and thick and dark tortillas by indigenous people. Ortega et al. [54] studied a network of 89 women who made tortillas by hand (without the use of a press) in Oaxaca, which indicates, however, the preference for non-pressed tortillas in that region.

The artisanal grinding process that transforms nixtamalized maize into dough requires a pre-Hispanic utensil called "metate" (from the nahuatl *metatl*). Some people say that the taste of tortillas made with the use of a metate is different from those made with an electric mill. However, the high physical effort involved in the use of a metate led some researchers in the 1940s to consider indigenous women as "metate slaves" [55]. Nowadays, the metate is commonly used in rural areas for the final kneading of dough already ground in the mill. In our study, seven women used it (Table 3). Doña Micaela, from Cucuchucho community, told us about the usefulness of metate:

(...) "it is for the last kneading pass before throwing the tortillas on the comal..."

In the rural areas studied here, 87.5% of the HCTE took their dough to one of the community mills. There are usually between three to four mills in each community and the grinding prices vary widely, since they depend on individual agreements between the millers and the tortilla makers. The remaining 12.5% of the HCTE own a small electric mill. In peri-urban areas, of the HCTE that do not buy dough and prepare their own nixtamal, all use the mills belonging to the nearby industrial tortilla shops. In urban areas, that proportion rises to 60%. However, the remaining 40% of urban HCTE invested in their own mill in order to have control over the delicate grinding process.

Technology should not be at odds with the production of a quality final product, especially if technological tools are well designed, taking into consideration the needs and cultural preferences of users and consumers. Some authors have demonstrated the preference of rural households for native maize (especially for tortillas) and their resistance to change [38,42]. Other authors, mainly with an agricultural engineering background are in favor of changing the original ingredients of tortillas to make them more accessible and nutritious for the urban population, even promoting tortillas made from white sorghum [56]. Since the 1980s, some authors have advocated for gas to be used in all rural households [57], while others have shown evidence that as long as firewood is accessible, rural households will always prefer it over using firewood to prepare food products other than tortillas [58,59]. Some mechanized pressing processes have also incorporated a porous material similar to that of metates in order to preserve the traditional quality of tortillas.

USE OF FLOURS: The National Company for Popular Subsistence (Compañía Nacional de Subsistencias Populares - CONASUPO) was a Mexican parastatal company created in 1961 with the objective of increasing the levels of food intake (especially maize) of the most vulnerable sectors of society, through generalized and distributive subsidies. This program reduced by more than half the price of products, such as nixtamalized corn flour, making them more accessible, which, in turn, increased their consumption, especially in times of grain shortages. Today, although the price of nixtamal nixtamalized corn flour (three brands of nixtamalized corn flour have been identified in Michoacán: MASECA, MINSA and AgroMINSA. They all fulfill the same function, although some people point out small differences between them) has increased significantly (Maseca: 12 MXN/kg, Minsa: 10.5 MXN/kg, AgroMinsa: 10.75 MXN/kg (data obtained in March 2019 at the Morelia Supply Center (Appendix A)), its use is still widespread. It is estimated that in the rural areas of the LPB, machine-made tortillas use 30% of nixtamalized corn flour [38]. Many consumers know that machine-made tortillas include nixtamalized corn flour and other products (whitener, preservatives, enhancers, and softeners) (see Appendix A). The sanitary specifications and commercial information of nixtamalized dough, tortillas, and other products derived from corn are regulated by the official Mexican norm NOM-187-SSA1/SCFI-2002. The Mexican Corn Tortilla Foundation (Fundación Tortilla de Maíz Mexicana or FTMM), integrated by government sectors, international organizations, academics and other members of the civil society, has requested a review of the official Mexican norm on tortillas regarding the allowed additives, labeling standards, control of artificial flavorings and colors, etc. The stated purpose of these changes is to avoid risks to human health and promote a fairer market for traditional maize tortillas [60].

HCT appear, in the imagination of consumers, as a symbol of the value of artisanal and natural products. In general, consumers of HCT are more demanding in terms of taste and texture and feel that machine-made tortillas cannot meet their requirements. Our results indicate, however, that 50% of rural

HCTE in the study area admit adding nixtamalized corn flour and/or wheat flour to the nixtamalized dough used to make HCT. This percentage decreases to 20% in peri-urban areas, while only two of the 15 HCTE studied in urban areas admit adding wheat or nixtamalized corn flour.

We found that, according to the interviewees, the use of wheat or nixtamalized corn flour for making HCT in rural areas can be explained by four main reasons:

(1) To fix a badly cooked nixtamal. If the nixtamal is overcooked, the dough loses consistency and becomes rubbery and watery. This can be fixed by adding wheat flour, which gives elasticity to the dough, and nixtamalized corn flour as a binder product.
(2) To fix a bad grind. If the stones of the mill have been recently cut, the dough can become *quebrada* or *martajada*, that is, with whole pieces of maize that have not been ground completely. In such a case, it is necessary to use a metate to finish the grinding. If, on the contrary, the stones of the mill are old and their marks have become blurred, the dough can turn out sticky, an effect similar to the one produced by overcooking the nixtamal.
(3) To increase the volume of the dough and make more tortillas. Due to the perishable nature of nixtamalized dough, it is important to calculate the volume of tortillas that will be possible to sell, so that the dough does not get spoiled. Sometimes, the opposite happens, more tortillas are sold than expected, and nixtamalized corn flour is mixed with the dough to make more tortillas and make a larger profit.

> *"What MASECA flour does is for the dough to yield more tortillas instantly if the prepared nixtamal is not enough"*

(4) To whiten the dough. The color of maize grains changes according to the variety. Tortilla establishments know that consumers have a greater preference for white tortillas, a fact that has been corroborated by other studies [61]. This drives tortilla makers to add wheat flour (totally white) to the dough and, in this way, whiten the tortillas made from it.

In addition to the reasons described above, which are closely linked to the particularities of the tortilla making process and the demands of the consumers, some people use nixtamalized corn flour to save time and make the elaboration of tortillas easier. This applies not so much to common tortillas, but to other products (mainly fried products) such as quesadillas and/or tostadas. Some of the interviewees indicated that they have become accustomed to using both corn and wheat flour because they have been doing so constantly since the implementation of the subsidy programs carried out by CONASUPO. It would be necessary to ask if, as Mitchell [15] suggests, these habits are the intended result of policies that aimed to alleviate a structural problem, such as food shortages, by implementing a temporary solution, such as subsidizing imported food products. Or if these habits are the result of short-term plans that did not take into consideration the possible consequences of the application of government policies. The paradoxes of state paternalism.

In urban environments, adding flours to the nixtamalized dough is more typical of industrial tortilla shops, since, according to some HCTE, nixtamalized corn flour is much more expensive than maize grain (12 MXN/kg vs. 5–8 MXN/kg). As indicated before, many urban consumers look for a differentiated product and are willing to pay more for tortillas made from 100% maize. This coincides with what Appendini [62] points out in one of his works when she talks about the phenomenon of industrialization through the generalized use of flour.

3.3. Commercialization and Consumption of Tortillas

VOLUME, PLACES, AND SALE PRICES: In urban areas, HCTE sell four times more tortillas than in rural areas, and twice as many as in peri-urban areas (Table 4). The figure in brackets in the third column corresponds to the average sales volume in urban HCTE, with the inclusion of an outlier establishment. The sales volume of the outlier establishment is similar to that of urban industrial tortilla shops, selling approximately 1000 kg of tortillas per day. This urban HCTE, however, produce various types of comal tortillas (in addition to other maize products such as tostadas or tortilla chips) and employ more than 15 workers. A higher sales volume allows them to diversify their products and marketing niches. This increase in sales volume as we approach urban areas goes together with increasing sophistication of HCTE, which start to be semi-permanent businesses. One passes from a rural context where a high number of women sell tortillas from their homes or go out to sell their tortillas in the community or in other communities, to established urban businesses with regular customers and fixed work schedules. The peri-urban environment shows a mixed picture, with cases of people selling tortillas from their homes or from informal street stalls, and cases of modest but settled tortilla shops.

Table 4. Average values associated with the sale of tortillas in urban, peri-urban, and rural HCTE.

	Rural	Peri-Urban	Urban
Sales volume (kg/week)	71	107	260 (741)
Efficiency Grain/Dough-Tortilla	G-T: 1.74	G-T: 1.76 D-T: 0.73	G-T: 1.27 D-T: 0.77
Sales price (MXN/kg)	18.5	16	Blue: 29.3 White: 17.8

G: Grain, D: Dough, T: Tortilla.

Regarding the efficiency in the conversion of grain to tortilla, we hardly see any difference between rural and peri-urban areas (1.74 and 1.76, respectively). These are remarkably high values compared with the average of the Mexican tortilla industry, which varies between 1.3 and 1.5 [63]. Greater conversion efficiency was identified in rural environments for HCT to which nixtamalized corn or wheat flour is added (1.83), compared to tortillas to which no flour is added (1.68). The lowest efficiency was observed in urban HCTE (1.27), which contradicts the widespread belief that hybrid maize yields more tortillas than native maize. We did not find any difference between them, which coincides with the results of other studies in the region [43]. The conversion of dough to tortilla does not vary significantly, it is around 75% efficient in both urban and peri-urban HCTE.

The price of each kg of tortillas is slightly higher in rural areas (Table 4), with no significant differences between the price of those made with native, hybrid, blue, or organic maize. The most affordable price is observed in peri-urban HCTE. In the urban area, there are significant differences between the price of white and blue tortillas, the latter being 1.65 times higher (Table 4). It should be noted that HCTE that sell blue tortillas are located in areas of the city that have high socioeconomic levels [64], which confirms that this type of product tends to be sold as gourmet food accessible only to certain urban strata. In the three study areas, the price of HCT is higher than that of machine-made tortillas, which is set at 14.71 MNX/kg for Michoacán [65].

These results raise the question of what alternatives are available for the lowest socioeconomic strata of urban areas to access native maize tortillas or organic maize products, and what is the relationship with the high levels of obesity found in those areas. It would be very interesting to conduct a study on these issues and assess their relationship with food sovereignty.

CONSUMPTION AND WORK: The sale and consumption of HCT are associated with different factors, depending on the area in which they are studied, and reflect variable and heterogeneous conditions. In rural areas, they reflect the marginalization of agricultural work, as well as the importance of tortillas for the livelihood of rural families. In the rural environment, a high proportion of maize and tortillas is used for self-consumption, and the sale of HCT to be a way to obtain an additional economic income to access other necessary goods. It has been reported that maize tortillas also function as exchange currency among residents of rural communities [66]. In rural areas, the producers and the consumers of the corn-tortilla system coexist, exchange goods, dialogue and establish links between each other that go beyond mere commercial relationships. As other studies focusing on Mexican rural areas have pointed out, maize is the central element in the organization of the work carried out by rural inhabitants, their eating habits and the perception of their quality of life [23,67,68]. Our study corroborated this fact through the testimony of Doña Marta:

"Here in Uricho, we go to bed with maize and get up with maize"

As one approaches urban environments, the purpose of making and selling tortillas turns to economic gain and business consolidation, with the selling of tortillas as the only productive activity of each HCTE. In peri-urban environments, however, different situations coexist. Here, strategies associated with the rural environment (purchasing grains from local farmers because they are neighbors, exchange of food for specific jobs, etc.) coexist with strategies resulting from the pressure exerted by the greater supply of tortillas that can be found in the nearby city (lower selling prices, etc.). In these areas, HCTE are usually informal and unstable. These are areas where some poverty indicators are higher than in rural areas [64].

We must emphasize that in many cases, and in the three environments studied here, the production and sale of HCT do not generate enough income to be able to live solely from this productive activity. However, there are at least two reasons why this activity is maintained in cases in which it does not generate a continuous economic benefit: (1) First, economic need. As part of their daily domestic work, which includes making tortillas for the family, some women make more tortillas to sell them and have some money of their own. This activity does not generate a large or continuous income, but it helps relieve some economic difficulties. (2) The second reason is the need of feeling useful. This feeling is not only about receiving a financial reward, but also about the satisfaction of maintaining a clientele that values your work. Doña Socorro has a small stall in the Plaza de Tenencia Morelos, in the peri-urban area of Morelia, and she explains it this way: *"The most satisfactory part of my work is the dignity that earning a living gives to me."*

Regarding consumption and nutritional issues, obesity and chronic diseases, such as type 2 diabetes among others, are strongly linked to high glycemic index diet [69]. Some authors have found that maize tortillas, produced from nixtamalized maize, present high levels of dietary fiber and resistant starch content associated with whole grains with lower glycemic index [70,71]. That corroborates the health benefits of nixtamalized tortilla consumption by improving food security.

3.4. Perceptions about the Quality of HCT

In some urban environments, it is usual to refer to comal tortillas as "traditional" or "artisanal" tortillas. These words confer a connotation of higher quality which, at the same time, gives the product an added value. Although quality is a social construction [46], and in many cases subjective, HCT are highly appreciated in towns and cities. A large market for them has usually existed in the country, providing a livelihood for "tortilleras" (women who make tortillas for sale) [52]. Lerner and Appendini [72] reported that many consumers look for "real and/or traditional" HCT rather than tortillas from commercial tortilla shops.

When we asked the people who run HCTE in rural areas about their perceptions of the terms "quality product", "artisanal product" and "natural product", they acknowledged not being familiar with these concepts and associated them indistinctly with flavor, texture, durability or the use of nixtamalized dough (without the addition of flour). They did not consider their product to be special or valuable. Most consider it a way to help them earn a livelihood, to survive. Doña Beatriz, from Cucuchucho, told us: *"Here, we only distinguish two types of tortillas, machine-made tortillas and handmade tortillas."* Doña Beatriz also revealed the secret of a good tortilla, which coincided with what most of the people interviewed in rural areas told us: The secret is the nixtamal and tortillas are made from "pure maize". Some people also mentioned that good tortillas should be "passed through the metate".

In peri-urban areas, people are more familiar with marketing terminology, associating "quality product" with flavor and cooking and the term "natural product" with tortillas made with nixtamalized dough, however, the term "artisanal product" confused 100% of respondents. Not all urban establishments identified the terminology chosen in the study to define or classify their product, but, paradoxically, the three urban establishments that sell blue tortillas in Morelia did. Much of the scientific literature about tortillas focuses on the chemical and physical properties of a "good tortilla," the most important being the breaking point, extensibility distance, and durability [46,73,74]. It is no surprise then that 75% of urban HCTE associate the term "quality product" with these properties. In this regard, Don Abdiel, the owner of a HCTE in Morelia declared: *"A soft tortilla does not break when bitten or bent, and swells again when reheated as if it were freshly made."*

People who make and sell tortillas also have perceptions about their future and the continuity of this occupation. Generational replacement in the HCT business is in danger for 80% of the women in the rural areas, given that this activity is associated with lack of education, lack of opportunities, and poverty. These women also consider it a highly physically demanding job, although the hardest part is the uncertainty of being able to sell the product. None of the interviewed women in the rural sector wants this job for their children. However, these perceptions change drastically when the business is prosperous and innovative. In such cases, women say they would like their sons or daughters to continue with the activity. Some of the interviewed are proactive and see this activity as a life choice, revaluing its benefits, and they say that selling tortillas is a dignifying job.

4. Conclusions

Michoacán is one of the most culturally representative states of traditional Mexico. Proof of this are its agriculture, its richness in maize native races, and its gastronomy. However, the data obtained in the present study suggests that there is a gradient of loss of culinary traditions from rural to urban environments. Even rural areas have strayed away from traditional methods and ingredients.

The most contrasting characteristics observed suggest that in rural areas, HMCTE are conceived as livelihoods, while in urban areas HMCTE are established businesses with a prospect of continuity in the medium term. On the other hand, in many urban HCTEs, women are not necessarily in charge of the steps involved in the whole tortilla making process: Dough can be purchased, maize milled in industrial mills or as women, that are part of a manufacturing chain, are only in charge of one particular task. This phenomenon implies a loss of the ancestral knowledge of tortilla making as a whole.

The limited availability of native maize in urban environments sometimes makes it impossible to access native maize products and/or makes them more expensive. That is why almost 100% of HCT in these areas are made with hybrid maize from Sinaloa or Jalisco. In rural areas, native maize is mostly produced for self-consumption. Half of the rural HCTE use native maize for making the HCT they sell. Paradoxically, 50% of the HCT sold in rural areas contain MASECA and/or wheat flour, something that goes against what those same people defined as "quality tortillas". The consumption of blue tortillas (which guarantees that they are made with native maize) is a luxury in urban areas, and a sign of backwardness in rural areas. There is no generational replacement perspective for tortilleras who make HCT due to the hardness of the work and its social connotations, associated with poverty. Tortilleras are one of the poorest socio-economic sectors, especially those from rural areas. These women will continue with this activity in the short future because of their limited access to other income-generating activities. For poor and marginal rural families, selling tortillas is an important source of income and daily currency, but generational replacement seems uncertain because nobody wants to condemn their daughters to poverty and/or a very physically demanding work.

Given that maize, nixtamalized corn flour, and fuelwood are the primary inputs needed to make HCT, the price of these determine its economic viability. Peri-urban and urban HCTE are more vulnerable to price increase and lack nixtamalized corn flour stocking because they depend almost 100% on gas and imported maize from other regions. Thus, rural HCTE were local maize and fuelwood are secured are less vulnerable and profitable. Firewood represents a relatively lower cost than gas for the ones that would need to purchase it.

There is a growing demand for HCT made of native maize and traditional nixtamal, however, in the peri-urban and urban sectors, innovation for making the whole process, with its four steps, easier and more efficient is urgent. Using appropriate technological tools for cultivating maize and cooking tortillas, such as *Patsari* or other improved cookstoves, could promote a more efficient and sustainable system of maize and tortilla production. Given the growing market of organic food, farmers could also find a secure market for the native maize they produce. This would allow indigenous landraces to be preserved and artisanal tortillas to be consumed in urban environments, preserving gastronomical traditions, securing local food security, and a providing tangible environmental, health, and economic benefits for rural and urban HCTEs.

Author Contributions: Conceptualization, E.A. and M.A.; data curation, E.A.; formal analysis, E.A.; investigation, E.A. and M.A.; methodology, E.A.; supervision, M.A.; validation, E.A.; writing—original draft, E.A.; writing—review & editing, E.A and M.A.

Funding: This research has not received external funding.

Acknowledgments: To all HCTEs' owners interviewed both in Morelia and in the LPB communities. To Carmen Patricio, resident of the LPB and expert connoisseur of the maize-tortilla system. To Ana Clara Pla and Omar Masera for their invaluable help and contributions. First author also thanks Centro de Investigaciones en Geografía Ambiental at Universidad Nacional Autónoma de México and DGAPA for her postdoctoral fellowship (2018–2019).

Conflicts of Interest: The authors declare no conflicts of interest.

Appendix A

Figure A1. Prices of maize-related products for sale at The Morelia Supply Center (MSC) dispenser.

References

1. FAO; IFAD; UNICEF; WFP; WHO. *The State of Food Security and Nutrition in the World 2018. Building Climate Resilience for Food Security and Nutrition*; FAO: Rome, Italy, 2018.
2. Ericksen, P.J. What Is the Vulnerability of a Food System to Global Environmental Change? *Ecol. Soc.* **2008**, *13*, 14. [CrossRef]
3. FAO. *Committee on World Food Security*; FAO: Rome, Italy, 2012.
4. Blas, A.; Garrido, A.; Unver, O.; Willaarts, B. A comparison of the Mediterranean diet and current food consumption patterns in Spain from a nutritional and water perspective. *Sci. Total Environ.* **2019**, *664*, 1020–1029. [CrossRef]

5. FAO. *El Sistema Alimentario en México—Oportunidades Para el Campo Mexicano en la Agenda 2030 de Desarrollo Sostenible*; FAO: Ciudad de México, Mexico, 2019.
6. Drewnowski, A.; Popkin, B.M. The Nutrition Transition: New Trends in the Global Diet. *Nutr. Rev.* **1997**, *55*, 31–43. [CrossRef] [PubMed]
7. Dhurandhar, E.J. The food-insecurity obesity paradox: A resource scarcity hypothesis. *Physiol. Behav.* **2016**, *162*, 88–92. [CrossRef] [PubMed]
8. Popkin, B.M. Rural areas drive the global weight gain. *Nature* **2019**, *569*, 200–201. [CrossRef] [PubMed]
9. Entrena-Duran, F. *Food Production and Eating Habits from around the World. A Multidisciplinary Aproach*, 1st ed.; Entrena-Duran, F., Ed.; Nova Science Publishers: New York, NY, USA, 2015; ISBN 9781634825405.
10. Bridle-Fitzpatrick, S. Food deserts or food swamps?: A mixed-methods study of local food environments in a Mexican city. *Soc. Sci. Med.* **2015**, *142*, 202–213. [CrossRef]
11. Cooksey-Stowers, K.; Schwartz, M.B.; Brownell, K.D. Food swamps predict obesity rates better than food deserts in the United States. *Int. J. Environ. Res. Public Health* **2017**, *14*, 1366. [CrossRef]
12. OCDE. *Obesity Update 2017*; OECD: Paris, France, 2017.
13. Unar-Munguía, M.; Monterubio Flores, E.; Colchero, M.A. Apparent consumption of caloric sweeteners increased after the implementation of NAFTA in Mexico. *Food Policy* **2019**, *84*, 103–110. [CrossRef]
14. Khoury, C.K.; Bjorkman, A.D.; Dempewolf, H.; Ramirez-Villegas, J.; Guarino, L.; Jarvis, A.; Rieseberg, L.H.; Struik, P.C. Increasing homogeneity in global food supplies and the implications for food security. *Proc. Natl. Acad. Sci. USA* **2014**, *111*, 4001–4006. [CrossRef]
15. Mitchell, K.E. *State-Society Relations in Mexico: Clientelism, Neoliberal State Reform, and the Case of Conasupo*; Routledge: New York, NY, USA, 2001; ISBN 9781138726505.
16. Pilcher, J. The Globalization of Mexican Cuisine. *Hist. Compass* **2008**, *6*, 529–551. [CrossRef]
17. Montanari, M. *Food Is Culture*; Columbia University Press: New York, NY, USA, 2006.
18. UNESCO. *Traditional Mexican Cuisine-Ancestral, Ongoing Community Culture, the Michoacan Paradigm: Nomination File No. 00400 for Inscription on the Representative List of the Intangible Cultural Heritage in 2010.* [English] [Online: United Nations Educational, Scientific and Cultural Organization]. Available online: https://ich.unesco.org/en/RL/traditional-mexican-cuisine-ancestral-ongoing-community-culture-the-michoacan-paradigm-00400 (accessed on 20 June 2019).
19. Sammells, C.A. Haute Traditional Cuisines: How UNESCO's List of Intangible Heritage Links the Cosmopolitan to the Local. In *Edible Identities: Food as Cultural Heritage*; Brulotte, R.L., di Giovine, M.A., Eds.; Ashgate Publishing: Burlington, VT, USA, 2014; pp. 141–158. ISBN 9781315578781.
20. Astier, M.; Pérez-Agiz, E.; Orozco-Ramírez, Q.; Patricio Chávez, M.d.C.; Moreno-Calles, A.I. Sistema agrícolas, conocimiento tradicional y agrobiodiversidad: El maíz en la cuenca del Lago de Pátzcuaro. In *Conocimiento Tradicional, Innovación y Reapropiación Social*; Argueta Villamar, A., Gomez Salazar, M., Navia Antezana, J., Eds.; Siglo XXI: México D.F., Mexico, 2012; pp. 146–172.
21. Peralta de Legarreta, A. *Cultura Gastronómica en la Mesoamérica Prehispánica*, 1st ed.; Siglo XXI Editores: Ciudad de México, Mexico, 2018; ISBN 9786070309403.
22. Vavílov, N.I. Mexico and Central America as a basic center of origin of cultivated plants in the new world. *Appl. Bot. Genet. Plant Breed.* **1931**, *26*, 207–238.
23. Cárdenas Marcelo, A.L.; Vizcarra Bordi, I.; Espinoza-Ortega, A.; Calderón, A.E. Tortillas artesanales mazahuas y biodiversidad del maíz nativo. Reflexiones desde el ecofeminismo de la subsistencia. *Soc. y Ambient.* **2019**, *7*, 265–291. [CrossRef]
24. Sabán de la Portilla, C.; Oroazco Ramírez, Q.; Astier, M. Análisis ambiental, social y económico del abasto de maíz y transformación en tortillas artesanales en la cuenca del lago Pátzcuaro, estado de Michoacán, México. *Agroecología* **2016**, *11*, 77–93.
25. SADER. *Reporte del Mercado del Maíz*; SADER: Ciudad de Mexico, Mexico, 2019.
26. Alarcón-Cháires, P. *Ecología y Transformación Campesina en la Meseta P'urhépecha: Una Tipología Socio-Ecológica de Productores Rurales de Nahuatzen, Michoacán*; S. de D.C. y E.U., Ed.; Universidad Michoacana de San Nicolás de Hidalgo: Morelia, México, 2001.
27. Orozco-Ramírez, Q. El sistema alimentario del maíz en Pátzcuaro Michoacán [Maize food system in Patzcuaro, Michoacan]. Master's Thesis, Universidad Nacional Autónoma de México, Ciudad de México, Mexico, 2007.

28. Cheetham, D. Corn, Colanders, and Cooking: Early Maize Processing in the Maya Lowlands and Its Implications. In *Pre-Columbian Foodways. Interdisciplinary Approaches to Food, Culture, and Markets in Ancient Mesoamerica*; Staller, J.E., Carrasco, M., Eds.; Springer: New York, NY, USA, 2010; pp. 345–368. ISBN 9781441904706.
29. Durán, H. *Apetito por la vida. Hijos del maíz*; Hojasanta: Ciudad de Mexico, Mexico, 2017.
30. de Dios Figueroa Cárdenas, J.; Acero Godínez, M.G.; Vasco Méndez, N.L.; Lozano Guzmán, A.; Flores Acosta, L.M. Nutritional quality of nixtamal tortillas fortified with vitamins and soy proteins. *Int. J. Food Sci. Nutr.* **2003**, *54*, 189–200. [CrossRef] [PubMed]
31. Paredes López, O.; Guevara Lara, F.; Bello Pérez, L.A. La nixtamalización y el valor nutritivo del maíz. In *Ciencias*; Universidad Nacional Autónoma de México: Ciudad de México, Mexico, 2009; pp. 60–70.
32. Katz, S.H.; Hediger, M.L.; Valleroy, L.A. Traditional maize processing techniques in the new world. *Science* **1974**, *184*, 765–773. [CrossRef] [PubMed]
33. Zazueta, C.; Ramos, G.; Fernández-Muñoz, J.L.; Rodríguez, M.E.; Acevedo-Hernández, G.; Pless, R.C. A radioisotopic study of the entry of calcium ion into the maize kernel during nixtamalization. *Cereal Chem.* **2002**, *79*, 500–503. [CrossRef]
34. Orozco-Ramírez, Q.; Astier, M. Socio-economic and environmental changes related to maize richness in Mexico's central highlands. *Agric. Human Values* **2017**, *34*, 377–391. [CrossRef]
35. INEGI. *Censo General de Población y Vivienda*; INEGI: Ciudad de México, Mexico, 2010.
36. Güiza, F.; Simmons, P.; Pola-Villaseñor, S.; McCall, M. Relaciones de poder y fallo institucional. Vulnerabilidad a desastres en dos ciudades medias. In *Procesos Periurbanos: Desequilibrios Territoriales, Desigualdad Social, Ambientales y Pobreza*; Vieyra, A., Méndez-Lemus, Y., Hernández-Guerrero, J.A., Eds.; CIGA-UNAM: Morelia, México, 2018; pp. 89–114. ISBN 9786073008877.
37. Ortíz Moreno, J.A.; Vieyra Medrano, A. Periurbanización y sus efectos en el ambiente y la calidad de vida: Análisis en dos localidades socieconómicamente contrastantes de Morelia, Michoacán. In *Procesos Periurbanos: Desequilibrios Territoriales, Desigualdad Social, Ambientales y Pobreza*; Vieyra, A., Méndez-Lemus, Y., Hernández-Guerrero, J.A., Eds.; CIGA-UNAM: Morelia, México, 2018; pp. 61–88. ISBN 9786073008877.
38. Astier, M.; Odenthal, G.; Patricio, C.; Orozco-Ramírez, Q. Handmade tortilla production in the basins of lakes Pátzcuaro and Zirahuén, Mexico. *J. Maps* **2019**, *15*, 52–57. [CrossRef]
39. Astier, M.; Barrera-Bassols, N.; Odenthal, J.; Ramirez, M.I.; Orozco, Q.; Mijangos-Cort'Es, J.O. Participatory identification and mapping of maize diversity in the Pátzcuaro-Zirahuén Basins, Michoacán, Mexico. *J. Maps* **2010**, *6*, 1–6. [CrossRef]
40. Morales Vallejo, P. *Tamaño Necesario de la Muestra: ¿Cuántos Sujetos Necesitamos?* UPC: Madrid, Spain, 2012.
41. FAO. *Developing Sustainable Food Value Chains—Guiding Principles*; FAO: Rome, Italy, 2015.
42. Orozco-Ramírez, Q.; Odenthal, J.; Astier, M. Maize diversity in Patzcuaro, Michoacan, Mexico, and its relationship with environmental and social factors. *Agrociencia* **2017**, *51*, 867–884.
43. Orozco-Ramírez, Q.; Barrera-Bassols, N.; Astier, M.; Masera, O.R. El sistema maíz-tortilla en Michoacán. In *Ciencia y Paciencia Campesina. El maíz en Michoacán*; Seefoó Luján, J.L., Baer, N.M.K., Eds.; El Colegio de Michoacán y Gobierno del Estado de Michoacán: Zamora, Michoacán, Mexico, 2010; p. 287.
44. AgroSíntesis. *Propone Sinaloa imponer arancel al maíz Amarillo*; Agro Síntesis: Ciudad de México, Mexico, 2019; p. 30.
45. Castañeda Zabala, Y.; González Merino, A.; Chauvet Sánchez, M.; Ávila Castañeda, J.F. The Maize Seed Industry in Jalisco. Social Actors in Conflict (in Spanish). *Sociológica* **2014**, *29*, 241–279.
46. Appendini, K.; Quijada, M.G. Consumption strategies in Mexican rural households: Pursuing food security with quality. *Agric. Hum. Values* **2016**, *33*, 439–454. [CrossRef]
47. Boué, C.; López Ridaura, S.; Rodríguez Sánchez, L.M.; Hellin, J.; Fuentes Ponce, M. Local dynamics of native maize value chains in a peri-urban zone in Mexico: The case of San Juan Atzacualoya in the state of Mexico. *J. Rural Stud.* **2018**, *64*, 28–38. [CrossRef]
48. SAGARPA. *Atlas Agoalimentario 2016*, 1st ed.; SIAP, Ed.; SIAP: Ciudad de México, Mexico, 2016.
49. Gallegos-Hernández, B.P.; Pérez-Villarreal, H.H.; Barahona, I.; Mayett-Moreno, Y. Analysis of the intrinsic signals, extrinsic signals and the expected quality of the organic tortilla to assess its purchasing intentions. *Cogent Bus. Manag.* **2018**, *5*, 1–23. [CrossRef]
50. Masera Astier, O.X.; Astier, M. *La red Tsiri: Una Experiencia de Sistemas Alimentarios Locales Sustentables*; Leisa: Lima, Perú, 2014; pp. 22–24.

51. Menozzi, D.; Halawany-Darson, R.; Mora, C.; Giraud, G. Motives towards traceable food choice: A comparison between French and Italian consumers. *Food Control* **2015**, *49*, 40–48. [CrossRef]
52. Berrueta, V.M.; Serrano-Medrano, M.; García-Bustamante, C.; Astier, M.; Masera, O.R. Promoting sustainable local development of rural communities and mitigating climate change: The case of Mexico's Patsari improved cookstove project. *Clim. Chang.* **2017**, *140*, 63–77. [CrossRef]
53. Novelo, V.; García, A. *Complementos del Seminario de Problemas Científicos y Filosóficos. La Tortilla: Alimento, Trabajo y Tecnología*; Universidad Nacional Autónoma de México: Ciudad de México, Mexico, 1987.
54. Ortega, T.; Vázquez, V.; Flores, D.; Nuñez, J.F. Agrobiodiversity, gender and food sovereignty in Tlaxiaco, Oaxaca. *Rev. Mex. Cienc. Agrícolas* **2017**, *18*, 3673–3682.
55. Keremitsis, D. Del metate al molino: La mujer mexicana de 1910 a 1940. *Hist. Mex.* **1983**, *33*, 285–302.
56. Astier, M.; Barrera-Bassols, N. *Catálogo de maices criollos de las Cuencas de Pátzcuaro y Zirahuén*, 1st ed.; GIRA, INE, INIFAP, SEDAGRO, UNAM: Ciudad de México, México, 2007; p. 56.
57. Soares Da Silva, A.; Smith, K.R. Clean household air for the Americas. *BMJ* **2019**, *364*, 1330. [CrossRef]
58. Galván-Miyoshi, Y. Integración de indicadores en la evaluación de sustentabilidad: De los índices agregados a la representación multicriterio. In *Evaluación de Sustentabilidad. Un Enfoque Dinámico y Multidimensional*; Astier, M., Masera, O., Galván-Miyoshi, Y., Eds.; SEAE-GIRA-ECOSUR-CIEco-UNAM-GIRA-Mundiprensa: Valencia, Spain, 2008; pp. 95–117.
59. Masera, O.R.; Saatkamp, B.D.; Kammen, D.M. From linear fuel switching to multiple cooking strategies: A critique and alternative to the energy ladder model. *World Dev.* **2000**, *28*, 2083–2103. [CrossRef]
60. La Campiña. Demandan Revisión de Norma de la Tortilla Para Prevenir Riesgos a la Salud. [Spanish] [Online: La Campiña]. 2019. Available online: https://revistalacampina.mx/2019/06/19/demandan-revision-de-norma-de-la-tortilla-para-prevenir-riesgos-a-la-salud/ (accessed on 19 June 2019).
61. Espejel-García, M.V.; Mora-Flores, J.S.; García-Salazar, J.A.; Pérez-Elizalde, S.; García-Mata, R. Caracterización del consumidor de tortilla en el Estado de México. *Agric. Sociedead y Desarro.* **2016**, *13*, 371–384. [CrossRef]
62. Appendini, K. *La Integración Regional de la Cadena Maíz-Tortilla*; CEDUA-COLMEX: Mexico D.F., Mexico, 2010.
63. SE. *Análisis de la Cadena de valor Maíz-Tortilla: Situación Actual y Actores de Competencia Local*; Secretaría de Economía: Mexico D.F., Mexico, 2012.
64. INEGI. Mapa interactivo de Morelia. [Spanish] [Online: Instituto Nacional de Estadística y Geografía, IMPLAN Morelia -Instituto Municipal de Planeación-]. 2010. Available online: https://www.sigemorelia.mx/ (accessed on 28 June 2019).
65. SNIIM. Información Mensual de Precios Diarios de Tortilla en Tortillerías y Autoservicios de México. [Spanish] [Online: Sistema Nacional de Información de Mercados]. 2019. Available online: http://www.economia-sniim.gob.mx/TortillaMesPorDia.asp?Cons=D&prod=1&dqMesMes=6&dqAnioMes=2019&preEdo=Amb&Formato=Nor&submit=Ver+Resultados (accessed on 22 June 2019).
66. Ortega Ortega, T.; Núñez Espinoza, J.F.; Vázquez Garcia, V.; Vizcarra Bordid, I.; Sesiae, P.M.; Flores Sánchez, D. Women and community organizing. The tortilla makers of Tlaxiaco, Oaxaca, Mexico. *EUTOPÍA* **2018**, *13*, 33–52. [CrossRef]
67. Díaz Hernández, B.M.; Ochoa Fernández, M.P.; Ramos Maza, M.T.; Cancino Córdova, S. *Trabajo, Mercado y Género: Mujeres Chiapanecas Productoras de Tostadas de Maíz*, 1st ed.; Universidad de Ciencias y Artes de Chiapas, Centro de Estudios Superiores de México y Centroamérica, El Colegio de la Frontera Sur, Universidad Autónoma de Chiapas: Ciudad de México, Mexico, 2015.
68. Jenatton, M.; Morales, H. Civilized cola and peasant pozol: Young people's social representations of a traditional maize beverage and soft drinks within food systems of Chiapas, Mexico. *Agroecol. Sustain. Food Syst.* **2019**, 1–35. [CrossRef]
69. Radulian, G.; Rusu, E.; Dragomir, A.; Posea, M. Metabolic effects of low glycaemic index diet. *Nutr. J.* **2009**, *8*, 1–8. [CrossRef] [PubMed]
70. Popkin, B.M.; Reardon, T. Obesity and the food system transformation in Latin America. *Obes. Rev.* **2018**, *19*, 1028–1064. [CrossRef]
71. Bello-Perez, L.A.; Osorio-Díaz, P.; Agama-Acevedo, E.; Gonzalez-Soto, R.A. Functional and Beneficial Properties of Corn Tortilla. In *Functional Properties of Traditional Foods*; Kristbergsson, K., Ötles, S., Eds.; Springer US: Boston, MA, USA, 2016; pp. 139–155. ISBN 978-1-4899-7662-8.
72. Lerner, A.M.; Appendini, K. Dimensions of Peri-Urban Maize Production in the Toluca-Atlacomulco Valley, Mexico. *J. Lat. Am. Geogr.* **2012**, *10*, 87–106. [CrossRef]

73. Rangel-Meza, E.; Orozco, A.M.; Miranda-Colín, S.; Vázquez-Carrillo, G.; Cuevas-Sánchez, J.; Merino-Castillo, J. Alkaline cooking, preparation and quality of corn tortilla from Ecatlán, Puebla, México. *Agrociencia* **2004**, *38*, 53–61.
74. Valderrama-Bravo, C.; Domínguez-Pacheco, A.; Hernández-Aguilar, C.; Zepeda-Bautista, R.; Del Real-López, A.; Pahua-Ramos, M.E.; Arellano-Vázquez, J.L.; Moreno-Martínez, E. Physical and chemical characterization of masa and tortillas from parental lines, crosses, and one hybrid. *Int. Agrophys.* **2017**, *31*, 129–138. [CrossRef]

International Journal of
Environmental Research and Public Health

MDPI

Article

Smart Approaches to Food Waste Final Disposal

Franco Cecchi [1] and Cristina Cavinato [2,*]

1 Department of Biotechnology, University of Verona, Strada le Grazie 15, 37134 Verona, Italy
2 Department of Environmental Sciences, Informatics and Statistics, University Ca' Foscari of Venice, Campus Scientifico via Torino 155, 30172 Mestre, Venice, Italy
* Correspondence: cavinato@unive.it; Tel.: +39-041-234-8909

Received: 27 June 2019; Accepted: 8 August 2019; Published: 10 August 2019

check for updates

Abstract: Food waste, among the organic wastes, is one of the most promising substrates to be used as a renewable resource. Wide availability of food waste and the high greenhouse gas impacts derived from its inappropriate disposal, boost research through food waste valorization. Several innovative technologies are applied nowadays, mainly focused on bioenergy and bioresource recovery, within a circular economy approach. Nevertheless, food waste treatment should be evaluated in terms of sustainability and considering the availability of an optimized separate collection and a suitable treatment facility. Anaerobic codigestion of waste-activated sludge with food waste is a way to fully utilize available anaerobic digestion plants, increasing biogas production, energy, and nutrient recovery and reducing greenhouse gas (GHG) emissions. Codigestion implementation in Europe is explored and discussed in this paper, taking into account different food waste collection approaches in relation to anaerobic digestion treatment and confirming the sustainability of the anaerobic process based on case studies. Household food waste disposal implementation is also analyzed, and the results show that such a waste management system is able to reduce GHG emissions due to transport reduction and increase wastewater treatment performance.

Keywords: anaerobic digestion; codigestion; food waste; organic waste; energy and resource recovery; food waste disposal

1. Introduction

Food waste anaerobic biological treatment is becoming an important issue within renewable energy recovery systems, even considering the circular economy approach developed in recent years. Three important keynote speeches about the great importance of the anaerobic codigestion process (AcoD), especially sludge and food waste cotreatment, were performed in 2014, where three world conferences related to anaerobic digestion (AD) took place: the 13th Anaerobic Digestion World Congress (AD13, Santiago de Compostela, Spain), the 11th Latin America Anaerobic Digestion Congress (DAAL XI, L'Avana, Cuba), and the 2nd International Conference on Sustainable Solid Waste Management (Athens, Greece). During the AD13 Conference, Prof. Juan Mata-Alvarez and co-authors presented a review of achievements and perspectives about anaerobic digestion and codigestion (AcoD) [1]. The authors observed that the anaerobic codigestion topic was the most relevant within anaerobic digestion research, and in fact 50% of the overall papers published between 2010 and 2013 were about codigestion: the most frequent substrates used were animal manure (54%), sewage sludge (22%), and the organic fraction of municipal solid waste (11%). Considering the type of substrates used, different sectors are involved: agricultural, public services, and industrial. The integration of these sectors requires an increased sharing of competencies, structures, and needs between them. Typically, anaerobic codigestion has been implemented to improve digester yields in terms of energy production from renewable sources, but it is clear that process stability improves with respect to AD of one single type of substrate thanks to nutrient balance. Nevertheless, Mata-Alvarez et al. observed that the full-scale application of AcoD of sewage sludge (SS) with other substrates has not been reported as would be expected, despite the successful integration of SS and the organic fraction of municipal solid waste (OFMSW) reported by Cecchi et al. [2]. Agricultural waste AD had not yet been considered: in fact, it was discussed for the first time at the IV International Symposium on Anaerobic Digestion (ISAD) of solid waste held in Copenhagen (2005) and was successively included in the topic of the 5th ISAD conference on solid waste and energy crops held in Hammamet (2008). According to this scenario, in the keynote speech held during the 2nd International Conference on Sustainable Solid Waste Management (Athens 2014), the role of anaerobic digestion of food waste was presented as a territorial and environmental process [3]. The role of anaerobic digestion of food wastes considering two proposed strategies based on the codigestion approach was discussed: one was anaerobic digestion applied as a service for the agricultural and farming sector, and the other was as a service for citizens (food waste, diapers, and wastewater treatment integration). The union of these two strategies was an environmentally and territorially friendly process that aimed to produce renewable energy and fertilizer material with low greenhouse gas emissions and nutrient recovery.

The 11th Latin America Anaerobic Digestion Congress was an opportunity for Prof. Polanco [4] to talk about the AcoD of food waste together with SS, in an integrated treatment approach implementing a household food waste disposal (FWD). A FWD is an electric device placed under the sink that is used to directly grind food waste into the sewer system together with wastewater and transport it directly to a wastewater treatment plant (WWTP). It was invented in 1927 and was mostly installed in the USA (late 50s) as a hygienic way to dispose of domestic organic wet wastes. In 2008, the FWD installation in USA households was 60% and in Canada 10%, whereas in Australia and New Zealand it was 12% and 30%, respectively [5]: in all of these countries, there were no legislative limits to FWD installation, but local institutions could avoid their use by considering the loading capacity of WWTPs and sewer conditions. In Europe, FWDs' potential as a waste management strategy has not been fully considered. FWD diffusion has mainly been constrained by controversial opinions and policies that have avoided the direct discharge of food waste into the sewer system. The legislation is made by each member state, so local legislation is adapted to several local aspects, such as the capacity of WWTPs, marketing opportunities for sludge and biogas, cultural attitudes, etc. In the United Kingdom, the installation rate is 5%, the highest among member states, while in other states, the application is still banned (except for Sweden, but only if the unit is connected to a tank) [6]. Recently, Perez et al. [4] discussed and tested the advantages of FWD implementation that gives higher biogas production in an AcoD plant and suggested the feasibility of this approach in terms of energy recovery: by implementing FWDs in half of households, full energy self-sufficiency was reached in the WWTP.

In a study carried out by Iacovidou et al. [7] on the situation of WWTPs and biogas production in the United States, the importance of proper integration of AcoD processes as a way to recover energy and even reduce greenhouse gas (GHG) impacts was highlighted: in fact, the authors reported that wastewater treatment was the eighth largest cause of anthropogenic sources of CH_4 emissions (in 2012, 12.8 million metric tons of CO_2 equivalent were produced). Among 14,780 WWTPs, only 1485 digested the sludge produced, but less than 10% of these plants recovered energy (heat and/or electricity) from biogas utilization. The USA is interested in implementing AcoD to increase the efficiency of biogas production and implement combustion or gas-upgrading systems in order to recover heat, electricity, and fuel. Full-scale implementation of food waste and water treatment was observed even in Germany: Krupp et al. [8] investigated the feasibility of codigestion of sewage sludge and food waste in two wastewater treatment plants from both technical and ecological points of view, and they concluded that the codigestion system was advantageous if compared to composting but that food waste must be pretreated properly and hydraulic retention times should be correctly applied.

Recently, Nghiem and colleagues [9] reviewed bottlenecks and possibilities of full-scale codigestion plants treating wastewater sludge and food waste, and they observed that two of the most important issues are the quality of collected food waste and food waste pretreatment technologies. This aspect was confirmed even in Tyagi et al. [10], a paper in which they reviewed the prospective benefits and challenges of food waste anaerobic digestion, highlighting the importance of pretreatment.

Considering the pioneering works of AcoD in Europe [2,11] on integrated food waste and wastewater treatment and the increasing codigestion studies and applications [9], it is possible to consider this approach a promising option both from a process and technological point of view. In order to sustain the advantage of this integration, two successful AcoD (sewage sludge and food waste) processes implemented on a full scale in Italy are discussed below, together with a discussion on the feasibility of using food waste disposals as alternatives to truck collection. The experimental data were complete with some specific details on industrial applications.

2. Concept of the Smart Wastewater Treatment Plant with Simultaneous Treatment of Wastewater and Food Waste

A smart wastewater treatment plant must represent a service for a city and territory, where citizens and the environment are the beneficiaries. Within this vision, a WWTP treats both wastewater and the organic fraction of municipal solid waste (OFMSW) obtained within separate collections system. The disposal of the OFMSW in the WWTP could be carried out through different ways: using trucks or through the sewer system after the use of an under-the-sink food waste disposal. Bernstad et al. [12] investigated from a Life Cycle Assessment (LCA) point of view the impact of different collection systems such as paper bags, under-the-sink grinders, vacuum systems, etc., and in all cases, they considered anaerobic digestion and energy and nutrient recovery. They observed that the direct discharge of food waste into the sewer decreased the loss of nutrients and materials that can cause eutrophication, and the codigestion balanced the negative impact with energy recovery and low carbon losses.

Regardless of the transfer modality, even if it could obviously change the treatment plant's configuration, what is of interest is the concept that a wet material such as food waste, which contains about 90% water, is treated inside a WWTP suitable for receiving and valorizing it. In fact, thinking about other disposal techniques, an incineration plant is not suitable for the treatment of organic waste because of its limited calorific value: even direct treatment in composting is not optimal, considering the high moisture content of the waste, which requires a substantial amount of structuring material to increase pile porosity. In a smart WWTP, wastewater is treated in a process with Biological Nutrient Removal (nitrogen and phosphorus, BNR) as well as organic carbon. The food waste that comes in the same plant, thanks to its high biodegradability can (after a fermentation step) ensure the effective biological removal of nitrogen and phosphorus, increasing the carbon to nitrogen ratio of the incoming waste. This process scheme was proposed in 1994 (see Figure 1) and realized on a full scale in 1999 (Treviso WWTP): the results of a long-time exercise will be illustrated later on.

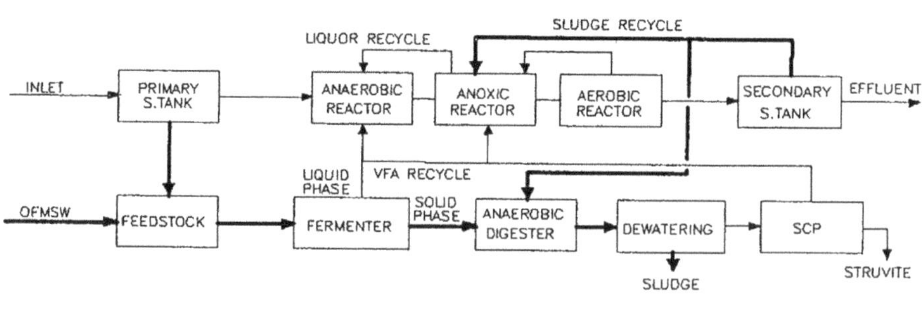

Figure 1. Integrated wastewater and waste treatment scheme [11].

In this context, it is possible to increase depurative efficiency (when needed) by extracting liquid fermentation products from biowaste and using them to enhance biological nutrient removal. This strategy avoids the purchase of expensive external carbon sources and complies with discharge limits in sensitive areas (defined by environmental legislation). The solid fraction of OFMSW obtained after the fermentation and separation steps is mixed with sewage sludge produced during the BNR phase and sent to the anaerobic digestion process. This is the codigestion approach of two different matrixes: the result is increased biogas production if compared to sludge alone and a consequent warranty of energy production in facing the thermal energy requirements of a digester. Koch et al. [13] observed that the codigestion of food waste holds promises not only due to a higher methane yield, but in particular due to the accelerated methane production rate, especially when the mixture of raw sludge with food waste is up to a volatile solids ratio of 35%. Moreover, the biomass mixed culture is modified and allows for a higher loading of waste, with a consequently higher yield. The thermal range adopted could be mesophilic (optimum at 35 °C) or thermophilic (optimum at 55 °C): the thermophilic range has some advantages from a thermodynamic point of view, as it in fact has a higher conversion efficiency of complex molecules and a high pathogen removal potential. The solid part of digestate obtained after thermophilic anaerobic digestion treatment is suitable as an agriculture amendment, closing the environmental cycle. The liquid fraction of digestate, rich in nitrogen and phosphorus, remains inside the WWTP and is treated in a crystallization process to obtain struvite ($(NH_4)MgPO_4 \cdot 6(H_2O)$), a slow release fertilizer: phosphorus is not a renewable resource, so its recovery is an environmental necessity.

This integrated approach could be considered smart because it is able to guarantee the best treatment results: electrical energy production from a renewable source that can be used directly by the WWTP, agronomic valorization of food waste and sewage sludge, and finally, phosphorus recovery. The unsuccessful applications of AD with organic waste are mainly linked to the quality of the waste, which means the presence of inert material such as plastics, iron, etc. There is a strong necessity to adopt an efficient separate collection system. A door-to-door collection system gives characteristics of the best quality in terms of inert material content and meets the quality requested for the AD process [14]: it is possible to assure the high quality of food waste also by simply implementing an under-the-sink food waste disposal connected to a sewer system.

The smart approach development could even be called a "biorefinery approach": the cotreatment of different organic wastes through biological technologies could lead to the production of biobased and biodegradable polyesters, such as polyhydroxyalkanoates (PHAs), by using mixed cultures and wastes. This technology is still being studied on a pilot scale because the volumetric productivity is still lower than industrial processes based on pure substrate and microbial cultures. Reis et al. [15] showed that this technology has significant yield, but some technical/scientific aspects need to be overcome in order to upscale the approach in a sustainable way. Some more specific details about food wastes and sewage sludge used as feedstock for an urban biorefinery with the aim of producing biofuels and added-value bioproducts were reported by Battista et al. [16].

Other interesting examples of the "biorefinery approach" are ongoing studies [17] on the anaerobic digestion of separate collected diapers, from which it is possible to recover a volatile fatty acid (VFA) stream usable for PHA production. Other authors have shown that it is possible to accumulate PHAs in sequencing batches (SBR), treating anaerobic digestion supernatants and recovering VFAs from alkaline fermentation with simultaneous nutrient removal [18,19]. The recovery of VFAs has been of great interest recently thanks to their value as chemical building blocks, and food waste fermentation is the most productive substrate giving the best yields [20].

3. Material and Methods

3.1. Integrated Waste and Wastewater Plant Description

Two cases studies are described below: they represent the first Italian examples of this smart WWTP approach to development. The first project was realized in Treviso, followed by Camposampiero (Padua, PD) and Rovereto (Trento, TN). The first and last WWTPs were monitored by the authors, and then a deeper presentation and discussion was carried out for these two plants.

3.1.1. The Treviso Full-Scale Food Waste Pretreatment and Anaerobic Codigestion Plant

The Treviso WWTP has been operative since 1999: it has 70,000 people equivalents (PEs) of wastewater treatment capacity with a BNR treatment process. The sludge produced is thickened at 3% of total solids and is cotreated with the OFMSW (up to 10 ton day^{-1}) in a 2200-m^3 anaerobic digester. The organic fraction from separate collection was pretreated (within the period considered in this paper) with a low-power system (Patent RN 2004A000038, 2004): the waste was first shredded and sieved with a rotary drum after the first removal of ferromagnetic materials, while those with a low magnetic permeability were removed in a second stage. The substrate thus obtained was reduced in size in a shredder with knives with a span of 15 mm. The biomass was then sent to a wet separator (mixer/separator), where the total solid content was lowered to 7%–8% using the sewage sludge coming from the wastewater treatment line. Here, light and heavy wastes were removed by flotation and gravity, respectively. The mixture sludge/OFMSW treated was sent to the digester by a grinder pump along with the rest of the thickened sewage sludge. The biogas produced was stored and used in a cogeneration unit (190 $kW_{electric\ energy}$) for heat and power production (CHP). The codigestion section operated under both mesophilic and thermophilic temperatures (this lasted for about 4 months) and under widely variable operating conditions, but almost never at full capacity (about 200 ton month^{-1}).

3.1.2. The Rovereto (TN) Full-Scale Food Waste Pretreatment and Anaerobic Codigestion Plant

Rovereto (TN) is a city with about 40,000 inhabitants with a production of 12.5 ton day^{-1} of OFMSW and a recycling rate of 65% [21]. The municipal wastewaters are mixed together with the wastewater produced in the industrial area and are treated in a WWTP with a design capacity of 90,000 PEs but with an effective capacity of approximately 72,000 PEs, arranged in two lines. The wastewater treatment process consists of pretreatment, primary sedimentation, alternate cycles of an activated sludge process [22], secondary sedimentation, and disinfection. The sludge produced by the primary and secondary sedimentation is thickened dynamically and sent to two mesophilic anaerobic digesters working in parallel (5000 m^3 total working volume). The biogas produced is collected in a gas meter and converted into electricity and heat by two high-efficiency gas turbines, while the digested sludge is dewatered to approximately 23% Total Solid (TS) by centrifugation. In 2014, the Rovereto WWTP completed construction on a pretreatment section for OFMSW preparation, implementing the concept of joint processing of biowaste–sewage and anaerobic codigestion. The organic waste is pretreated with a hammer mill to remove inert and to produce a liquid flow to be pumped in the two digesters. The treatment unit is a Wackerbauer TM75, which provides the grinding/pulping of the organic substance and the separation of plastic and other residues. The operation of the equipment requires the addition of water in a ratio of about 1:1 (1 m^3 water ton $_{OFMSW}$$^{-1}$). The slurry obtained has a TS content equal to 5.2% and Total Volatile Solid (TVS) of 82% of TS. This stream is treated in a hydrocyclone for the separation of sands/plastics and stored in two tanks before being fed directly to the digesters, distributing the load over 16 h per day. The residual waste flows are sent to a landfill (15%–20% wet weight). The plant was monitored in the periods before (fed with sludge only) and after the implementation of the codigestion process.

3.1.3. The Camposampiero (PD) Full-Scale Food Waste Pretreatment and Anaerobic Codigestion Plant

Camposampiero WWTP has been operative since 2005: it has 35,000 PEs of wastewater treatment capacity and a wet anaerobic codigestion process working at 55 °C (biphase Linde process, 1500 m^3 first phase and 3300 m^3 s phase). The design data considered 16,000 ton year^{-1} of OFMSW treated together with 25,000 ton year^{-1} of manure and 7800 ton year^{-1} of thickened sludge (8% total solids). As for the pretreatment of organic waste, after a primary shredding, the material passes through an iron removal and a milling step, and then it reaches a wet-pulper. Here, the material is diluted with water or leachate from industrial organic waste. The heavy aggregates (glass, stones, etc.) are removed from the bottom of the pulper while the liquid pulp is fed to a sieve drum for the removal of light coarse materials and plastic, before reaching the first stage of the anaerobic codigestion process. The excess sludge produced by the WWTP and manure is fed directly to the thermophilic digester. In Table 1, there are data reported from the literature on AcoD process performances in order to compare them to the case studies monitored.

Table 1. Camposampiero anaerobic digestion process performance [23]. OFMSW: organic fraction of municipal solid waste.

Parameter	AcoD of Sludge, Manure, and OFMSW at Thermophilic Range
Hydraulic Retention Time (HRT), day	22
Organic Loading Rate (OLR), kg TVS m^3 day^{-1}	3.52
Specific Gas Production (SGP), Nm^3 kg TVS^{-1}	0.67
Gas Production Rate (GPR), Nm^3 m^{-3} day^{-1}	1.46
CH_4, %	58–60

4. Results and Discussion

4.1. The Treviso and Rovereto (TN) Full-Scale Anaerobic Codigestion Plants

Over eight years of Treviso plant monitoring activity, it was possible to observe different operative process conditions: During the first year, the digester was fed with only sewage sludge and the biogas production was not sufficient to reach the mesophilic temperature (35 °C). Starting in August 2001, the OFMSW started to be fed into the anaerobic digestion process, with a consequent increase in biogas production and temperature. Mesophilic conditions were maintained during all of the years after until August 2007, when the thermophilic temperature (55 °C) was tested. The monthly biogas production trend shows a production that ranged from a minimum of 6000 m^3 $month^{-1}$ to a maximum of 21,000 m^3 $month^{-1}$ (Figure 2).

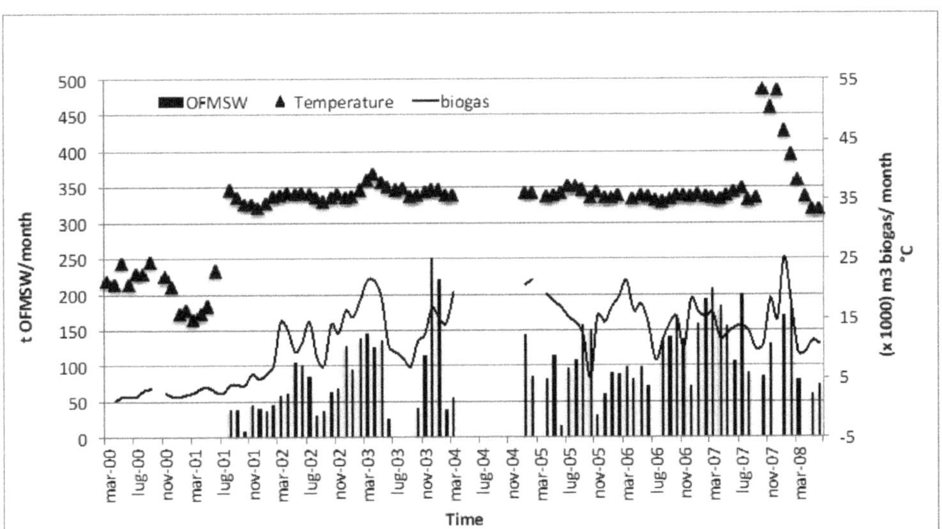

Figure 2. Amount of organic waste treated per month, average biogas production per month, working temperature behavior (Treviso wastewater treatment plant (WWTP)).

The average monthly gas production was 14,097 m^3 $month^{-1}$, and the seven-year average specific gas production (SGP) was 0.3 m^3 kg TVS_{feed}^{-1} instead of the 0.12 typical of AD in secondary sewage sludge alone: this was double the biogas production even with a limited quantity of OFMSW.

The amount of OFMSW fed into codigestion changed with time and was related to waste characteristics, to the behavior of the anaerobic process, and to the annual limit of treatment authorized at 1300 ton year^{-1}. A mass balance analysis (Figure 3) of the Treviso selection system allowed for an evaluation of the selection efficiency and the content of rejected waste of the street separate collection system. The balance was done considering 9 ton day^{-1} of organic waste disposed.

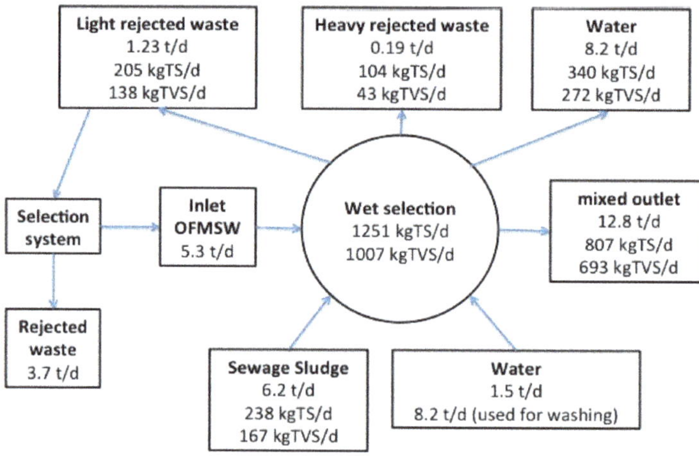

Figure 3. OFMSW mass balance during mechanical selection and wet refinery of Treviso plant (considering 9 ton day^{-1} of food waste treated).

Forty-one percent of the 9 ton day^{-1} treated was rejected during the mechanical waste selection, and as a consequence 59% (5.3 ton day^{-1}) of organic waste was sent to the wet mixer refiner. In this section, water (1.5 ton day^{-1} plus 8.2 t d^{-1} for washing) and activated sludge (6.2 ton day^{-1}) were mixed together in order to decrease the solid content and to allow for heavy and light fraction separation. The mixture obtained had 65% of the inlet TS and 69% of the TVS, and in that way, 27% of TVS was in the wastewater obtained and was recirculated in the WWTP, while 8% was discharged with the heavy fraction. Depending on the season, the quality could change, with a decreased amount of rejected waste. During these seven years of mesophilic working temperatures, the process was monitored in terms of stability parameters, showing no overload conditions and increasing process stability: total alkalinity ranged between 1600 and 3300 mg $CaCO_3$ L^{-1}, with an average of 2328 mg $CaCO_3$ L^{-1}, while the pH was between 6.8 and 7.5, with an average value of 7.15 (Figure 4).

In the case of the Rovereto WWTP (Figure 5), it was possible to observe how the high biodegradability of the organic waste used as a cosubstrate ensured a doubling of biogas production and the consequent doubling of electrical energy production, driving the system toward energy autonomy, while the production of sludge to be disposed of increased only by 10%.

In Figure 5, the monthly average values of biogas production together with total alkalinity during one year of monitoring are reported. It is possible to observe that the increase in organic loading due to codigestion of the OFMSW and SS did not affect the total alkalinity, which was maintained on an almost stable level, with an average value of 3842 mg $CaCO_3$ L^{-1}. The energy yield passed from 4.5 to 7 MWh per day, while the energy demand for pretreatment was only 0.5 MWh per day. These figures allowed for covering the energy demand of some 35 kWh per person equivalent per year [24].

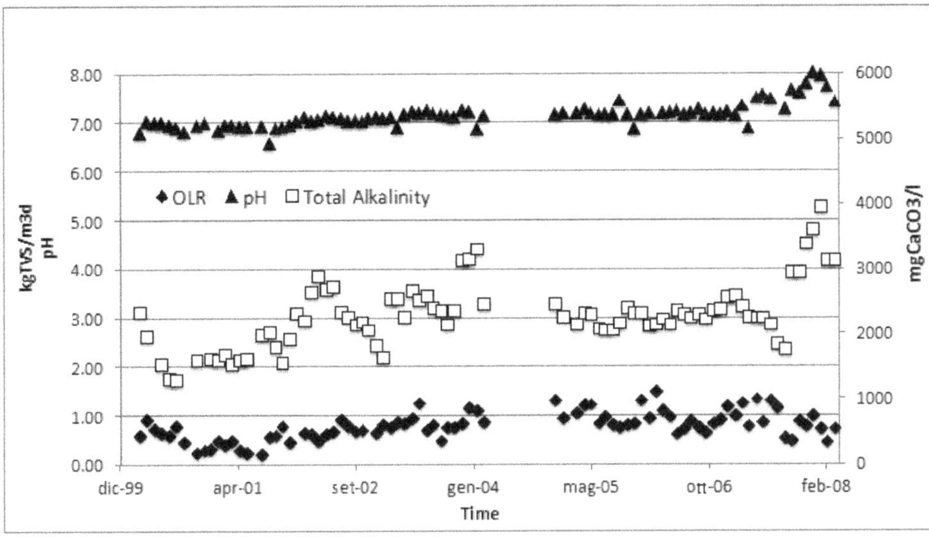

Figure 4. Temperature, total alkalinity, and organic loading rate behavior during eight years of anaerobic codigestion process monitoring of the Treviso anaerobic digestion (AD) process.

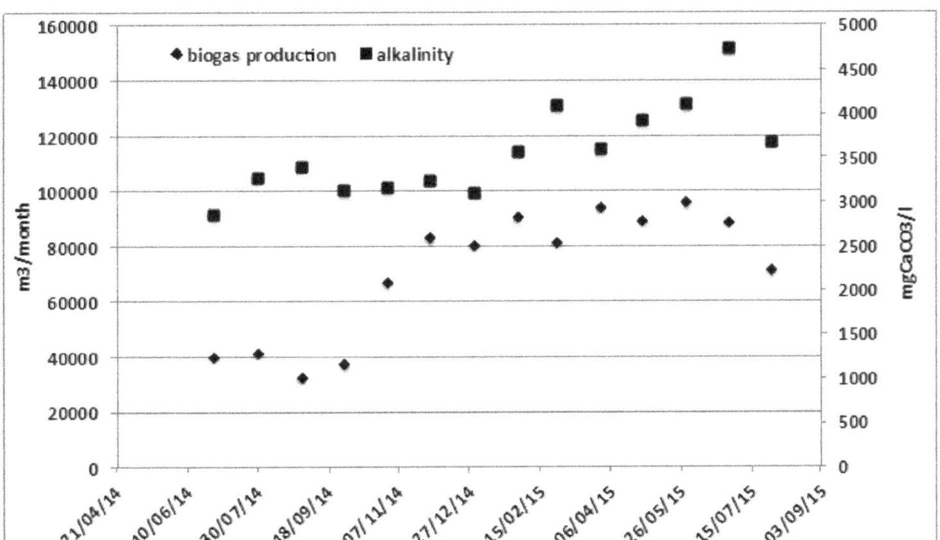

Figure 5. Rovereto AD biogas production changing from mono- to codigestion.

The average values of the monitored period are given in Table 2.

Table 2. Treviso (seven years) and Rovereto (one year, [24]) anaerobic digestion process performance monitoring. TS: total solids; TVS: total volatile solids.

Parameters	Unit	Treviso	Rovereto
Temperature	°C	35.7	35–37
pH		7.15	7.42
Total alkalinity	Mg $CaCO_3$ L^{-1}	2328	3842
Total solids	g L^{-1}	30.1	23.98
Total volatile solids	g L^{-1}	15.8	16.34
TS/TVS	%	52.4	68
Biogas production	m^3 $month^{-1}$	14,097	86,070
Gas production rate (GPR)	m^3 $m^3_{reactor}$ day^{-1}	0.21	0.43
Specific gas production (SGP)	m^3 kg TVS^{-1}	0.3	0.5
Hydraulic retention time (HRT)	day	25–30	30–40
Organic loading rate (OLR)	kg TVS $m^3_{reactor}$ d^{-1}	0.87	1.38
OFMSW collected	ton $month^{-1}$	107.7	

4.2. Under-The-Sink Food Waste Disposal

In terms of using the sewer system to carry the OFMSW to the WWTP, under-the-sink food disposals are of great interest, as Prof. Polanco highlighted in his keynote speech, but the use of this technique does not have particular applications in Italy. The situation is not so different in Europe: significant experiences with this transportation modality of food waste to the treatment plant have been studied. They have demonstrated the incongruity of impediments to the use of this technique through illustrations of data from some experiments carried out on a full scale. In particular, it has been noted that there are more than 110 million devices installed worldwide. Despite this fact, there are objections to the use of under-the-sink disposals concerning clogging and settling phenomena in the sewer system, the pollution of water bodies, overloading, and increased sludge production in wastewater treatment plants. A technical study emphasized that there is no problem with clogging and no need for extraordinary measures to be implemented in a sewer system as a cause of the use of under-the-sink disposals [25], there are no particular problems caused by solid loads increasing in the wastewater [26], and there is no problem with clogging, fouling, and sedimentation [27]: the sedimentation problems are related to the improper use of the equipment, as the material in the sewer system is only partially hydrolyzed and not fermented, and there are no dangerous methanogenic phenomena. Concerning objections about impacts to water bodies, it has been reported that possible problems can occur during periods of rain when the sewers are not provided with overflow systems [25]; however, these problems could be solved with the implementation of equalization tanks for overflow [27]. Concerning the impact on wastewater treatment plants, it is obvious that the increase of organic carbon and ammonia introduced with the food waste requires a larger amount of oxygen for the oxidation step, with a consequent increase of costs; however, there is an important reduction in costs for municipal solid waste disposal [25]. The anaerobic sludge stabilization performance increases significantly [26], with a consequent increase in biogas production [27]. Finally, the increased amount of rapidly biodegradable organic carbon load improves the biological nutrient removal [28,29]. Other aspects related to FWD impacts on energy consumption, water consumption, and GHG emissions have been reviewed by several authors [30–35]. The energy consumption of an FWD is variable and depends on the model, frequency, and duration of use, so it is difficult to calculate the real consumption, but in most of the studies reviewed, this value resulted in insignificance: Bolzonella et al. [36] estimated an additional energy use of 4.3 kWh and 8.5 kWh per year, meaning 0.55 € and 1.10 € per year. Water consumption depends on how much tap water is needed to produce a homogenized mixture to be flushed into the sewer: Iacovidou et al. [30] reviewed consumption ranging between 1 and 6.6 L/capita/day, which is 0.3% to 3.5% of total consumption. The GHG emissions impacts of FWDs have not been considered too much: in fact, there have been few papers on this topic, and Iacovidou et al. [26] reviewed one study in which it was estimated that the use of an FWD and anaerobic digestion generates −16.8 kg CO_2 equiv per 100 kg of food waste compared to −1.4 kg and +74.3 kg CO_2 equiv per 100 kg for composting and landfilling, respectively, but they took into account biogas production, carbon sequestration, and fertilizer offsets. In two life cycle assessment (LCA) studies, FWD ranked second after home composting, with 42.2 kg CO_2 equivalent per 100 kg, and it was the best if coupled with a WWTP (compared to municipal solid waste collection followed by composting, landfilling, etc.), with 44 kg CO_2 equivalent per 100 kg of food waste. Bernstad Saraiva et al. [31] compared two systems in the collection of food waste in households: (a) the use of food waste disposals (FWDs) in kitchen sinks and (b) the collection of food waste in paper bags for further treatment. For both cases, they considered anaerobic digestion and the use of digestate as a fertilizer. They evaluated GHG emissions from the collection and treatment of 1 ton of food waste (dry matter) and found (according to the performed assessment) lower emissions from the FWD system compared to the reference system (−990 and −770 kg CO_2 equivalent ton^{-1} food waste dry matter, respectively). Other authors [35] have assessed different solutions for diverting food waste away from incineration toward biogasification, focusing on the feasibility of implementing FWDs in Aarhus City (Sweden) compared to transport by truck, and they concluded that the separate collection of organic waste for combined biogas and fertilizer production is the most flexible, robust, and least risky economic solution compared to FWD implementation.

In order to clarify the benefits of FWDs, Battistoni et al. [37] studied the effect before and after the implementation of an under-the-sink food waste disposal, monitoring both the sewer system and a small WWTP's performance. The experimentation considered 95 people served and a school canteen (industrial FWD) with a 60-person equivalent capacity: the total "penetration market factor" was about 67% of the resident population. The sewage system had a retention time of 1.5 h: therefore, the time was not sufficient to trigger the fermentation process [36]. The small WWTP had a treatment capacity of 250 PE and a max flow rate of 6.9 m^3 h^{-1}, and the biological process was an intermittent controlled aeration [38,39]. The authors reported no significant solid sedimentation into pipes: therefore the hydraulic overload expected, due to the need for water for the FWD to function, did not give substantial changes, with a flow rate that was slightly leveled (maximum flow rate of 4 m^3 h^{-1}). The quality of wastewater obviously changed, and the Total Suspended Solid (TSS), Chemical Oxygen Demand (COD), and Total Nitrogen (TN) content increased 30%, 44%, and 19%, respectively, with the COD/TN ratio changing from 9.9 up to 12. In addition, the COD/TSS ratio increased from 1.4 up to 2.6. In general, they observed that the FWD technology did not overload the WWTP in any dry or wet period, and the increase in rapidly biodegradable COD could optimize the use of nitrogen-bound oxygen, which meant saving energy in the aeration system. From an economic point of view, the implementation of FWDs avoids the management costs related to source collection and transportation (from 191.400 € y^{-1} to zero) and to treatment and disposal (usually composting technology, from 47.100 € y^{-1} to 6.900 € y^{-1}). Considering the capital costs of collection organization, the use of FWDs could be beneficial after 4–5 years of operation. This experience could be compared to a large-scale case study in Surhammar (Sweden): Evans et al. [40] described a "15-year" application of under-the-sink FWDs and their effects on the sewer system (plus a cost analysis). The results obtained showed that when 50% of households had an FWD installed, this meant that food waste was separated at the source and well managed. Thus, when the system reached an equilibrium with the new loading conditions, the cost effects on wastewater treatment were neutral, and with additional biogas recovery, FWDs made a positive financial contribution.

5. Conclusions

Smart approaches to food waste final and sustainable disposal are those able to reduce first of all the environmental impacts related to waste management: in this context, the application of an under-the-sink food waste disposal avoids transportation pollution and management costs. The second smart aspect is energy recovery through the anaerobic codigestion of food waste and sewage sludge, together with a reduction in CO_2 emissions (if compared to composting technologies). Considering the long-term successful applications of anaerobic codigestion processes within the Italian territory, it is possible to conclude that from technological and environmental points of view, there are no further impediments for new applications of these approaches. Some important evaluations must be carried out related to the application of an effective separate collection, the choice of appropriate food waste pretreatments, and an economical evaluation of the whole process, considering incentives for biogas exploitation and the revenue achievable from these processes from other new products. Food waste is a renewable resource, and through its valorization, it is possible to transform wastewater treatment plants into small biorefineries able to manage waste systems in a sustainable way, with interesting economic revenue for citizens.

Author Contributions: C.F.: conceptualization, review, and supervision; C.C.: data curation, writing—review editing.

Funding: This research received no external funding.

Conflicts of Interest: The authors declare no conflicts of interest.

Abbreviations

AcoD	anaerobic codigestion
AD	anaerobic digestion
BNR	biological nutrient removal
COD	chemical oxygen demand
FWD	food waste disposer
GHG	greenhouse gas
GPR	gas production rate
HRT	Hydraulic retention time
LCA	Life cycle assessment
OFMSW	organic fraction of municipal solid waste
OLR	Organic loading rate
PE	people equivalent
PHA	polyhydroxyalcanoate
rbCOD	rapidly available chemical oxygen demand
SBR	sequencing batch reactor
SGP	specific gas production
SS	sewage sludge
TN	total nitrogen
TS	total solid
TSS	total suspended solid
TVS	total volatile solid
VFA	volatile fatty acid
WWTP	wastewater treatment plant

References

1. Mata-Alvarez, J.; Dosta, J.; Romero-Guza, M.S.; Fonoll, X.; Peces, M.; Astals, S. A critical review on anaerobic co-digestion achievements between 2010 and 2013. *Renew. Sustain. Energy Rev.* **2014**, *36*, 412–427. [CrossRef]
2. Cecchi, F.; Traverso, P.G.; Perin, G.; Vallini, G. Comparison of codigestion performance of two differently collected organic fractions of municipal solid waste with sewage sludge. *Environ. Technol. Lett.* **1988**, *9*, 391–400. [CrossRef]
3. Cecchi, F.; Cavinato, C. Anaerobic digestion of bio-waste: A mini-review focusing on territorial and environmental aspects. *Waste Manag. Res.* **2015**, *33*, 429–438. [CrossRef] [PubMed]
4. Pérez-Elvira, S.I.; Cano, R.; Polanco, M.; Sousa, T.S.O.; Fdz-Polanco, F. Domestic food waste and sewage sludge combined treatment implementing household food waste disposers. In Proceedings of the XI Latin American Workshop and Symposium on Anaerobic Digestion, XI DAAL 2014, La Havana, Cuba, 24–27 November 2014; ISBN 9789592614703.

5. U.S. Environmental Protection Agency. *Examining the Use of Food Waste Disposer, STRIVE Report Series (11)*; EPA: Washington, DC, USA, 2008.
6. Iacovidou, E.; Ohandia, D.G.; Gronow, J.; Voulvoulis, N. The household use of food waste disposal units as a waste management option: A review. *Crit. Rev. Environ. Sci. Technol.* **2012**, *42*, 1485–1508. [CrossRef]
7. Shen, Y.; Lville, J.L.; Urgun-Demirtas, M.; Mintz, M.M.; Snyder, S.W. An overview of biogas production and utilization at full-scale wastewater reatment plants (WWTPs) in the United States: Challenges and opportunities towards energy-neutral WWTPs. *Renew. Sustain. Energy Rev.* **2015**, *50*, 346–362. [CrossRef]
8. Krupp, M.; Schubert, J.; Widmann, R. Feasibility study for co-digestion of sewage sludge with OFMSW on two wastewater treatment plants in Germany. *Waste Manag.* **2005**, *25*, 393–399. [CrossRef]
9. Nghiem, L.D.; Koch, K.; Bolzonella, D.; Drewes, J.E. Full scale co-digestion of wastewater sludge and food waste: Bottlenecks and possibilities. *Renew. Sustain. Energy Rev.* **2017**, *72*, 354–362. [CrossRef]
10. Tyagi, V.K.; Fdez-Güelfo, L.A.; Zhou, Y.; Garcia, L.I.R.; Ng, W.J. Anaerobic co-digestion of organic fraction of municipal solid waste (OFMSW): Progress and challenges. *Renew. Sustain. Energy Rev.* **2018**, *93*, 380–399. [CrossRef]
11. Cecchi, F.; Battistoni, P.; Pavan, P.; Fava, G.; Mata-Alvarez, J. Anaerobic digestion of OFMSW and BNR processes: A possible integration–Preliminary results. *Water Sci. Technol.* **1994**, *30*, 65–72. [CrossRef]
12. Bernstad, A.; Jansen, J.L. Separate collection of household food waste for anaerobic degradation—Comparison of different techniques from a systems perspective. *Waste Manag.* **2012**, *32*, 806–815. [CrossRef]
13. Koch, K.; Helmreich, B.; Drewes, J.E. Co-digestion of food waste in municipal wastewater treatment plants: Effect of different mixtures on methane yield and hydrolysis rate constant. *Appl. Energy* **2015**, *137*, 250–255. [CrossRef]
14. Cavinato, C.; Bolzonella, D.; Pavan, P.; Fatone, F.; Cecchi, F. Mesophilic and thermophilic anaerobic co-digestion of waste activated sludge and source sorted biowaste in pilot- and full-scale reactors. *Renew. Energy* **2013**, *55*, 260–265. [CrossRef]
15. Reis, M. Current status and future prospects on the PHA production by mixed cultures and wastes. In Proceedings of the European Symposium on Biopolimers, Rome, Italy, 16–18 September 2015.
16. Battista, F.; Frison, N.; Pavan, P.; Cavinato, C.; Gottardo, M.; Fatone, F.; Eusebi, A.L.; Majone, M.; Zeppilli, M.; Valentino, F.; et al. Food wastes and sewage sludge as feedstock for an urban biorefinery producing biofuels and added-value bioproducts. *J. Chem. Technol. Biotechnol.* **2019**, in press. [CrossRef]
17. Bolzonella, D.; Brida, M.; Brida, G.; Giurin, G.U.; Mattioli, A.; Cecchi, F. Anaerobic digestion of separately collected diapers: Closing the loop in a circular bio-economy perspective. In Proceedings of the 14° Anaerobic Digestion World Congress, Vina del Mar, Cile, 15–18 November 2015.
18. Frison, N.; Katsou, E.; Malamis, S.; Oehmen, A.; Fatone, F. Nutrient removal via nitrite from reject water and polyhydroxyalkanoate (PHA) storage during nitrifying conditions. *J. Chem. Technol. Biotechnol.* **2015**, *90*, 1802–1810. [CrossRef]
19. Fatone, F.; Frison, N.; Katsou, E.; Malamis, S. Integrating the Selection of PHA Storing Biomass and Nitrogen Removal via Nitrite for the Treatment of the Sludge Reject Water. In Wastewater and Biosolids Treatment and Reuse: Bridging Modeling and Experimental Studies. Available online: http://dc.engconfintl.org/wbtr_i/15 (accessed on 3 May 2019).
20. Strazzera, G.; Battista, F.; Garcia, N.H.; Frison, N.; Bolzonella, D. Volatile fatty acids production from food wastes for biorefinery platforms: A review. *J. Environ. Manag.* **2018**, *226*, 278–288. [CrossRef] [PubMed]
21. ISPRA. *Rapporto Rifiuti Urbani–Dati di Sintesi*; ISPRA: Roma, Italy, 2014; ISBN 978-88-448-0665-1.
22. Nardelli, P.; Gatti, G.; Eusebi, A.L.; Battistoni, P.; Cecchi, F. Full-Scale Application of the Alternating Oxic/Anoxic Process: An Overview. *Ind. Eng. Chem. Res.* **2009**, *48*, 3526–3532. [CrossRef]
23. Gatto, O.; Dindo, A.; Scarpa, C. Digestione anaerobica a liquido e a secco: Un confronto su scala industriale. In Proceedings of the SEP Pollution, Padova, Italy, 14–15 March 2006.
24. Mattioli, A.; Gatti, G.B.; Mattuzzi, G.P.; Cecchi, F.; Bolzonella, D. Co-digestion of the organic fraction of municipal solid waste and sludge improves the energy balance of wastewater treatment plants: Rovereto case study. *Renew. Energy* **2017**, *113*, 980–988. [CrossRef]
25. Department of Environmental Protection, The City of New York. The Impact of Food Waste Disposers in Combined Sewer Areas of New York City 1997. Available online: http://www.nyc.gov/html/dep/pdf/grinders.pdf (accessed on 26 June 2019).

26. Nilsson, P.; Hallin, P.; Johansson, J.; Karlén, L.; Lilja, G.; Petersson, B.; Petterson, J. *Waste Management at the Source Utilising Food Waste Disposers in Home: A Case Study in the Town of Staffanstrop—Final Report;* Department of Environmental Engineering, Lund Institute of Technology, The University of Lund: Lund, Sweden, 1990.
27. De Koning, J.; Van der Graaf, J.H.J.M. *Kitchen Food Waste Disposers. Effects on Sewer Systems and Wastewater Treat- Ment. Technical Report;* Department of Water Management, Environmental and Sanitary Engineering, Technische Universiteit Delft: Delft, The Netherlands, 1996.
28. Pavan, P.; Battistoni, P.; Traverso, P.; Musacco, A.; Cecchi, F. Effect of addition of anaerobic fermented OFMSW on BNR process: Preliminary results. *Water Sci. Technol.* **1998**, *38*, 327–334. [CrossRef]
29. Bolzonella, D.; Innocenti, L.; Cecchi, F. BNR wastewater treatments and sewage sludge anaerobic mesophilic digestion performances. *Water Sci. Technol.* **2002**, *46*, 199–208. [CrossRef]
30. Iacovidou, E.; Ohandja, D.G.; Voulvoulis, N. Food waste disposal units in UK households: The need for policy intervention. *Sci. Total Environ.* **2012**, *423*, 1–7. [CrossRef]
31. Maalouf, A.; El-Fadel, M. Effect of a food waste disposer policy on solid waste and wastewater management with economic implications of environmental externalities. *Waste Manag.* **2017**, *69*, 455–462. [CrossRef] [PubMed]
32. Maalouf, A.; El-Fadel, M. Carbon footprint of integrated waste management systems with implications of food waste diversion into the wastewater stream. *Resour. Conserv. Recycl.* **2018**, *133*, 263–277. [CrossRef]
33. Bernstad Saraiva, A.; Davidsson, A.; Bissmont, M. Lifecycle assessment of a system for food waste disposers to tank–A full-scale system evaluation. *Waste Manag.* **2016**, *54*, 169–177. [CrossRef] [PubMed]
34. Lijo, L.; Malamis, S.; Gonzalez-García, S.; Moreira, M.T.; Fatone, F.; Katsou, E. Decentralised schemes for integrated management of wastewater and domestic organic waste: The case of a small community. *J. Environ. Manag.* **2017**, *203*, 732–740. [CrossRef] [PubMed]
35. Thomsen, M.; Romeo, D.; Caro, D.; Seghetta, M.; Cong, R.G. Environmental-Economic Analysis of Integrated OrganicWaste andWastewater Management Systems: A Case Study from Aarhus City (Denmark). *Sustainability* **2018**, *10*, 3742. [CrossRef]
36. Bolzonella, D.; Pavan, P.; Battistoni, P.; Cecchi, F. The under sink garbage grinder: A friendly technology for the environment. *Environ. Technol.* **2003**, *24*, 349–359. [CrossRef] [PubMed]
37. Battistoni, P.; Fatone, F.; Passacantando, D.; Bolzonella, D. Application of food waste disposers and alternate cycles process in small-decentralized towns: A case study. *Water Res.* **2007**, *41*, 893–903. [CrossRef] [PubMed]
38. Battistoni, P.; Boccadoro, R.; Bolzonella, D.; Marinelli, M. An alternate oxic–anoxic process automatically controlled: Theory and practice in a real treatment plant network. *Water Sci. Technol.* **2003**, *48*, 337–344. [CrossRef] [PubMed]
39. Battistoni, P.; de Angelis, A.; Boccadoro, R.; Bolzonella, D. An automatically controlled alternate oxic–anoxic process: A feasible way to perform high nitrogen biological removal also during wet weather periods. *Ind. Eng. Chem. Res.* **2003**, *42*, 509–515. [CrossRef]
40. Evans, T.D.; Andersson, P.; Wievegg, A.; Carlsson, I. Surahammar—A case study of the impacts of installing food waste disposers in fifty percent of households. *Water Environ. J.* **2010**, *24*, 309–319. [CrossRef]

International Journal of
Environmental Research and Public Health

MDPI

Article

Climate Change and Consumer's Attitude toward Insect Food

Hsiao-Ping Chang [1,2], Chun-Chieh Ma [3] and Han-Shen Chen [1,2,*]

1 Department of Health Diet and Industry Management, Chung Shan Medical University, No. 110, Sec. 1, Jianguo N. Rd., Taichung City 40201, Taiwan; pamela22@csmu.edu.tw
2 Department of Medical Management, Chung Shan Medical University Hospital, No. 110, Sec. 1, Jianguo N. Rd., Taichung City 40201, Taiwan
3 Department of Public Administration and Management, National University of Tainan, Taiwan No. 33, Sec. 2, Shu-Lin St., Tainan 70005, Taiwan; ccma@mail.nutn.edu.tw
* Correspondence: allen975@csmu.edu.tw; Tel.: +886-4-2473-0022 (ext. 12225)

Received: 15 March 2019; Accepted: 6 May 2019; Published: 8 May 2019

check for updates

Abstract: Given the influence of rising environmental awareness, food systems and security are receiving increasing international attention. Previous studies have discussed the acceptance of insect foods but have been primarily conducted in a European context. Hence, their results cannot be applied to Taiwanese consumers. Regarding this, our study is centered on the theory of planned behavior and considers environmental concern and food neophobia to discuss the effects of consumer attitudes, subjective norms, and perceived behavioral control on the purchase intention toward insect food. We used purposive sampling to survey questionnaire answers face-to-face in Taichung city, Taiwan. We distributed 408 surveys of which 77.45% were used in this study. The results revealed that consumer attitudes, perceived behavioral control, and food neophobia significantly influence purchase intention, whereas subjective norms and environmental concern did not demonstrate significant relationships with purchase intention. According to these results, we suggest that businesses emphasize consumers' product experience or reduce levels of food neophobia to increase consumer interest in insect foods and improve the acceptability of such foods, thereby increasing purchase intention.

Keywords: vulnerability of food systems; food neophobia; environmental concern; global environmental change; behavior change

---

1. Introduction

According to a report by the Food and Agriculture Organization (FAO) [1] of the United Nations, the global population will reach 9 billion by 2050. Furthermore, in the light of the myriad of environmental concerns we are faced with, including global climate change, shortages in global arable land, and so on, in response to this rapid population growth, food volumes must increase. Comparing edible insects with average traditional livestock, raising 1 kg of edible insects on average only requires 2 kg of feed, whereas raising 1 kg of beef requires 8 kg of feed. Furthermore, they also help to reduce environmental burdens, emit far less greenhouse gases, and require much less land and water than mammals and birds [2]. The main reason insects can be used as meat substitutes is that they can be used at lower economic and environmental costs. The amount of land used per kg of insect protein is 50–90% lower than that of traditional livestock, 40–80% less per kilogram of edible food, and their GHGEs (greenhouse gas emissions) are 1000–2700 grams lower. The study confirmed that there are five edible insects (*Tenebrio molitor*, *Acheta domesticus*, *Locusta migratoria*, *Pachnoda marginata*, and *Lapatia dubia*) with greenhouse gas and ammonia emissions equivalent to pigs, but far less than the emissions of cattle. In addition, the amount of carbon dioxide produced per kilogram is lower than that of traditional livestock [3].

Past literature has evidenced that insects have important effects in the nutritional history of Africa, Asia, and Latin America [4–6], such as the Yukpa people in Colombia and Venezuela who use certain insects in their meat dishes. There are also studies exploring *zonocerus variegatus*, bee larvae, and pupae in southern Nigeria, which are all rich in protein [5,7]. Researchers have shown that insects are a good source of protein and micronutrients [5,6,8]. Zielińska, Baraniak, Karaś, Rybczyńska, and Jakubczyk [9] pointed out that insects have high levels of monounsaturated fatty acids and polyunsaturated fatty acids, which satisfy human requirements for amino acids. Moreover, Poma, Cuykx, Amato, Calaprice, Focant, and Covaci [10] noted that compared to common animal products (meats, fish, eggs), insects (locusts) can even serve as a replacement for common proteins. That study encouraged the consumption of insect foods. Existing literature refers to insect foods as insect-based food [11], insect food [12], and insects as food [13]. This indicates that insect foods do not yet have a standard name, although a global population of 2 billion people considers insects a part of their traditional diet and more than 1900 types of insects are used by humans for food [1]. In remote rural and biodiversity-rich tropical countries, insects have been an important source of protein and micronutrients for thousands of years [6], with the world's most common edible insects accounting for 31% of beetles, followed by caterpillars, bees, wasps, ants, etc. [2]. In Europe, edible insect foods are relatively novel foods [8], and companies already sell edible insect products, such as the French company Micronutris, which sells the cricket *sigillatus*, tablette-tenebrio's chocolate, aldente-aenebrio pasta and crackers, etc., [14] while Finnish chain bakery Fazer sells 'Fazer Cricket Bread' made from glutinous flour [15]. Not only that, the European Union's non-profit organization, IPIFF (International Platform for Insects as Food and Feed, IPIFF), promotes the use of insects and insect-derived products as a source of human consumption and animal feed, and believes that it can be cultured with fewer resources (such as land, water, feed, energy), while generating lower greenhouse gas emissions and pollutants. It is likely to be the most common source of protein for aquaculture and livestock animals, and it is believed that insect proteins will become a generally accepted dietary component of Western society. Not only Western countries however, as palm weevil and cricket have also been cultivated on a commercial scale in Thailand [16], and similar commercial farming models are being developed in Africa [17]. In Taiwan, although it started late, there is an ecological rehabilitation of the Fuyang enterprises that have both green energy and alternative food. In Europe, foods such as insect flour and insect crackers exist. In Asia, Thailand has a well-developed insect-eating culture, and insect dishes including fried silkworm pupae and fried crickets are eaten in Taiwan. This study's discussion on this topic means that insect foods, as food products which have insect ingredients, are on the market and can be presented in any form. For example: cricket biscuits, cricket bread, fried insects (such as grasshoppers, pupae, meal-worms) and so on.

Tan, Tibboel, and Stieger [18] demonstrated that when a novel food, such as insects, is mixed with a familiar product, this can generate positive expectations and improve the sensory experience and attractiveness of the novel food, thereby increasing consumer approval and helping reduce barriers to consumption. In Hartmann, Shi, Giusto, and Siegrist's [19] study comparing the cultures of Germany and China regarding insect eating, Chinese people were found to evaluate insect foods according to their taste, nutritional value, familiarity, and social acceptance. That study indicates that if insects could be combined with a familiar food, this could possibly encourage their use as a source of food in the West and reduce consumers' negative attitudes toward insect foods. Research related to consumers and edible insect food products is not very common, especially in Taiwan, but what does exist includes the manufacturing methods of products [20], consumer knowledge [12], and food suitability [20].

Numerous studies have examined the link between insect foods and purchase intention [12,21,22]. To predict and explain human behaviors, social psychologists have proposed several models. The most frequently utilized among these is the theory of planned behavior (TPB), which is commonly considered as being effective in predicting general behavior. The TPB argues that the following three factors, in combination, are responsible for forming behavioral intentions—perceived behavioral control, attitude toward the behavior in question, and subjective norms [23]. Based on the theory of planned behavior, Ajzen [24] pointed out that perceived behavioral control means "an individual's perceived ease or difficulty in performing a particular behavior". An attitude toward a behavior is considered to be the "degree to which a person has a favorable or unfavorable evaluation or appraisal of the behavior in question". "Subjective norms" can also be divided into individual behavioral norms and social norms. Personal behavioral norms mainly come from the influence of important reference objects formed by parents, friends, peers, or experts. Social norms refer to the influence of pressure from other social groups to perform or not perform the behavior. This means that an individual who possesses a reasonably high level of behavioral control for a particular behavior will have a higher chance of show-casing a firm intention to engage in that behavior. "Behavioral intention" indicates a person's willingness to conduct a certain behavior, with the assumption that this willingness must necessarily exist before the behavior itself is executed [21]. Throughout its development, the TPB has been widely applied by scholars to study purchase intention [25–29]. Menozzi et al. [30] used the TPB to investigate the intentions and expected behaviors of young Italian consumers on the consumption of flour and chocolate biscuits containing 10% crickets. Their study found that the primary obstacles to the consumption of insect-containing flour products is the aversion generated from seeing nearby insects, nonconformance to local food culture, and the lack of related products in supermarkets. According to Menozzi et al. [30] in relation to insect foods, modern consumers maintain a certain degree of reservations toward purchasing insect food products. This may be the result of consumers' aversion to insects, incompatible local food cultures, or consumers not having seen such products before.

Previous studies have noted that when compared with familiar foods, a consumer's willingness to choose new foods, or those that have not been previously eaten, may be affected by neophobia [31–33]. Pliner and Hobden [31] referred to food neophobia as the fear of food that has either not been eaten or seen in an eating situation, and describe an unwillingness to approach and eat unfamiliar food. People generate a mixture of conflicting curiosity and fear toward foods they have not previously seen [34]. Pliner and Hobden [31] developed the food neophobia scale (FNS) to measure aversion to new foods, and previous studies have utilized the FNS to evaluate consumer willingness to eat or choose new foods [18,31,35,36], familiarity and experience with foreign foods [31,33], willingness to explore food flavors [35,37], and expectations from new foods [31,36]. Some scholars have applied a food technology neophobia scale to measure consumer attitudes toward new and traditional food technologies [38], university students' satisfaction regarding food-related life [39], and consumer neophobia to red wine [40]. In addition, the degree of an individual's food neophobia influences their purchase of food products and frequency of purchase [41], as well as their eating habits [42]. Along the same lines, La Barbera et al. [21] noted that consumer food neophobia and aversions affect the willingness to eat insect foods. The present study postulates that due to varying levels of food neophobia among Taiwanese consumers, consumers will have varying reactions, which may influence their willingness to purchase insect food products. For these reasons, the present study includes food neophobia as a research variable.

Whether consumers are willing to purchase a product does not depend simply on their preference for the product. Rather, environmental awareness tactics influence consumer behaviors, such as emphasizing low carbon foods and beverages, encouraging consumers to support local foods, supporting the purchase of in-season, organic, and fair-trade products, and focusing on product labeling [43,44]. Junior et al. [45] observed that individual views on environmental concern affect ones willingness to purchase green products. As a result, this study uses environmental concern as a research variable to study whether consumer willingness to purchase insect foods is influenced by environmental concern.

In summary, this study uses the TPB as a theoretical core for investigating the effects of consumer attitudes, subjective norms, and perceived control on the willingness to purchase insect foods. It also adds the two elements of environmental concern and food neophobia. Through this, the study aims to provide an understanding of the likelihood of insects being used as food in Taiwan. In the event that the insect food-product industry develops, these results can serve as a reference for the development of marketing strategies.

2. Materials and Methods

2.1. Research Framework

This study is centered on the TPB and includes two variables—environmental concern and food neophobia. It investigates Taiwanese consumer knowledge of, and attitudes toward, insect foods, and forecasts consumer purchase intention. The research framework of this study is shown in Figure 1.

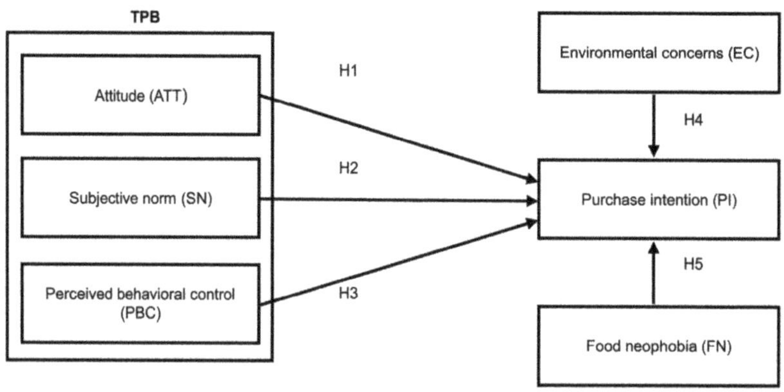

Figure 1. Study hypotheses and conceptual framework. H1, attitude positively affects purchase intention; H2, subjective norms positively affect purchase intention; H3, perceived behavioral control positively affects purchase intention; H4, environmental concern positively affects purchase intention; H5, food neophobia positively affects purchase intention.

2.2. Literature Review

2.2.1. Purchase Intention

Purchase intention is defined as the likelihood of a consumer purchasing a given product. A higher purchase intention indicates an increase in purchase probability [46–50]. Marketers have long argued that purchase intention can be used to accurately forecast purchase behavior [51–53]. Moreover, previous studies have explored product knowledge [54–57], food neophobia [41,58–60], and purchase intention. In this study we define the definition of purchase intention as a customer's intention to purchase insect food.

2.2.2. Attitude

Ajzen [23] believes that attitude is a product of an individual's behavioral beliefs and outcome evaluations [61]. Studies have shown that consumer attitudes toward organic foods [62], green foods [63], and green products [26,64–66] affect purchase intention. Furthermore, Asif, Xuhui, Nasiri, and Ayyub [67] observed that attitude and health consciousness may more accurately forecast purchase intention toward organic products. As such, this study postulates that consumer attitudes toward insect foods will influence their purchase decisions. On this basis, we propose hypothesis 1:

Hypothesis 1 (H1). *Consumer attitude has a significant and positive influence on purchase intention toward insect foods.*

2.2.3. Subjective Norms

Ajzen [23] argued that subjective norms are social pressures (e.g., opinions of relatives, close friends, or colleagues) that influence decision making and whether to take action [68,69]. Additionally, Scalco et al.'s [62] study showed that subjective norms significantly influence purchase intention toward organic foods. On the other hand, this study speculates that consumers are influenced by their relatives and friends when making decisions regarding whether to purchase insect foods. On this basis, hypothesis 2 is proposed:

Hypothesis 2 (H2). *Consumers' subjective norms positively and significantly influence their purchase intention toward insect foods.*

2.2.4. Perceived Behavioral Control

Ajzen [24] argued that perceived behavioral control is the ease or difficulty with which individuals perceive the accomplishment of a certain behavior. This control is affected by external factors, which influence these behaviors, meaning that individuals may be subject to obstacles from past experiences and expectations, including their understanding of their own abilities (ability), urgent needs (resources), and convenience (opportunity). Yadav and Pathak [70] observed that perceived behavioral control significantly influences consumers' purchase intentions with regard to organic foods. With this, we propose hypothesis 3:

Hypothesis 3 (H3). *Consumers' perceived behavioral control positively and significantly influences their purchase intention toward insect foods.*

2.2.5. Environmental Concern

Minton and Rose [71] define environmental concern as the general attitude toward preserving the environment. Newton et al. [72] assert that environmental concern does not directly affect purchase intention but rather allows consumers to understand the environmental result produced by purchasing a product. Furthermore, Arisal and Atalar [73] note that collectivists are more attentive toward environmental topics and find that environmental concern affects individual purchase intention. Collectivists involve emotions such as empathy and indebtedness [74]. Because of their focus on society rather than the individual, public benefits are often considered altruistic [75,76]. Some examples of public benefits include improved environmental outcomes and enhanced animal welfare [77]. The new ecological paradigm measures individual perspectives toward environmental attitudes [78–80]. Lee [81] notes that environmental concern is the second major index for forecasting green purchasing behaviors. In summary, environmental concern is correlated with purchase intention, and on this basis, we propose hypothesis 4:

Hypothesis 4 (H4). *Consumers' environmental concerns positively and significantly affect their purchase intention toward insect foods.*

2.2.6. Food Neophobia

Studies have shown that individuals with high degrees of food neophobia purchase some types of foods (e.g., poultry and fish) at a lower frequency and others (e.g., pork) at a higher frequency [41]. A study by Jaeger et al. [42] confirms that the eating habits of adult New Zealanders are subject to the varying influences of different degrees of food neophobia. Adults with high degrees of food neophobia have lower intake frequencies and selection norms for foods such as fruits, vegetables, protein, seasonings, beverages, and dairy products when compared to those with low degrees of food neophobia. According to this analysis and the definitions and topics of this study, food neophobia is defined as the degree to which consumers are unwilling to consume insect foods when they experience dislike or fear [31,32]. This study also uses the FNS to measure the degree of food neophobia, which reveals that when consumers have higher degrees of food neophobia, stronger negative emotions result, refusals to eat certain foods become more intense, and purchase intention decreases. Previous studies have also stated that in some Western countries, people's dislike toward and refusal to eat insects is a kind of food neophobia [30]. As such, this study argues that because insect foods are a relatively unfamiliar food product for Taiwanese consumers, they may experience fear or disgust when they see insect foods. On this basis, we present hypothesis 5:

Hypothesis 5 (H5). *Consumers' food neophobia negatively and significantly affects their purchase intention toward insect foods.*

2.3. Questionnaire Design

The questionnaire design of this study comprises six sections and is presented in Appendix A. Sections one to three deal with the scale for the TPB. In these sections, questions related to attitude and subjective norms refer to the study by Menozzi et al. [30] and investigate the perception of consumers toward the consumption of insect food products using three questions, and whether consumers are influenced by reference objects and purchase insect food as a result using another three questions. Next, the questionnaire addresses perceived behavioral control, referring to the studies by Ajzen [23] and Menozzi et al. [30] to investigate the ease with which consumers purchase insect foods. This section includes four items, with a total of ten questions. The fourth section addresses consumers' environmental concern, referring to the research by Dunlap and Van Liere [78] to establish a scale for environmental concern and investigate consumer perceptions and attitudes toward the environment. The fifth section addresses consumer food neophobia with regard to insect foods and applies a scale developed by consulting the work of Pliner and Hobden [31], Siegrist et al. [41], and Jaeger et al. [42]. It includes six questions that measure respondents' individual levels of food neophobia. The sixth section addresses consumer willingness to purchase insect foods. The studies by Chen and Cheng [82], Chen et al. [83], and Singh and Verma [84] were consulted to develop a scale for purchase intention and determine the likelihood that consumers will purchase insect foods. In lists 1–6 (Appendix A), a 7-point Likert scale is used, allowing respondents to provide ratings ranging from "strongly disagree" (1) to "strongly agree" (7). The seventh section addresses basic information about respondents, including their gender, profession, education level, and average monthly income.

2.4. Sample Size and Composition

In this study, we used purposive sampling to survey questionnaire answers face-to-face in Taichung city, Taiwan. We distributed 408 survey forms and 100% of these were returned. After eliminating 92 invalid survey forms, 316 valid survey forms remained for a valid return rate of 77.45%. Regarding the respondents' demographics, slightly fewer were women than men (41.1% versus 58.9%). The largest age group among the respondents were those aged 31–40 years (38.4%), followed by 41–50 years (31.5%). The most common highest level of educational achievement was senior high school (42.5%), followed by university (30.6%). Over half of the respondents worked in the service industry (38.6%) or the traditional manufacturing industry (22.8%). The average monthly income among the respondents was most commonly between NT$40,001 and NT$50,000 (47.3%), followed by between NT$30,001 and NT$40,000 (22.9%).

2.5. Statistical Analysis

This study adopted structural equation modeling (SEM) to examine the structural relationship between attitude, subjective norm, perceived behavioral control, environmental concern, food neophobia and purchase intention, and model fit. The SEM is an effective model test and improvement method that enables theoretical models to be tested and can explain the causal relationships among the variables in hypotheses which are related to the models based on statistical dependence. The analysis used the SPSS version 21.0 statistical software package (IBM Corp.: New York, NY, USA) and Amos version 21.0 (IBM Corp.: New York, NY, USA).

3. Results

3.1. Measurement Model: Reliability and Validity

Reliability and convergent validity analysis results for each construct are presented in Table 1. Fornell and Larcker [85] stated that a Cronbach's α coefficient of greater than 0.7 indicates high reliability, whereas a coefficient lower than 0.35 indicates low reliability. The Cronbach's alpha of each dimension was greater than 0.80, indicating good reliability [86]. Hair et al. [87] suggested that the composite reliability (CR) of latent variables should be greater than 0.70. A high CR of latent variables for an examined variable indicates that the examined variable is valid for use in measuring the latent variable. The CR of the variables in this study ranged from 0.713 to 0.940, indicating that this model had good internal consistency. The average variance extracted (AVE) for each factor was between 0.560 and 0.839, which is higher than the recommended benchmark of 0.5 [85]. However, the value of factor loading (0.624–0.952) was higher than the recommended level of 0.6 [88]. Means, standard deviations, and correlations among the constructs are presented in Table 2. Significant positive correlations were found to exist between attitude and purchase intention ($r = 0.39, p < 0.01$), subjective norm and purchase intention ($r = 0.40, p < 0.01$), perceived behavioral control and purchase intention ($r = 0.31, p < 0.01$), and environmental concern and purchase intention ($r = 0.27, p < 0.01$). These results indicate that the higher the attitude, subjective norm, perceived behavioral control, and environmental concern of consumers, the stronger their purchase intention to insect food. By contrast, food neophobia exhibited a significant negative correlation with purchase intention ($r = -0.40, p < 0.01$), indicating that the higher the food neophobia of consumers, the weaker their purchase intention to insect food.

Table 1. Results of factor loading, reliability, and validity.

Items	Factor Loading	Cronbach's α	AVE	CR
Attitude		0.917	0.795	0.921
Eating insect food is pleasant.	0.952			
Eating insect food is relevant.	0.791			
Eating insect food is tasty.	0.924			
Subjective Norm		0.874	0.778	0.875
I would buy insect food because: doctors/nutritionists are in favor.	0.862			
I would buy insect food because: environmental groups are in favor.	0.902			
Perceived Behavioral Control		0.674	0.560	0.713
It would be very easy for me to buy insect food.	0.855			
It is mostly up to me whether to buy insect food.	0.624			
Environmental Concern		0.898	0.706	0.905
Space and resources on the Earth are limited.	0.902			
Humans must strive for harmonic coexistence with nature for survival.	0.950			
Humans have overly exploited nature.	0.749			
The balance of nature is delicate and fragile.	0.740			
Food Neophobia		0.873	0.710	0.880
I dare to try new and different foods.	0.892			

Table 1. *Cont.*

Items	Factor Loading	Cronbach's α	AVE	CR
Insect food looks too strange	0.799			
I like food from different countries.	0.835			
I will try insect food on specific occasions.	0.746			
I will be afraid to eat food that I have never eaten before.	0.725			
I will eat almost all types of food.	0.758			
Purchase Intention		0.939	0.839	0.940
I would consider buying insect food.	0.934			
I am willing to recommend others to buy insect food.	0.906			
I intend to eat insect food in the future.	0.908			

Note: CR: Composite reliability; AVE: Average variance extracted.

Table 2. Means, standard deviations, and correlations of constructs.

Construct	Mean	S.D.	1	2	3	4	5	6
1. Attitude	4.23	0.62	1.00					
2. Subjective Norm	3.17	0.73	0.21 **	1.00				
3. Perceived Behavioral Control	4.06	0.68	0.23 **	0.32 **	1.00			
4. Environmental Concern	5.68	1.26	0.36 **	0.31 **	0.32 **	1.00		
5. Food Neophobia	4.51	0.74	0.47 **	0.38 **	0.30 **	0.37 **	1.00	
6. Purchase Intention	4.84	0.46	0.39 **	0.40 **	0.31 **	0.27 **	−0.40 **	1.00

Note: ** $p < 0.01$.

3.2. Structural Model: Goodness-Of-Fit Statistics and Hypothesis Testing

AMOS version 21.0 was first used to conduct confirmatory factor analysis (CFA). Five latent constructs were contained within the measurement model (Figure 1). As shown in Table 3, the revised model exhibited an appropriate fit after CFA (χ^2/df = 2.174, goodness-of-fit index = 0.908, root mean square error of approximation = 0.069, comparative fit index = 0.932, normalized fit index = 0.946, adjusted goodness-of-fit index = 0.873).

Table 3. Results of the fit indicators of the evaluation model.

Fit Index	Ideal Value	Result	Conclusion
χ^2/df	<3	2.174	Acceptable
GFI	>0.9 (good fit) 0.8–0.89 (acceptable fit)	0.908	Good fit
AGFI	>0.9 (good fit) 0.8–0.89 (acceptable fit)	0.873	Acceptable
NFI	>0.9	0.946	Acceptable
CFI	>0.9	0.932	Acceptable
RMSEA	≤0.05 (close fit) 0.05–0.08 (fair fit) 0.08–0.10 (mediocre fit) >0.10 (poor fit)	0.069	Fair fit

Note: GFI: goodness-of-fit index; AGFI: adjusted goodness-of-fit index; NFI: normalized fit index; CFI: comparative fit index; RMSEA: root mean square error of approximation.

4. Discussion

SEM was utilized in the research, and a maximum likelihood estimation was employed to measure the associations among attitudes, subjective norms, perceived behavioral control, environmental concern, food neophobia, and purchase intention. Figure 2 demonstrates standardized path coefficients arising from examining the proposed structural model. The results of this study reveal that consumer attitudes significantly and positively affect purchase intention toward insect foods ($\beta = 0.772$, $p < 0.001$), thus supporting H1. This also indicates that when consumers believe that an insect food tastes good or is acceptable, their attitudes toward insect foods are more positive and they are more likely to purchase them. As such, consumer attitudes do influence purchase intention and this result is consistent with previous literature. For example, Yazdanpanah and Forouzani [25] stated that student attitudes toward organic foods are an important factor for predicting purchase intention. Another study showed that attitudes toward the use of smartwatches not only strongly influence willingness to use such products, but also strengthen the purchase intention of potential consumers [89].

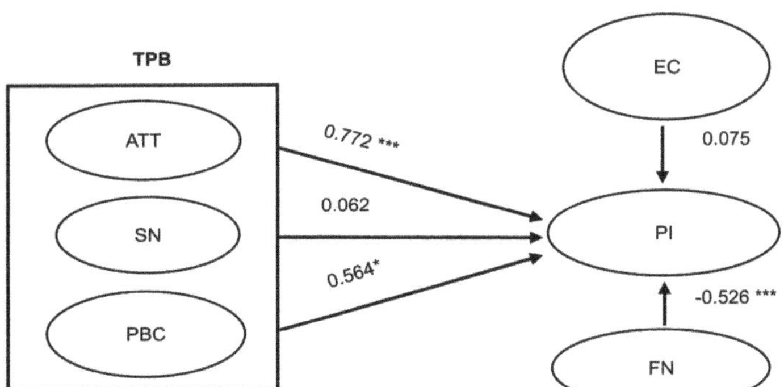

Figure 2. Results of structural equation modeling. * $p < 0.05$; *** $p < 0.001$; GFI = 0.908; AGFI = 0.873; CFI = 0.932; TLI = 0.733; RMSEA = 0.069.

However, the results of this study found that consumers' subjective norms toward insect foods are not significantly correlated with purchase intention ($\beta = 0.062$, $p > 0.05$) and therefore, H2 is not supported. This indicates that consumers will not purchase insect food as a result of promotions by famous people or environmental groups. The results of this study are consistent with those of Tan, Ooi, and Goh [90], who state that consumers' subjective norms related to energy-saving home appliances are not significantly correlated with purchase intention. Yadav and Pathak [70] nonetheless observed that young consumers' willingness to purchase green products can be predicted through subjective norms. This study postulates that this is probably because insect foods are not consistent with Taiwanese food culture, and therefore consumers are not likely to try such foods on a whim. With regard to home appliances, it is likely that the variation in the previous research results is because there are few brands that are respected by consumers, meaning consumers do not consult the opinions of others on this topic.

Consumers' perceived behavioral control is positively and significantly correlated with purchase intention ($\beta = 0.564$, $p < 0.05$), thus supporting H3. This indicates that when a consumer believes they can easily obtain insect foods, their purchase intention will increase. According to TPB, perceived behavioral control influences customers' intention and behavior because customers use perceived behavioral control to judge the likelihood of successfully performing the behavior or accomplishing the task [91]. Perceived control can increase as a result of a convenience channel to use, a channel which the customer is comfortable with or has experience using, or a channel which provides clear and transparent information about the process from purchase to delivery. If a channel provides a customer higher perceived control, then their purchase intention also increases [92]. On the other hand, the perceived difficulty of using a channel can result from obstacles and challenges in searching for products or interacting with the vendor, which can easily prevent the customer from making a purchase [93]. The results of this study are the same as those of previous literature, which state that consumers' perceived behavioral control with regard to green skincare products positively and significantly affects their purchase intention [94]. Perceived behavioral control is also an important factor in determining consumers' intentions to purchase local pork products [95]. In other words, accessibility is an important factor related to purchasing products. With regard to environmental concern, this study reveals that it is not significantly correlated with purchase intention ($\beta = 0.075$, $p > 0.05$) and therefore, H4 is not supported. This indicates that consumers do not purchase insect foods as a result of environmental influences. The results of this study are consistent with those of Asif et al. [67], who conclude that environmental concern does not significantly influence consumers' intentions to purchase organic food products. However, Prakash and Pathak [96] state that consumers' environmental concern positively and significantly affects their intention to purchase products with environment-friendly packaging. This study postulates that although insect foods, organic foods, and products with environment-friendly packaging are helpful for the environment, conceptually, consumers do not consider insect foods as environment-friendly, and therefore environmental concern is not a factor influencing purchase intention. Furthermore, with regard to consumer perception of organic foods, consumers likely view such products from a health perspective and therefore, environmental concern is not included as an influencing factor.

Finally, consumers' food neophobia was significantly and negatively correlated with their intention to purchase insect foods ($\beta = -0.526$, $p < 0.001$), thus supporting H5. This indicates that when consumers have higher degrees of food neophobia with regard to insect foods, they will not purchase them. This is consistent with the results of Piha et al. [12], La Barbera et al. [21], and Tan et al. [22], which state that consumers with high degrees of food neophobia are unwilling to try new functional foods [97] and this, in turn, reduces purchase intention. But Piha et al. [12] suggested that differences in cultural regions also affect consumers' purchasing intention to buy insect food. When consumers know more about edible insect food, they will have a higher willingness to buy it. Food neophobia is also correlated to consumer interest in food [41] and can affect consumers' hedonic assessment of food. Jaeger et al. [42] identified food neophobia as the consumer preference for certain kinds of food, and an important impediment to changing food habits and resolving food-related health problems. As such, resolving consumer food neophobia toward certain types of foods is helpful for improving consumer acceptance of such foods. Per these findings, H1, H3, and H5 are supported but H2 and H4 are not. A summary of the verification of the hypotheses made in this study is shown in Table 4.

Table 4. Summary of hypothesis verification.

Hypothesis	Content	Verification
H1	Attitude positively affects insect-food purchase intention	Supported
H2	Subjective norms positively affect insect-food purchase intention	Not supported
H3	Perceived behavioral control positively affects insect-food purchase intention	Supported
H4	Environmental concerns positively affect insect-food purchase intention	Not supported
H5	Food neophobia will negatively affect purchase intention toward insect food	Supported

5. Conclusions and Limitations

5.1. Conclusions

The results of this study reveal that for consumers, attitude, perceived behavioral control, and food neophobia have significant effects on their purchase intention. Scholars have identified that when consumers have positive feelings about their experiences, their attitudes toward products are more positive [22,98,99]. Therefore, this study shows that edible insect companies can enhance consumer knowledge and experience of edible insect products. For foods that consumers are not familiar with or that they have never eaten before, or edible insect foods that they are afraid of trying, edible insect companies can educate consumers about edible insects, as well as edible insects with high protein and nutrients. Consumers can also reduce fear or uncertainty about eating edible insect products through product experience. Furthermore, edible insects can be considered as sustainable food because insects are simply far less demanding than beef, e.g., 7 times less feed, 50 times less water, and 100 times fewer greenhouse gases [100]. The cultivation of edible insects can reduce the environmental burden and contribute to environmental protection more than the conservation of traditional livestock [101]. In addition, edible insect products are also sold in other countries. For example, the French company Micronutris sells the cricket *sigillatus*, tablette-tenebrio's chocolate, and aldente-aenebrio pasta and crackers [102], and the Fazer chain bakery are going to start selling 'insect bread' and offer a product called Fazer Cricket Bread [103], and so on. Fazer is one of the largest corporations in the Finnish food industry. Edible insect companies can cook insect foods in a familiar cooking style [13] or introduce insect foods into common products [90]. These methods can increase consumer interest and acceptability of insect foods, thereby increasing the intention to purchase.

5.2. Limitations of the Research and Future Research

In the context of global appeals for environmental issues, modern consumers are attentive to environmental, ecological, and sustainable development topics. However, insect foods are environmentally friendly and sustainable products, like organic foods and green products, and although relevant magazine articles have reported this fact, this has yet to be widely accepted by consumers. As such, it is recommended that future studies compare consumer attitudes towards different environmentally friendly products (e.g., organic foods and insect foods), including environmental awareness, behaviors, and values, and compare the variation among consumer knowledge. Additionally, it is recommended that future studies analyze the knowledge and perspectives of consumers belonging to various age groups or different cultural backgrounds (e.g., Thai and Indian consumers). One of the study's main limitations is the use of a purposive sampling technique, which does not allow the findings to be generalized to the overall population of the customers' attitudes toward insect food in Taiwan. Future research could also investigate the effect of other customer-related constructs, such as living in different regions, between urban and non-urban responses. In addition, the cost of purchasing insect foods is another future research limitation.

Author Contributions: Analysis and interpretation of data, H.-P.C.; Project administration, H.-S.C.; Writing—review & editing, C.-C.M.

Funding: This research did not receive any specific grant from funding agencies in the public, commercial, or not-for-profit sectors.

Acknowledgments: First, we would like to express our deepest gratitude to Yu-Ju Chou and Hsiao-Han Chen, who provided the statistics used in the thesis. Second, we would like to express our heartfelt thanks to all the experts who have taken the time to review this article and provide valuable comments.

Conflicts of Interest: The authors declare no conflict of interest.

Appendix A. Questionnaire

Basic Description of Insect Food

The insect foods discussed in this study are foods that have insect ingredients on the market and can be presented in any form. For example: cricket biscuits, cricket bread, fried insects (such as grasshoppers, pupae, meal-worms) and so on.

Figure A1. Insect food display photo.

Questions for attitude, subjective norm, perceived behavioral control, environmental concern, food neophobia and purchase intention. All question items are answered using a 7-point Likert scale (from strongly agree to strongly disagree).

1. **Attitude**

    (1) Eating insect food is pleasant.
    (2) Eating insect food is relevant.
    (3) Eating insect food is tasty.

2. **Subjective Norm**

    (1) I would buy insect food because: doctors/nutritionists are in favor.
    (2) I would buy insect food because: environmental groups are in favor.

3. **Perceived Behavioral Control**

    (1) It would be very easy for me to buy insect food.
    (2) It is mostly up to me whether to buy insect food.

4. **Environmental Concern**

    (1) Space and resources on the Earth are limited.
    (2) Humans must strive for harmonic coexistence with nature for survival.
    (3) Humans have overly exploited nature.
    (4) The balance of nature is delicate and fragile.

5. **Food Neophobia**

    (1) I dare to try new and different foods.
    (2) Insect food looks too strange.
    (3) I like food from different countries.
    (4) I will try insect food on specific occasions.
    (5) I will be afraid to eat food that I have never eaten before.
    (6) I will eat almost all types of food.

6. **Purchase intention**

(1) I would consider buying insect food.
(2) I am willing to recommend others to buy insect food.
(3) Intend to eat insect food in the future.

References

1. Food and Agriculture Organization of the United Nations, FAO. Insects for Food and Feed. Available online: http://www.fao.org/edible-insects/en/ (accessed on 25 December 2017).
2. Van Huis, A.; Van Itterbeeck, J.; Klunder, H.; Mertens, E.; Halloran, A.; Muir, G.; Vantomme, P. *Edible Insects: Future Prospects for Food and Feed Security*; No. 171; Food and Agriculture Organization of the United Nations (FAO): Rome, Italy, 2013.
3. Oonincx, D.G.; van Itterbeeck, J.; Heetkamp, M.J.; van den Brand, H.; van Loon, J.J.; van Huis, A. An exploration on greenhouse gas and ammonia production by insect species suitable for animal or human consumption. *PLoS ONE* **2010**, *5*, e14445. [CrossRef]
4. Bodenheimer, F.S. *Insect as Human Food*; Juuk, W., Ed.; The Hague: Dordrecht, The Netherlands, 1951; p. 352.
5. Banjo, A.D.; Lawal, O.A.; Songonuga, E.A. The nutritional value of fourteen species of edible insects in southwestern Nigeria. *Afr. J. Biotechnol.* **2006**, *5*, 298–301.
6. Durst, P.B.; Johnson, D.V.; Leslie, R.N.; Shono, K. Edible forest insects: Humans bites back. In Proceedings of the Workshop on Asia-Pacific Resources and Their Potential for Development, Chiang Mai, Thailand, 19–21 February 2008.
7. Fasoranti, J.O.; Ajiboye, D.O. Some edible insects of Kwara state, Nigeria. *Am. Entomol.* **1993**, *39*, 113–116. [CrossRef]
8. Payne, C.L.R.; Scarborough, P.; Rayner, M.; Nonaka, K. Are edible insects more or less 'healthy' than commonly consumed meats? A comparison using two nutrient profiling models developed to combat over-and undernutrition. *Eur. J. Clin. Nutr.* **2016**, *70*, 285–291. [CrossRef] [PubMed]
9. Zielińska, E.; Baraniak, B.; Karaś, M.; Rybczyńska, K.; Jakubczyk, A. Selected species of edible insects as a source of nutrient composition. *Food Res. Int.* **2015**, *77*, 460–466. [CrossRef]
10. Poma, G.; Cuykx, M.; Amato, E.; Calaprice, C.; Focant, J.F.; Covaci, A. Evaluation of hazardous chemicals in edible insects and insect-based food intended for human consumption. *Food Chem. Toxicol.* **2017**, *100*, 70–79. [CrossRef] [PubMed]
11. Hamerman, E.J. Cooking and disgust sensitivity influence preference for attending insect-based food events. *Appetite* **2016**, *96*, 319–326. [CrossRef] [PubMed]
12. Piha, S.; Pohjanheimo, T.; Lähteenmäki-Uutela, A.; Křečková, Z.; Otterbring, T. The effects of consumer knowledge on the willingness to buy insect food: An exploratory cross-regional study in Northern and Central Europe. *Food Qual. Prefer.* **2018**, *70*, 1–10. [CrossRef]
13. Tan, H.S.G.; Fischer, A.R.H.; Tinchan, P.; Stieger, M.; Steenbekkers, L.P.A.; van Trijp, H.C.M. Insects as food: Exploring cultural exposure and individual experience as determinants of acceptance. *Food Qual. Prefer.* **2015**, *42*, 78–89. [CrossRef]
14. Micronutris. 2018. Available online: https://www.micronutris.com/fr/accueil (accessed on 25 February 2018).
15. Fazer. Finland Rolls Out Bread Made from Crushed Crickets. 2017. Available online: http://www.bbc.com/news/world-europe-42101700 (accessed on 25 February 2018).
16. Hanboonsong, Y.; Jamjanya, T.; Durst, P.B. *Six-Legged Livestock: Edible Insect Farming, Collection and Marketing in Thailand*; Food and Agriculture Organization of the United Nations, Regional Office for Asia and the Pacific: Bangkok, Thailand, 2013.
17. Muafor, F.J.; Gnetegha, A.A.; Le Gall, P.; Levang, P. *Exploitation, Trade and Farming of Palm Weevil Grubs in Cameroon*; CIFOR: Bogor, Indonesia, 2015; Volume 178.
18. Tan, H.S.G.; Tibboel, C.J.; Stieger, M. Why do unusual novel foods like insects lack sensory appeal? Investigating the underlying sensory perceptions. *Food Qual. Prefer.* **2017**, *60*, 48–58. [CrossRef]
19. Hartmann, C.; Shi, J.; Giusto, A.; Siegrist, M. The psychology of eating insects: A cross-cultural comparison between Germany and China. *Food Qual. Prefer.* **2015**, *44*, 148–156. [CrossRef]
20. Tan, H.S.G.; Fischer, A.R.H.; van Trijp, H.C.M.; Stieger, M. Tasty but nasty? Exploring the role of sensory-liking and food appropriateness in the willingness to eat unusual novel foods like insects. *Food Qual. Prefer.* **2016**, *48*, 293–302. [CrossRef]

21. La Barbera, F.; Verneau, F.; Amato, M.; Grunert, K. Understanding Westerners' disgust for the eating of insects: The role of food neophobia and implicit associations. *Food Qual. Prefer.* **2018**, *64*, 120–125. [CrossRef]
22. Tan, H.S.G.; Verbaan, Y.T.; Stieger, M. How will better products improve the sensory-liking and willingness to buy insect-based foods? *Food Res. Int.* **2017**, *92*, 95–105. [CrossRef] [PubMed]
23. Ajzen, I. The theory of planned behavior. *Organ. Behav. Hum. Decis. Process.* **1991**, *50*, 179–211. [CrossRef]
24. Ajzen, I. From intentions to actions: A theory of planned behavior. In *Action Control*; Springer: Berlin/Heidelberg, Germany, 1985; pp. 11–39.
25. Yazdanpanah, M.; Forouzani, M. Application of the Theory of Planned Behaviour to predict Iranian students' intention to purchase organic food. *J. Clean. Prod.* **2015**, *107*, 342–352. [CrossRef]
26. Paul, J.; Modi, A.; Patel, J. Predicting green product consumption using theory of planned behavior and reasoned action. *J. Retail. Consum. Serv.* **2016**, *29*, 123–134. [CrossRef]
27. Zhang, Y.; Jing, L.; Bai, Q.; Shao, W.; Feng, Y.; Yin, S.; Zhang, M. Application of an integrated framework to examine Chinese consumers' purchase intention toward genetically modified food. *Food Qual. Prefer.* **2018**, *65*, 118–128. [CrossRef]
28. Maichum, K.; Parichatnon, S.; Peng, K.C. Application of the extended theory of planned behavior model to investigate purchase intention of green products among Thai consumers. *Sustainability* **2016**, *8*, 1077. [CrossRef]
29. Zhang, L.; Chen, L.; Wu, Z.; Zhang, S.; Song, H. Investigating young consumers' purchasing intention of green housing in China. *Sustainability* **2018**, *10*, 1044. [CrossRef]
30. Menozzi, D.; Sogari, G.; Veneziani, M.; Simoni, E.; Mora, C. Eating novel foods: An application of the Theory of Planned Behaviour to predict the consumption of an insect-based product. *Food Qual. Prefer.* **2017**, *59*, 27–34. [CrossRef]
31. Pliner, P.; Hobden, K. Development of a scale to measure the trait of food neophobia in humans. *Appetite* **1992**, *19*, 105–120. [CrossRef]
32. Pliner, P.; Pelchat, M.; Grabski, M. Reduction of neophobia in humans by exposure to novel foods. *Appetite* **1993**, *20*, 111–123. [CrossRef]
33. Choe, J.Y.; Cho, M.S. Food neophobia and willingness to try non-traditional foods for Koreans. *Food Qual. Prefer.* **2011**, *22*, 671–677. [CrossRef]
34. Rozin, P.; Vollmecke, T.A. Food likes and dislikes. *Annu. Rev. Nutr.* **1986**, *6*, 433–456. [CrossRef]
35. Raudenbush, B.; Schroth, F.; Reilley, S.; Frank, R.A. Food neophobia, odor evaluation and exploratory sniffing behavior. *Appetite* **1998**, *31*, 171–183. [CrossRef]
36. Raudenbush, B.; Frank, R.A. Assessing food neophobia: The role of stimulus familiarity. *Appetite* **1999**, *32*, 261–271. [CrossRef]
37. Knaapila, A.; Laaksonen, O.; Virtanen, M.; Yang, B.; Lagström, H.; Sandell, M. Pleasantness, familiarity, and identification of spice odors are interrelated and enhanced by consumption of herbs and food neophilia. *Appetite* **2017**, *109*, 190–200. [CrossRef]
38. Vidigal, M.C.; Minim, V.P.; Simiqueli, A.A.; Souza, P.H.; Balbino, D.F.; Minim, L.A. Food Technology Neophobia and Consumer Attitudes toward Foods Produced by New and Conventional Technologies: A Case Study in Brazil. *LWT-Food Sci. Technol.* **2015**, *60*, 832–840. [CrossRef]
39. Schnettler, B.; Grunert, K.G.; Miranda-Zapata, E.; Orellana, L.; Sepúlveda, J.; Lobos, G.; Hueche, C.; Höger, Y. Testing the Abbreviated Food Technology Neophobia Scale and its relation to satisfaction with food-related life in university students. *Food Res. Int.* **2017**, *96*, 198–205. [CrossRef]
40. Ristic, R.; Johnson, T.E.; Meiselman, H.L.; Hoek, A.C.; Bastian, S.E.P. Towards development of a Wine Neophobia Scale (WNS): Measuring consumer wine neophobia using an adaptation of the Food Neophobia Scale (FNS). *Food Qual. Prefer.* **2016**, *49*, 161–167. [CrossRef]
41. Siegrist, M.; Hartmann, C.; Keller, C. Antecedents of food neophobia and its association with eating behavior and food choices. *Food Qual. Prefer.* **2013**, *30*, 293–298. [CrossRef]
42. Jaeger, S.R.; Rasmussen, M.A.; Prescott, J. Relationships between food neophobia and food intake and preferences: Findings from a sample of New Zealand adults. *Appetite* **2017**, *116*, 410–422. [CrossRef]
43. Owen, L.; Seaman, H.; Prince, S. *Public Understanding of Sustainable Consumption of Food. A Report to the Department for Environment, Food and Rural Affairs by Opinion Leader*; Department for Environment, Food and Rural Affairs: London, UK, 2007.

44. Lea, E.; Worsley, A. Australian consumers' food-related environmental beliefs and behaviours. *Appetite* **2008**, *50*, 207–214. [CrossRef]
45. Junior, S.S.B.; da Silva, D.; Gabriel, M.L.D.S.; de Oliveira Braga, W.R. The effects of environmental concern on purchaseof green products in retail. *Procedia Soc. Behav. Sci.* **2015**, *170*, 99–108. [CrossRef]
46. Zeithaml, V.A. Consumer perceptions of price, quality, and value: A means-end model and synthesis of evidence. *J. Mark.* **1988**, *52*, 2–22. [CrossRef]
47. Dodds, W.B.; Monroe, K.B.; Grewal, D. Effects of price, brand, and store information on buyers' product evaluations. *J. Mark. Res.* **1991**, *28*, 307–319.
48. Schiffman, L.G.; Kanuk, L.L. *Comportamiento del Consumidor*; Pearson Educación: London, UK, 2005.
49. Reinders, M.J.; Banovi', M.; Guerrero, L.; Krystallis, A. Consumer perceptions of farmed fish. *Br. Food J.* **2016**, *118*, 2581–2597. [CrossRef]
50. De Angelis, M.; Adıgüzel, F.; Amatulli, C. The role of design similarity in consumers' evaluation of new green products: An investigation of luxury fashion brands. *J. Clean. Prod.* **2017**, *141*, 1515–1527. [CrossRef]
51. Morwitz, V.G.; Schmittlein, D. Using segmentation to improve sales forecasts based on purchase intent: Which "intenders" actually buy? *J. Mark. Res.* **1992**, *29*, 391–405.
52. Sun, L.; Zheng, X.; Su, M.; Keller, L.R. Intention–behavior discrepancy of foreign Versus domestic brands in emerging markets: The relevance of consumer prior knowledge. *J. Int. Mark.* **2017**, *25*, 91–109. [CrossRef]
53. Wang, H.J. Determinants of consumers' purchase behaviour towards green brands. *Serv. Ind. J.* **2017**, *37*, 896–918. [CrossRef]
54. Yiridoe, E.K.; Bonti-Ankomah, S.; Martin, R.C. Comparison of consumer perceptions and preference toward organic versus conventionally produced foods: A review and update of the literature. *Renew. Agric. Food Syst.* **2005**, *20*, 193–205. [CrossRef]
55. Pieniak, Z.; Aertsens, J.; Verbeke, W. Subjective and objective knowledge as determinants of organic vegetables consumption. *Food Qual. Prefer.* **2010**, *21*, 581–588. [CrossRef]
56. Gomez, P.; Werle, C.O.C.; Corneille, O. The pitfall of nutrition facts label fluency: Easier-to-process nutrition information enhances purchase intentions for unhealthy food products. *Mark. Lett.* **2017**, *28*, 15–27. [CrossRef]
57. Lai, Y.H.; Liu, L.W. Organic products purchase behavior: A case study for Urabn consumers in central Taiwan. *Int. J. Supply Chain Manag.* **2017**, *6*, 222–225.
58. Arvola, A.; Lähteenmäki, L.; Tuorila, H. Predicting the intent to purchase unfamiliar and familiar cheeses: The effects of attitudes, expected liking and food neophobia. *Appetite* **1999**, *32*, 113–126. [CrossRef]
59. Labrecque, J.; Doyon, M.; Bellavance, F.; Kolodinsky, J. Acceptance of functional foods: A comparison of French, American, and French Canadian Consumers French. *Can. J. Agric. Econ. Rev.* **2006**, *54*, 647–661. [CrossRef]
60. Barrena, R.; Sánchez, M. Neophobia, personal consumer values and novel food acceptance. *Food Qual. Prefer.* **2013**, *27*, 72–84. [CrossRef]
61. Lee, C.; Green, R.T. Cross-cultural examination of the Fishbein behavioral intentions model. *J. Int. Bus. Stud.* **1991**, *22*, 289–305. [CrossRef]
62. Scalco, A.; Noventa, S.; Sartori, R.; Ceschi, A. Predicting organic food consumption: A meta-analytic structural equation model based on the theory of planned behavior. *Appetite* **2017**, *112*, 235–248. [CrossRef]
63. Ueasangkomsate, P.; Santiteerakul, S. A study of consumers' attitudes and intention to buy organic foods for sustainability. *Procedia Environ. Sci.* **2016**, *34*, 423–430. [CrossRef]
64. Varshneya, G.; Pandey, S.K.; Das, G. Impact of social influence and green consumption values on purchase intention of organic clothing: A study on collectivist developing economy. *Glob. Bus. Rev.* **2017**, *18*, 478–492. [CrossRef]
65. Yadav, R.; Pathak, G.S. Determinants of consumers' green purchase behavior in a developing nation: Applying and extending the theory of planned behavior. *Ecol. Econ.* **2017**, *134*, 114–122. [CrossRef]
66. Sreen, N.; Purbey, S.; Sadarangani, P. Impact of culture, behavior and gender on green purchase intention. *J. Retail. Consum. Serv.* **2018**, *41*, 177–189. [CrossRef]
67. Asif, M.; Xuhui, W.; Nasiri, A.; Ayyub, S. Determinant factors influencing organic food purchase intention and the moderating role of awareness: A comparative analysis. *Food Qual. Prefer.* **2018**, *63*, 144–150. [CrossRef]
68. Park, H.S. Relationships among attitudes and subjective norms: Testing the theory of reasoned action across cultures. *Commun. Stud.* **2000**, *51*, 162–175. [CrossRef]

69. Kim, Y.; Han, H. Intention to pay conventional–hotel prices at a green hotel–a modification of the theory of planned behavior. *J. Sustain. Tour.* **2010**, *18*, 997–1014. [CrossRef]
70. Yadav, R.; Pathak, G.S. Intention to purchase organic food among young consumers: Evidences from a developing nation. *Appetite* **2016**, *96*, 122–128. [CrossRef]
71. Minton, A.P.; Rose, R.L. The effects of environmental concern on environmentally friendly consumer behavior: An exploratory study. *J. Bus. Res.* **1997**, *40*, 37–48. [CrossRef]
72. Newton, J.D.; Tsarenko, Y.; Ferraro, C.; Sands, S. Environmental concern and environmental purchase intentions: The mediating role of learning strategy. *J. Bus. Res.* **2015**, *68*, 1974–1981. [CrossRef]
73. Arısal, İ.; Atalar, T. The exploring relationships between environmental concern, collectivism and ecological purchase intention. *Procedia Soc. Behav. Sci.* **2016**, *235*, 514–521. [CrossRef]
74. Aaker, J.L.; Williams, P. Empathy Versus Pride: The Influence of Emotional Appeals Across Cultures. *J. Consum. Res.* **1998**, *25*, 241–261. [CrossRef]
75. Kareklas, I.; Carlson, J.; Muehling, D.D. 'I Eat Organic for My Benefit and Yours': Egoistic and Altruistic Considerations for Purchasing Organic Food and Their Implications for Advertising Strategists. *J. Advert.* **2014**, *43*, 18–32. [CrossRef]
76. Magnusson, M.K.; Arvola, A.; Hursti, U.; Aberg, L.; Sjoden, P. Choice of organic food is related to perceived consequences for human health and to environmentally friendly behaviour. *Appetite* **2003**, *40*, 109–117. [CrossRef]
77. Bellows, A.C.; Onyango, B.; Diamond, A.; Hallman, W.K. Understanding Consumer Interest in Organics: Production Values vs. Purchasing Behavior. *J. Agric. Food Ind. Organ.* **2008**, *6*. [CrossRef]
78. Dunlap, R.E.; Van Liere, K.D. The "new environmental paradigm". *J. Environ. Educ.* **1978**, *9*, 10–19. [CrossRef]
79. Dunlap, R.E.; Van Liere, K.D.; Mertig, A.G.; Jones, R.E. New trends in measuring environmental attitudes: Measuring endorsement of the new ecological paradigm: A revised NEP scale. *J. Soc. Issues* **2000**, *56*, 425–442. [CrossRef]
80. Pienaar, E.F.; Lew, D.K.; Wallmo, K. The importance of survey content: Testing for the context dependency of the new ecological paradigm scale. *Soc. Sci. Res.* **2015**, *51*, 338–349. [CrossRef]
81. Lee, K. Opportunities for green marketing: Young consumers. *Mark. Intell. Plan.* **2008**, *26*, 573–586. [CrossRef]
82. Chen, C.F.; Chang, Y.Y. Airline brand equity, brand preference, and purchase intentions—The moderating effects of switching costs. *J. Air Transp. Manag.* **2008**, *14*, 40–42. [CrossRef]
83. Chen, J.; Teng, L.; Yu, Y.; Yu, X. The effect of online information sources on purchase intentions between consumers with high and low susceptibility to informational influence. *J. Bus. Res.* **2016**, *69*, 467–475. [CrossRef]
84. Singh, A.; Verma, P. Factors influencing Indian consumers' actual buying behaviour towards organic food products. *J. Clean. Prod.* **2017**, *167*, 473–483. [CrossRef]
85. Fornell, C.; Larcker, D.F. Structural equation models with unobservable variables and measurement error: Algebra and statistics. *J. Mark. Res.* **1981**, *18*, 382–388. [CrossRef]
86. Nunnally, J.C. *Psychometric Theory*; McGraw-Hill: New York, NY, USA, 1978.
87. Hair, J.F.; Anderson, R.E.; Tatham, R.L.; Black, W.C. *Multivariate Data Analysis*, 5th ed.; Prentice-Hall International: Englewood Cliffs, NJ, USA, 1998.
88. Chin, W.W.; Gopal, A.; Salisbury, W.D. Advancing the theory of adaptive structuration: The development of a scale to measure faithfulness of appropriation. *Inf. Syst. Res.* **1997**, *8*, 342–367. [CrossRef]
89. Hsiao, K.L.; Chen, C.C. What drives smartwatch purchase intention? Perspectives from hardware, software, design, and value. *Telemat. Inform.* **2018**, *35*, 103–113. [CrossRef]
90. Tan, C.S.; Ooi, H.Y.; Goh, Y.N. A moral extension of the theory of planned behavior to predict consumers' purchase intention for energy-efficient household appliances in Malaysia. *Energy Policy* **2017**, *107*, 459–471. [CrossRef]
91. Kidwell, B.; Jewell, R.D. An Examination of Percieved Behavioral Control: Internal and External Influences on Intention. *Psycol. Mark.* **2003**, *20*, 625–642. [CrossRef]
92. Chen, M.F. Consumer attitudes and purchase intentions in relation to organic foods in Taiwan: Moderating effects of food-related personality traits. *Food Qual. Prefer.* **2007**, *18*, 1008–1021. [CrossRef]
93. Hansen, T.; Solgaard, H.S. Strategic Pricing: Fundamental considerations and Future Perspectives. *Mark. Rev.* **2004**, *4*, 99–111. [CrossRef]

94. Hsu, C.L.; Chang, C.Y.; Yansritakul, C. Exploring purchase intention of green skincare products using the theory of planned behavior: Testing the moderating effects of country of origin and price sensitivity. *J. Retail. Consum. Serv.* **2017**, *34*, 145–152. [CrossRef]
95. Lorenz, B.A.; Hartmann, M.; Simons, J. Impacts from region-of-origin labeling on consumer product perception and purchasing intention—Causal relationships in a TPB based model. *Food Qual. Prefer.* **2015**, *45*, 149–157. [CrossRef]
96. Prakash, G.; Pathak, P. Intention to buy eco-friendly packaged products among young consumers of India: A study on developing nation. *J. Clean. Prod.* **2017**, *141*, 385–393. [CrossRef]
97. Stratton, L.M.; Vella, M.N.; Sheeshka, J.; Duncan, A.M. Food neophobia is related to factors associated with functional food consumption in older adults. *Food Qual. Prefer.* **2015**, *41*, 133–140. [CrossRef]
98. Peracchio, L.A.; Tybout, A.M. The moderating role of prior knowledge in schema-based product evaluation. *J. Consum. Res.* **1996**, *23*, 177–192. [CrossRef]
99. Banović, M.; Fontes, M.A.; Barreira, M.M.; Grunert, K.G. Impact of product familiarity on beef quality. perception. *Agribusiness* **2012**, *28*, 157–172. [CrossRef]
100. Micronutris. Meet the Edible Bugs. Available online: https://www.micronutris.com/en/our-insects (accessed on 22 April 2019).
101. Nanalyze Weekly. Edible Insects and Bugs. 2018. Available online: https://www.nanalyze.com/2018/04/8-startups-edible-insects-bugs/ (accessed on 22 April 2019).
102. Micronutris. Fine Edible Insects. Available online: https://www.micronutris.com/en/home (accessed on 22 April 2019).
103. News Talk. 2017. Available online: https://www.newstalk.com/news/finnish-bakery-chain-to-start-selling-insect-bread-519562 (accessed on 22 April 2019).

International Journal of
Environmental Research and Public Health

MDPI

Review

Consumer Attitudes Towards Environmental Concerns of Meat Consumption: A Systematic Review

Ruben Sanchez-Sabate [1,*] and Joan Sabaté [2]

1 Centro de Excelencia en Psicología Económica y del Consumo (CEPEC), Núcleo Científico y Tecnológico en Ciencias Sociales y Humanidades, Universidad de La Frontera, Temuco 4811230, Chile

2 School of Public Health. Loma Linda University, Loma Linda, CA 92350, USA; jsabate@llu.edu

* Correspondence: ruben.sanchez@ufrontera.cl

Received: 1 March 2019; Accepted: 29 March 2019; Published: 5 April 2019

check for updates

Abstract: Meat consumption is a major contributor to global warming. Given the worldwide growing demand of meat, and the severe impact of meat production on the planet, reducing animal protein consumption is a matter of food security and public health. Changing consumer food behavior is a challenge. Taste preferences, culinary traditions and social norms factor into food choices. Since behavioral change cannot occur without the subject's positive attitude based on reasons and motivations, a total of 34 papers on consumer attitudes and behavior towards meat consumption in relation to environmental concerns were examined. The results show that consumers aware of the meat impact on the planet, willing to stop or significantly reduce meat consumption for environmental reasons, and who have already changed their meat intake for ecological concerns are a small minority. However, environmental motives are already appealing significant proportions of Westerners to adopt certain meat curtailment strategies. Those who limit meat intake for environmental reasons are typically female, young, simply meat-reducer (not vegan/vegetarian), ecology-oriented, and would more likely live in Europe and Asia than in the U.S.

Keywords: consumer attitudes; meat avoiders; meat reducers; environmental concerns; global warming; climate change; sustainability; ecology; planetary health

---

1. Introduction

Worldwide demand for meat and other animal products is increasing due to rising incomes, growing populations and other sociocultural factors [1,2]. This trend is a global problem because meat production is a major responsible for global warming and environmental degradation [1,3–6]. The livestock industry pollutes freshwater with antibiotics, hormones and chemical substances among others, depletes freshwater availability, contributes to the loss of biodiversity, and is a major source of anthropogenic greenhouse gas emissions [1]. Consequently, finding ways to make diets more sustainable by reducing animal protein consumption has become a matter of food security and thus, a public health issue [7].

Changing consumer food behaviors is a challenge. They are the result of strongly held factors like taste preferences, culinary traditions and social norms [8]. Health behavior theorists have described the stages a person undergoes when trying to adopt healthy behaviors. They cite that behavioral change can only occur with the adoption of a positive attitude based on reasons and motivations [9]. It is therefore relevant to know if environmental reasons can prompt individuals to reduce or avoid meat consumption.

In Western societies, meat-based diets are the norm. Meat avoiders like vegans and vegetarians represent a small minority. For instance, in the United States and the United Kingdom, vegetarians account for significantly less than 5% of the population [10]. The motivations of converted vegans and vegetarians—those raised on a meat-based diet—have been described as non-static and related to health, economy, environment, society and culture, ethics and religion [11]. Vegetarians can be categorized in two large groups: health oriented and ethically motivated [12]. This is because the most prevalent motivations among vegetarians are health and animal welfare [13–17]. Environmental reasons, on the contrary, are important to a small fraction of vegetarians [11].

Another group of consumers to consider are those not ready to give up meat, but who have, or are willing to consider reducing meat consumption. These are known as meat-reducers or flexitarians. Contrary to vegans and vegetarians who have been studied for decades, meat-reducers have received scant attention [18].

The goal of this systematic review is to enhance our understanding of consumer attitudes on meat consumption in relation to environmental sustainability in order to support potential public health interventions oriented towards meat intake reduction. We looked into the three main stages of behavioral change process as proposed by Glanz et al.'s [9]: awareness (precontemplation), willingness (contemplation and preparation) and change (action, maintenance and termination). Having a general overview of the three stages should give public health professionals a general understanding of the role environmental reasons may play in the food eating behavior change process. Thus, this systematic review aims to answer the three following research questions: (1) Are people aware of the environmental impact of meat production and consumption? (2) Are people willing to stop or reduce meat consumption based on environmental concerns? and (3) Have ecological/environmental concerns been the motivation for people who have altered their meat consumption?

2. Materials and Methods

This systematic review was reported following the PRISMA (Preferred Reporting Items for Systematic Reviews and Meta-Analyses) guidelines [19] We performed a preliminary search in Google Scholar for articles that reported data on at least one of the following three topics: people's awareness of the environmental impact of meat production and consumption; people's willingness to stop or reduce meat consumption because of environmental concerns; and people who have already stopped or reduced meat consumption because of environmental reasons or motivations (diet change).

This initial search allowed us to identify a series of keywords that we later used to conduct a literature search of the Web of Science (WOS) Core Collection in March 2018. A separate query was conducted for each topic (awareness, willingness and diet change). Each query consisted of a series of search strings that combined no more than three terms each from one of the following categories: consumer related, meat related, and planet related. For example, one query looked like this: consumer attitudes AND meat AND climate change.

Thus, for "awareness" we used a series of search strings that combined the following terms: "consumer/people attitudes/perceptions" AND "meat"/"livestock" AND "climate change"/"GHG emissions"/"global near/2 warming"/"environment"/"water near/3 use"/"land near/3 use". Similar search strings were used for "willingness" and "change".

The screening process was completed by both authors independently to reduce bias. It comprised three stages for each one of the three topics considered. First, articles and abstracts were screened. Citations that met the eligibility criteria (Figure 1) were imported to the reference manager Zotero. Second, selected citations were read in full to make a final decision on their relevance for any of the three topics considered, and to locate new relevant articles that had not been found by the WOS search. Third, these first two steps were conducted for the new bibliography until no new eligible references were detected. The few articles considered pertinent by only one reviewer were included or discarded after a discussion between the two coauthors. The search for "awareness" yielded a total of 14 articles that met the eligibility criteria. The search for "willingness" yielded a total of 16 articles that met the eligibility criteria. And the search for "diet change" yielded a total of 17 articles that met the eligibility criteria. This systematic review rendered a total of 34 articles since some publications were relevant for more than one topic. Pertinent data from these articles was abstracted in tables with categories including: study design, sample characteristics, question or dependent variable and covariates effects, among other relevant information.

INCLUSION CRITERIA:

- Quantitative studies
- Government studies/reports.
- Full-text papers in English or Spanish published in peer-reviewed journals.
- Focus on:
    - Consumer Awareness of meat consumption and production environmental impact
    - Consumer Willingness to stop or reduce meat consumption because of its environmental impact
    - Consumer environmental reasons or motivations to have adopted a plant-based diet or have reduced / avoided meat consumption.

EXCLUSION CRITERIA:

- Qualitative studies
- Opinion papers, outlook, concept papers, books or book chapters.
- Not related to consumer attitudes or behavior (e.g. environmental impact of meat production).
- Studies that only consider one type of meat (e.g. beef, lamb, poultry…).

Figure 1. Eligibility criteria.

3. Results

3.1. People Awareness of the Environmental Impact of Meat Production and Consumption

The results from 14 articles that examined awareness of the negative impact meat production and consumption have on the environment are summarized in Table A1, presented in the Appendix A at the end of this dcocument. In short, the main findings are: (1) aware consumers are a minority; (2) consumers either underestimate or ignore the potential of either stopping or reducing meat production and consumption to reduce the anthropogenic impact on the environment; and (3) it is not clear for the consumer that a vegetarian diet is more environmental friendly than a diet including meat.

Consumer awareness of the meat environmental toll has been studied in Belgium, Finland, Germany, the Netherlands, Portugal and the United States using different methods. The percentages of aware participants ranged from 23% to 35% across studies [20–22]. One study in which subjects received prior information, the percentage jumped to 58% [23]. Another study required respondents to list concrete impacts of meat production on the planet: only 24% named "pollution" and 20% "erosion of natural resources" [24]. Another study showed a tendency toward a neutral opinion on the negative environmental impact of meat [25]. And regarding behaviors that damage the earth, one study showed that consumers rarely (less than 10%) thought of "meat eating" [26].

Consumer estimation of meat production and consumption toll on the environment was studied in Australia, Belgium, the Netherlands, Switzerland, the United Kingdom, and the U.S. Only two studies specifically queried participants on meat production. Less than half (38%) agreed that changing animal husbandry can counter climate change [20], but still its toll was underestimated relative to other activities like transport, even when prior information on meat and the environment was given [23]. All other studies focused on meat consumption reduction. Percentages of participants agreeing with it as a way to help the environment varied between 18% to 29% across studies [27–29]. Percentages of subjects that considered it an effective way to alleviate climate change varied from 5% to 64%. This big range can be explained by different methodological and geographical factors across studies. Still, reducing meat consumption was usually considered the least or second least effective when compared to other options [26,30,31]. Still consistent with this finding, the only longitudinal study found by the reviewers showed that participants gave slightly higher effectiveness to meat reduction in the follow-up survey four years later [32]. Finally, it is not clear to consumers that a vegetarian diet is more environmentally friendly than a diet with meat [25].

Not all studies report on covariate effects. From those which do, the gender variable is the most frequent one. Women are more conscious about the negative impact meat has on the environment [22,25], and thus, they perceive a higher effectiveness in reducing meat consumption to alleviate climate change than men [26,30–33]. One study found that the only important covariates were the frequency of meat intake and already established concerns about the environment. As meat intake went up, the perceived effectiveness of meat reduction went down. But the subjects who held a strong belief in human causation of climate change assigned a positive association between eating less meat and helping the planet. Other covariates like age and level of education presented no correlations [30]. Another study also showed no correlations of awareness with age, but surprisingly, neither with gender nor with meat consumption frequency [28].

3.2. People Willingness to Stop or Reduce Meat Consumption Because of Environmental Reasons or Motivations

The results from 15 papers plus a European Union Report (EUR) that examined people willingness to stop or reduce meat consumption for environmental reasons are summarized in Table A2 (see Appendix B). The main findings are: (1) those motivated by ecological concerns to reduce meat intake are a minority, and (2) meat curtailment is among the least preferred personal options to counter climate change.

When no prior information on the meat environmental toll was given, participants from Finland, Germany, The Netherlands, Switzerland and the U.S. willing to stop or reduce meat consumption because of environmental reasons ranged from 12.8% to 25.5% [22,25,33]. Reducing meat intake was usually the least chosen option to curb climate change [26,30]. Belief in the negative impact of meat on the planet associated positively with willingness to change meat consumption in three studies [26,30,33]. One study also revealed a positive association between consciousness, understood as cognitive and affective awareness of the environmental toll of meat, with willingness to reduce meat consumption [22]. Another study that specifically distinguished between belief and actual knowledge on the effectiveness of meat reduction for climate change mitigation, showed that while belief was positively associated with willingness, knowledge was not [26]. Only one study explicitly reported that education and age were not related to willingness [33].

Eight studies conducted throughout Belgium, Germany, The Netherlands, Portugal, Sweden and the U.S., and the EUR did provide information to the participants connecting meat production and consumption with the environment before the data collection. The results show disparate percentages of people willing or maybe willing to reduce meat consumption for environmental reasons. If simply asked for their willingness to make such a dietary change, participants "certainly willing" were a small minority (5–18%), while those "maybe willing" were 41% [20,25]. Regarding agreement with certain direct meat curtailment strategies, percentages varied widely (15–60%) depending on the strategy considered. Meat substitution for vegetables was significantly less popular than meat reduction, but the latter was still among the least preferred options unless compared with eating insects or meat substitutes [21,23,24]. In one study, participants did not find altering meat consumption easy to do [34].

The EUR [35] reported that about 50% of Europeans would be willing to replace most of the meat they eat with vegetables, and 80% of them would be willing to eat less meat but of certified origin. Considering some countries separately, the UK, the Netherlands, Denmark, Finland and Belgium present lower percentages of people willing to replace meat with vegetables (29–49%) and of people willing to consume less meat but of certified origin (62–73%) than countries like Portugal, Spain, Italy and Romania in which percentages range from 53% to 69% and 83% to 89% respectively.

Covariate effects are similar to those presented in the awareness section. Being female is usually a strong predictor of willingness to decrease meat consumption or choose meat-free menus [21,29,31,35,36]. Meat consumption frequency and positive attitudes to meat are negatively associated with willingness to eat it less [21,24,31,37]. Ethnicity and culture can strongly influence willingness. Turks living in the Netherlands were less willing to alter meat consumption than Chinese and Native Dutch [36]. Mediterranean Europeans responded more positively to replacing most of the meat with vegetables (56% average) and to reduce meat consumption (86%) than Northern Europeans (46% and 80%, respectively) [38]. Regarding income, one study presented a negative association between affluence and willingness [38]. Age and education, on the contrary, had in general no influence [31].

Finally, the effect of information on meat and the environment on willingness is less clear. In one study, it could be seen that prior information increased the percentage of people willing to eat less meat from 12% to 18% [25]. In two other studies, information did not alter the number of participants willing to choose meals with less or no meat [21,29]. However, one study reported that participants concerned for the environment and/or already aware before the experiment about the negative impact of meat, were more likely to support meat curtailment strategies [21]. Still, another study found that pro-environmental beliefs had no significant predictive value [29]. In any case, it should be kept in mind that each study provided participants with different types, degrees, and formats of information on the meat environmental toll and thus, generalizing results is not recommendable.

3.3. Meat Consumption Changes for Environmental Reasons

The results from 17 articles that examined motivations for limiting meat consumption are summarized in Table A3 (Appendix C). The main findings show that those who have already adopted a meatless diet or have already reduced its consumption are: (1) a small minority among samples from the general population, and a significantly bigger one among certain population groups; and (2) female, most likely young, partial meat limiters and reside in Europe.

The studies reviewed referred to people who follow a low or no-animal product diet in two different ways: (1) vegans and vegetarians; and (2) "meat avoiders", "animal product limiters", and similar expressions. This fact directly affected the wording of questions and sentences that participants had to answer or rate. Thus, some studies looked for reasons for being "vegan", "vegetarian" or something similar like "semi-vegetarian", while other studies searched for reasons for "avoiding meat", "reducing meat consumption" or any other wording that means the curtailment of animal products consumption. It is necessary to bring attention to this point because veganism and vegetarianism are not only a diet choice but an identity [39]. Deciding to become a vegetarian is a much more complex process than simply opting for reducing or avoiding meat consumption, or even adopting a plant-based diet.

Studies that specifically asked for reasons or motives for being vegan/vegetarian were all conducted in the U.S. Those who indicated environmental concerns were few (>3.2%) [39,40] in recent surveys with a general population of vegans/vegetarians. However, among specific population groups environmental vegan/vegetarians were significant minorities: 14% in the case of marathon runners [41], and 32.1% in the case of women physicians surveyed two decades ago [42]. Other research conducted in the U.S. and Finland showed that vegans, vegetarians and semi-vegetarians tend to agree with and give a moderate importance to the protective benefits of a vegetarian diet towards the environment [43,44].

Only a few consumers (4–19%) indicated environmental concerns for having reduced or avoided meat intake in studies conducted in Belgium, The Netherlands and the U.S. [23,45,46]. However, when specific population groups and certain meat curtailment strategies are considered the percentage of environmental meat reducers or avoiders increases. More than a 50% of a general population sample from The Netherlands reported to have "one meat-free day a week" and "smaller meat portions" at least once a month [47]. Other studies showed that meat avoiders/reducers gave a moderate importance to environmental concerns in their meat purchasing and consumption habits [48]. Those who considered ecology important were the 38.2% of a Dutch sample [45,49,50]. And 38.1% of university students from eleven Eurasian countries pointed to the environment as their major reason for meat avoidance [51].

Reported covariate effects across studies, and research on specific groups like vegans, portray those who limit meat consumption because of the environment as female, young, semi-vegetarian/meat reducer, ecology-oriented, and more likely living in Europe and Asia than in the U.S. Four studies that specifically asked participants to indicate their main reason for meat reduction or avoidance further reflect this profile [39,40,49,51]. Once more, women proved more likely to reduce meat intake because of the environment than men. This was true for Euromerican women [48,49], and for a multiethnic sample from The Netherlands [47,52]. Studies rarely reported age as a significant covariate. However, considering the one study that did [43], and the fact that this review found the highest percentage of meat avoiders because of the environment, in a survey of 3433 students attending different universities based in eleven Eurasian countries [51], it appears that young people may be the most motivated by ecology for already having reduced or stopped meat intake. The degree of involvement with food and sustainability, regardless of age, is another covariate that also correlated positively with environmental reasons for meat curtailment [46,47]. Ethnicity, as well, had a significant impact in one study conducted in The Netherlands [36].

Considering only studies published after 2010, vegans and meat limiters may be more likely to be influenced by environmental reasons than vegetarians. Samples from the U.S., Europe and Asia presented much lower percentages (9–21%) of vegetarians that consider sustainability an important factor that shapes their diet than semi-vegetarians (30–49%), light semi-vegetarians (34–44%) or meat limiters in general (41%) [48,49,51,52]. Two studies carried out in the U.S. before the year 2000, add to this pattern: 60.7% of all types of meat limiters including vegetarians [45] and 32.1% of self-described vegetarians indicated ecological concerns as current reason for their dietary choices [42]. An older study in the UK also showed that vegans and meat reducers are more likely to be influenced by environmental reasons than vegetarians [53]. Opposite results to this pattern, meaning that vegetarians reported to be more influenced by ecological concerns than vegans and meat reducers, appeared to a certain extent, in a study conducted in Finland [44]. In any case, more evidence is needed in order to draw conclusions on differences between vegans, vegetarians and meat reducers. Finally, two recent surveys of vegans living in the U.S. yielded very low percentages (2–3.2%) of consumers motivated by the environment [39,40], adding country location as another significant variable to consider.

4. Discussion

The reduction of meat production and consumption would alleviate the anthropogenic impact on the environment [1]. Individual choices for diets low in meat and high in vegetables are urgently needed according to the latest scientific evidence [7]. Previous studies have identified two main motivations that prompt people in the West to become vegan or vegetarian: animal welfare and health [14,16,45,54]. Ecological concerns, however, are only relevant to a minority of them [11]. In addition to vegans and vegetarians, there are a significant number of consumers who limit meat consumption. Known as meat-reducers or flexitarians, few studies have explored their motivations for reducing meat intake [18].

Review of the main findings shows that, in the so-called developed countries, those aware of the meat impact on the planet, and those willing to alter their meat consumption for environmental reasons, are a small minority. This result is in line with a previous review of awareness and willingness only [55].

Regarding change, the present review shows that people who altered their meat consumption patterns because of the environment represent also a small minority of the studied samples. Within this minority of people aware, willing, or who have already changed, women are a clear majority. Considering in addition that the reduction of meat consumption tends to be among the least preferred strategies to alleviate climate change when compared to other non-food activities like driving less, it looks like environmental reasons are not a major motive for reducing meat intake for the general Western population.

Giving information on the environmental toll of meat production could be a promising strategy to increase awareness and willingness. Studies that provided participants with such information before the test showed significantly higher percentages of people aware and willing. However, there are two other factors that could very well explain such increases. First, social desirability, i.e., survey respondents' tendency to give answers they believe will be viewed favorably by researchers or other participants. The second factor, which applies only to the studies reviewed on willingness, has to do with their different designs. Percentages of people willing to alter meat consumption when prior information is given vary from 5% to 80% in the papers reviewed. Such significant disparity could be explained by studies variations in: (1) methodology; (2) the assessed behavioral action state: some studies measured "belief" while others "intention" or "willingness"; (3) the definition of target behavior (it is not the same to aim for a plant-based diet than for eating meat-free meals regularly) and (4) the time frame to adopt the favorable behavior: for instance, having a meat-free meal x times per month or per week. Therefore, it remains unclear how beneficial the strategy of informing the consumer on the meat environmental toll will actually be for the reduction of its intake.

It is also necessary to pay attention to how the information on the meat impact on the environment is usually introduced. The papers reviewed present the environmental problem in a very rational and detached way. By this we mean that prior information given or questions addressed to participants are based on the common-sense supposition that the environment is separate from, and around, humans. As Lakoff [56] has argued, this is a false supposition because humans are an inseparable part of nature. Yet, this mode of thinking and understanding ("frame" in communication sciences parlance) is common in mass media and public policy communications [56], as well as how scientists word the questions they use and how study participants interpret them. Thus, it is necessary to explore how subjects would react to meat curtailment strategies when ecological concerns are presented to them in an emotional fashion. Research on this regard is promising as environmental messages that appealed to emotions and/or values reduced the intentions of participants to eat meat and affected their attitudes towards meat consumption [34,57,58]. However, research on the effects of emotional messages on people's attitudes and behaviors towards climate change in general has shown that fear-based appeals can backfire and lead to a decrease in participants' willingness to reduce their carbon footprints [59]. In addition, a longitudinal study conducted in the UK showed that levels of concern and motivation to behaviorally address climate change decrease as time passes from participants' exposure to climate change communications [60]. Therefore, more research on communication strategies to increase awareness and willingness to alter meat consumption among Westerners is needed.

Surprisingly, despite increased media attention in recent years to the environmental concerns linked to meat consumption, percentages of vegan, vegetarian and meat reducer participants who claim to follow such dietary patterns on environmental concerns have remained largely unchanged in studies conducted after 2010 compared to the few published before 2002 included in this review. This could be explained by the fact that scientific knowledge and even dietary recommendations for reducing meat consumption based on environmental reasons precede the time span (1987–2016) of the studies included in this systematic review [61]. Such knowledge evidently permeated to vegans and vegetarians long before the more recent mass media attention, probably because they have belief systems and/or sources of information outside the mainstream.

The studies reviewed have limitations that should be addressed in future research. The geographical limitation (the fact that the majority of studies were conducted in only a small number of countries of northern Europe and North America) is the most noticeable. The large survey carried out by the European Commission showed big differences in willingness between northern and southern European countries [35,38]. This gives reason to believe that research on awareness, willingness, and change regarding meat consumption in relation to planetary health can yield significantly different results when Mediterranean, Latin American, and the so-called developing countries are considered. Were this the case, such differences could be explained due to cultural and economic determinants.

There are also methodological limitations worth considering when designing future studies. The majority of the studies reviewed used convenience samples. Random samples are better in order to generalize results to the general populations. Another limitation is that we have found only one longitudinal study. Longitudinal studies could be of interest to identify the evolution of the influence environmental reasons may have on subjects throughout their lives. Cultural aspects may not have been sufficiently taken into account. One study noted large differences in willingness and diet change across ethnicities living in the same country [36]. Further research exploring willingness and change could benefit from an understanding of the cultural significance meat has in the culture/society to be studied. For this and the geographical limitation mentioned before, we consider the results of this systematic review hard to generalize cross-nationally.

Future research could incorporate covariates such as gastronomic and hedonistic dimensions of meat intake and people's cooking skills when examining willingness and change. Previous studies have already shown that people rarely want to give up meat for the pleasure it gives them [27,62,63]. Thus, it is probable that those who do not have the skills to cook palatable meat-free meals, may not reduce its consumption not because they do not want to, but because they do not know how to have an enjoyable food experience without meat. Another covariate to consider in future research is the participant's social networks. Since eating is a socially regulated behavior [64], such an important dietary change as altering meat consumption may be favored or impeded by, for instance, family and/or significant communities such as churches, vegetarian associations.

5. Conclusions

This systematic review reveals a lack of disposition by the general population in Western countries to stop eating meat on environmental reasons. Even for vegans/vegetarians, ecological concerns are more of another motive to further justify their dietary pattern than an original motivation to give up animal products altogether. However, the reviewed evidence also shows that environmental motives are already appealing to significant proportions of Western meat-eaters to adopt certain meat curtailment strategies like meat-free days. This appeal is more prevalent among women and people from certain cultures. Given that dietary habits are not static, and the fact that mass media attention to sustainable food systems and diets is increasing, it is feasible that ecological concerns become a trigger to at least minor reductions in meat consumption for a majority of the Western population, especially for those not motivated by health or animal welfare. Since a small reduction in meat intake among a large proportion of Westerners could mean a significant contribution to reducing the anthropogenic impact on the environment, mass media outlets, public health educators, nutritionists, policy makers, and the food industry may also consider environmental reasons to promote healthy and sustainable diets.

Author Contributions: Both authors contributed to the conceptualization of this systematic review. R.S.-S. wrote the original draft. Both authors reviewed and edited the original draft. Both authors have read and approved the final manuscript.

Funding: No funding received.

Conflicts of Interest: The authors declare no conflict of interest.

Appendix A

Table A1. People's awareness of the environmental impact of meat production and consumption.

Title					Outcome Measure: Perceived Environmental Impact		
Author(s), Year	**Design; Year Data Collected**	**Country; Sample**	**Main Research Question**	**Provided Information Prior to the Experiment**	**Question or Dependent Variable**	**Response or Finding**	**Effect of Covariates**
Campbell-Arvai, 2015 [29] *	Survey in dining halls; unspecified	U.S.; undergraduate students, convenience sample, N = 320, 46% men	Food-related environmental beliefs and behaviors	No info	(1) Eating less meat can help the environment. (2) Adopting a vegetarian diet can help the environment	(1) 29% agree; 20% unsure; 51% disagree (2) 22% agree; 13% unsure; 65% disagree Lowest level of agreement compared with other behaviors (e.g., using less packaging, grown locally)	n.a.
Clonan et al., 2015 [28] *	Postal survey; 2009	UK (Nottinghamshire); random sample from electoral registers, N = 842, 41% men	Meat consumption attitudes and sustainable meat purchase	No info	To help reduce the impact of climate change, it is better to eat less animal foods (meat, dairy products and eggs).	18% agree 46% unsure 36% disagree	Red and processed meat intake frequency, sustainable meat purchase frequency, gender, age, SES were not significant
Cordts et al., 2014 [25] *	Online experiment; 2013	Germany; quota sample, N = 590, 52% men	Consumer response to negative information on meat consumption	Variables measured before info provision (experimental manipulation)	(1) Farming animals and producing animal products (e.g., milk or meat) has a considerable negative environmental impact. (2) A vegetarian diet is more environmentally friendly than a diet including meat.	(1) M = 3.07, SD = 1.12 (1 = do not agree at all to 5 = fully agree) (2) M = 3.10, SD = 1.21 (1 = do not agree at all to 5 = fully agree)	(1) Women agreed more than men (M = 3.19; SD = 1.11; M = 2.95; SD = 1.12; $p \leq 0.01$) (2) Women agreed more than men (M = 3.23; SD = 1.19, M = 2.98; SD = 1.21; $p \leq 0.05$)

Table A1. *Cont.*

Title					Outcome Measure: Perceived Environmental Impact		
Author(s), Year	**Design; Year Data Collected**	**Country; Sample**	**Main Research Question**	**Provided Information Prior to the Experiment**	**Question or Dependent Variable**	**Response or Finding**	**Effect of Covariates**
De Boer et al., 2016 [30]	Nation-wide consumer surveys; 2014	Netherlands and the USA; representative sample N = 527 (The Netherlands). Weighted variables: gender, age, level of education, region, and a value-related test score on "mentality-environment". (efficiency of the weighting 89%, effective sample size 478) N = 556 (USA). Weighted variables: gender, age, and level of education (efficiency of the weighting 90%, effective sample size 500) Total = 1083	Consumer awareness of meat consumption environmental impact and their willingness to reduce meat consumption, among other research questions.	No prior info given.	"For each of the following lifestyle- changes, please let us know whether you think this is an effective way of combatting climate change". The options, which were presented in randomized order, were: "Eat less meat", "Buy local, seasonal, unprocessed foods (e.g., by going to farmer's markets)", "Buy (more) organic foods", "Drive less", "Save energy at home (e.g., turning thermostat down, using saving bulbs, air-drying laundry)", and "Install solar panels on my house".	Dutch: "eating less meat" option, second less effective 12% recognized the outstanding effectiveness of the less meat option in the eyes of climate experts 46% attributed effectiveness to the "eating less meat" option Americans: "eating less meat" option, the least effective 6% recognized the outstanding effectiveness of the less meat option in the eyes of climate experts 30% attributed effectiveness to the "eating less meat" option	Regular meat eaters assigned lower effectiveness ratings to the less meat and the organic food option, but not to the other options. Belief in human causation and personal importance were associated with assigning higher effectiveness ratings to all the options. The pattern of profile results remained unchanged when gender, age, and level of education were entered as covariates. The analysis revealed that these variables had small effects on the effectivity ratings. Females gave slightly higher ratings than males, especially to the food-related options
Study 1 de Boer, Schösler, et al., 2013 [20] *	Online survey; 2010	The Netherlands; quota sample, N = 1083, 50% men	Motivational explanations for responses to the meat-free meal idea	No info before questions	(1) Agriculture and animal husbandry together are one of the major causes of climate change. (2) If agriculture and animal husbandry change the way they work, they can counter climate change.	(1) 23% agree 36% unsure 41% disagree (2) 38% agree 37% unsure 25% disagree	n.a.

Table A1. *Cont.*

Title					Outcome Measure: Perceived Environmental Impact		
Author(s), Year	**Design; Year Data Collected**	**Country; Sample**	**Main Research Question**	**Provided Information Prior to the Experiment**	**Question or Dependent Variable**	**Response or Finding**	**Effect of Covariates**
Study 2 de Boer et al., 2014 [31] *	Online survey; 2010	The Netherlands; quota sample, $N = 1083$, 50% men	Consumer strategies to reduce meat consumption and its' association with their willingness to eat meatless meals	As an individual, you can make a big difference to nature and climate protection by choosing one (or more) meals without meat every week.	Did you know that?	64% yes, 36% no	More 'yes' responses for older and better educated people
De Groeve, et al., 2017 [21]	Online survey. Two samples. Data collected in 2015 (sample 1) and 2016 (sample 2)	Belgium; Ghent University Business Administration Students; $N = 429$	Assess students support for six less meat initiatives (LMIs) to be implemented in student restaurants.	No prior info given.	Students' knowledge about the negative impact of meat on the environment	4.66% reported "Very much" 24.4% rather much 36.6% not little, not much 24.4% Little 9.79% Very little	n.a.
Graca, Oliveira, et al., 2015 [24] *	Online survey; 2013	Portugal; convenience sample, $N = 410$, 30% men	Multiple correspondence analysis to identify clusters of meat-related associations	Info provided after the question	Participants responded to an open ended question about how meat consumption may impact nature and the environment	24% pollutes nature and the environment; 20% erosion, disruption, depletion of natural resources; 18% references to mass production, artificial methods; 14% impacts only if unregulated or in excess; 11% does not impact nature and the environment;	n.a.
Lea & Worsley, 2008 [27] *	Postal survey; 2004	Australia (Victoria); random sample, $N = 223$, 48% men	Food-related environmental beliefs and behaviors	No info	Consumers eating less meat' is important to help the environment	22% agree 22% unsure 56% disagree Lowest level of agreement compared with other behaviors (e.g., using less packaging, grown locally)	n.a.

Table A1. *Cont.*

Title					Outcome Measure: Perceived Environmental Impact		
Author(s), Year	**Design; Year Data Collected**	**Country; Sample**	**Main Research Question**	**Provided Information Prior to the Experiment**	**Question or Dependent Variable**	**Response or Finding**	**Effect of Covariates**
Pohjolainen et al., 2016 [22]	Postal survey; 2010	Finland; representative sample. $N = 1890$	The level of environmental consciousness among Finnish consumers concerning meat production and consumption	No prior info given.	Participants had to agree or disagree with the following three statements: (1) meat production strengthens climate change significantly more than plant production (2) meat production causes eutrophication significantly more than plant production (3) food production causes significant environmental problems	(1) 35.7% agree; 47% neutral; 17.3% disagree (2) 34.8% agree; 45% neutral;20.2% disagree (3) 35.6% agree; 37.7% neutral; 26.7% disagree	Consumers clustered in six groups depending on their awareness of meat-related environmental questions: Those aware (highly conscious and rather conscious), those resistant to the idea (Resistant), those who give neutral answers (highly unsure and rather unsure) and those "careless conscious". Among the groups highly and rather conscious, the majority is female (66.2% and 55.3%), two thirds aged between 46–75, 40% or more have tertiary education. When occupation is considered, in both groups more than 40% are not in labor force and blue-collar workers are slightly more represented than white-collar (26.1–21.1%/21.9–19.6%).

Table A1. *Cont.*

Title					Outcome Measure: Perceived Environmental Impact		
Author(s), Year	**Design; Year Data Collected**	**Country; Sample**	**Main Research Question**	**Provided Information Prior to the Experiment**	**Question or Dependent Variable**	**Response or Finding**	**Effect of Covariates**
Tobler et al., 2011 [33] follow-up study by Siegrist et al., 2015 [32] *	Postal survey; longitudinal study: 2010, follow-up 2014	Switzerland (German- and French-speaking regions); random panel sample, N2010 = 6189, N2014 = 2781, 48% men	Consumer willingness to adopt ecological food consumption	No info	Perceived environmental benefit of eating less meat (maximum of once or twice per week), (1 = very small to 6 = very large)	M = 3.75, SD = 1.71, reducing meat consumption was perceived as having the lowest environmental effect compared with other behaviors (e.g., avoiding excessive packaging or organic food). Longitudinal study; Increase across time (M2010 = 3.89, SD = 1.69; M2014 = 4.23, SD = 1.56; $p < 0.001$)	Women perceived meat reduction as more beneficial for the environment than men (M = 3.96, SD = 1.69; M = 3.52, SD = 1.70; $p < 0.001$) Larger improvement for women and higher educated participants; $p < 0.001$

Table A1. *Cont.*

Title					Outcome Measure: Perceived Environmental Impact		
Author(s), Year	**Design; Year Data Collected**	**Country; Sample**	**Main Research Question**	**Provided Information Prior to the Experiment**	**Question or Dependent Variable**	**Response or Finding**	**Effect of Covariates**
Truelove et al., 2012 [26]	Mixed methods. Online survey with open ended questions and behavior ratings.; 2008	USA; Undergraduate psychology majors (*N* = 112) (69 women and 43 men)	Students perceptions of the relative impact and effectiveness of certain behaviors on global warming.	No prior info given.	(1) Open-ended request to participants to list their own behaviors that cause global warming. (2) Respondents asked to rate the impact of 16 behaviors in contributing to GW. Rate went from 1 (Negligible impact) to 11 (Major impact). (3) Open-ended request to participants to list behaviors that reduce global warming. (4) Respondents asked to rate the impact of 20 behaviors that contribute to reduce GW. 1 (Extremely ineffective) to 11 (Extremely effective)	(1) Driving was mentioned by 90% participants. Eat meat only by less than 10% (2) Eat meat was rated with median of 3.83/11, just above behaviors like riding your bike and skiing. SD: 2.52 (3) Drive less and use alternate transportation was mentioned by almost 80% of the participants. Recycle by more than 45%. Reduce meat consumption by less than 5% (4) Reduce your meat consumption: 4.35/11 effectiveness. SD: 2.96	In answer (4), women scored higher than men.
Vanhonacker et al., 2013 [23] *	Online survey; 2011	Belgium (Flanders); convenience sample, *N* = 221, 36% men	Attitudes towards more sustainable food choices and consumer segmentation based on their self-evaluated ecological footprint	Explanation of the concept 'ecological footprint' Participants were informed about the contribution of animal production to Co2 emissions.	Participants had to score the contribution to greenhouse gas emissions for various industry sectors, including livestock production. Participants were asked how aware they were of the extent of this contribution.	Approx. M = 3.7 (no number, only bar chart presented) (1 = does not contribute at all to 5 = contributes very much) Livestock production was underestimated relative to other activities (e.g., transport, energy use) 58% reported awareness	n.a.

Notes n.a.: not assessed; M = arithmetic mean; SD = standard deviation; SES = socioeconomic status. *: As reported by [55].

Appendix B

Table A2. People willingness to stop or reduce meat consumption because of its environmental impact.

					Outcome Measure: Willingness to Reduce/Replace		
Author(s), Year	**Design; Year Data Collected**	**Country; Sample**	**Main Research Question**	**Provided Information Prior the Experiment**	**Question or Dependent Variable**	**Response or Finding**	**Effect of Covariates**
Campbell-Arvai et al., 2014 [29] *	Experimental between-subject design with control group; unspecified.	U.S.; convenience sample of students, $N = 319$, 46% men	Nudging intervention; food-related environmental beliefs and behaviors	Use of a default vegetarian meal option vs. provision of information on the menus.	Hypothetical choice of a lunch or dinner meal (with or without meat)	Offering a vegetarian option as default increased the probability that participants would choose a meat-free meal (OR = 4.10, $p < 0.001$), information on the menu did not significantly influence meal choice (OR = 1.09, $p = 534$).	Females were more likely to choose meat-free menus (OR = 0.49, $p = 0.02$), biospheric value orientation and pro-environmental beliefs were not significant
Cordts et al., 2014 [25] *	Online experiment; 2013	Germany; quota sample, $N = 590$, 52% men	Consumer response to negative information on meat consumption	Randomization to info about negative consequences of meat consumption for animal welfare/health/climate change/personal image; no control group.	Consumers' belief that they will reduce their meat consumption in the future (measured before and after info)	Before info: 12.8% After info: 18.8% (climate change) to 28.0% (animal welfare)	Condition climate change: Smaller effect in men compared with women (15.5% vs. 22.8%)

Table A2. *Cont.*

					Outcome Measure: Willingness to Reduce/Replace		
Author(s), Year	**Design; Year Data Collected**	**Country; Sample**	**Main Research Question**	**Provided Information Prior the Experiment**	**Question or Dependent Variable**	**Response or Finding**	**Effect of Covariates**
De Boer et al., 2018 [38]	Survey data obtained from EU Report; 2012	See EU Report	See EU Report	See EU Report	See EU Report	See EU Report.	(1) Willingness to replace meat (%yes) (2) Willingness to eat less but better meat (%yes) High-income zones Northern zone (1) 38% (2) 77% Western Central zone (1) 42% (2) 78% Medit. zone (1) 55% (2) 86% Medium-to-low income zones Northern zone (1) 54% (2) 83% Western Central zone (1) 63% (2) 86% Medit. zone (1) 57% (2) 86%
Study 1: de Boer, Schösler, et al., 2013 [20] *	Online survey; 2010	Netherlands; quota sample, $N = 1083$, 50% men	Motivational explanations for responses to the meat-free meal idea	As an individual, you can make a big difference to nature and climate protection by choosing one (or more) meals without meat every week.	Willingness to choose one or more meals without meat every week	5% certainly 41% maybe 21% doing so already 23% do not want to	Predictors for 'does not want to change' vs. 'maybe' (reference): skepticism about climate change (OR = 1.98, $p < 0.001$), value of care for nature (OR = 0.64, $p < 0.001$), level of education (OR = 0.90, $p < 0.05$) (based on standardized predictors)

Table A2. *Cont.*

Author(s), Year	Design; Year Data Collected	Country; Sample	Main Research Question	Provided Information Prior the Experiment	Outcome Measure: Willingness to Reduce/Replace		
					Question or Dependent Variable	**Response or Finding**	**Effect of Covariates**
Study 2: de Boer et al., 2014 [31] *	[the same]	[the same]	Consumers' strategies to reduce meat consumption and its' association with their willingness to eat meat-less meals.	[the same]	Willingness to choose one or more meals without meat every week	Same results as in study 1	Predictors for 'certainly' vs. 'maybe' (reference): Female gender (OR = 2.02, $p < 0.01$), familiarity with topic (OR = 2.67, $p < 0.001$), buying meat substitutes (OR = 1.39, $p < 0.001$), preference for plant-based proteins (OR = 1.34, $p < 0.01$) and number of meat-eating days (OR = 0.70, $p < 0.001$); education and age were n.s.
De Boer et al., 2016 [30]	Nation-wide consumer surveys; 2014	Netherlands and the USA; representative sample N = 527 (Netherlands) Weighted variables: gender, age, level of education, region, and a value-related test score on "mentality-environment". (efficiency of the weighting 89%, effective sample size 478) N = 556 (USA). Weighted variables: gender, age, and level of education (efficiency of the weighting 90%, effective sample size 500) Total = 1083	Consumers awareness of meat consumption environmental impact and their willingness to reduce meat consumption, among other research questions.	No prior info given.	Willingness to personally make lifestyle-changes (those already doing it at the time of experiment were instructed to choose the option "certainly willing")." The answer categories were "Certainly not willing" (1), "Likely not willing" (2), "Likely willing" (4), "Certainly willing" (5), and "Don't know" (recoded to 3).	Only a small group of participants of both countries were willing to change. Reducing meat consumption was the second less chosen behavior to curb climate change among the DUTCH (M = 3.58 SD = 0.36) and the least chosen among the U.S. (M = 3.01 SD = 1.44)	When participants believed eating less meat to be a highly effective behavior to curb climate change, the medians increased: Dutch M = 4.26 SD = 0.96; US M = 3.88 SD = 1.19.

Table A2. *Cont.*

					Outcome Measure: Willingness to Reduce/Replace		
Author(s), Year	**Design; Year Data Collected**	**Country; Sample**	**Main Research Question**	**Provided Information Prior the Experiment**	**Question or Dependent Variable**	**Response or Finding**	**Effect of Covariates**
De Groeve, et al., 2017 [21]	Online survey. Two samples. Data collected in 2015 (sample 1) and 2016 (sample 2)	Belgium; Ghent University Business Administration Students $N = 429$	Assess students support for six less meat initiatives (LMIs) to be implemented in student restaurants.	Each respondent had a 50% chance of receiving information about the climate impact of meat before assessing their support for the LMIs	Support for indirect and direct meat curtailment actions: DIRECT MEAT CURTAILMENT "Eating beef or mutton once a week at maximum." M "Reduce your portions of meat per meal (for example, 100 g instead of 120 g) P "Increase the supply of vegetarian main meals up to 50% of the meals." V "Switching to a 'contrarian week' in student restaurants whereby meals with meat are served one day a week, and vegetarian meals four days a week." C	DIRECT MEAT CURTAILMENT STRATEGIES: M Strongly disagree 20% Tend to disagree 27% Neutral 21% Tend to agree 25% Strongly agree 9% P Strongly disagree 9% Tend to disagree 15% Neutral 17% Tend to agree 41% Strongly agree 17% V Strongly disagree 12% Tend to disagree 26% Neutral 28% Tend to agree 24% Strongly agree 10% C Strongly disagree 35% Tend to disagree 33% Neutral 17% Tend to agree 11% Strongly agree 4%	A higher concern for environmental problems is correlated with more positive appraisals of all the LMIs (each $p < 0.001$). A higher KNIM [knowledge about the negative impact of meat **] is also significantly (but less strongly) associated with more positive appraisals of all LMIs, except for LMI-M. Higher appraisals of the direct strategies for meat curtailment (LMIs M, P, V and C) are highly significantly associated with sex and meat consumption frequencies: female students and students who eat meat (or fish) with their main meals less often are more willing to support these LMIs (in every case $p < 0.001$). Prior information about the climate impact of meat appears to have no effect on the support for the LMIs, except for LMI-C, where there is a significant negative effect of information ($U = 20{,}197$; $p = 0.024$) **: KNIM's four themes: environment, animal welfare, health, Global food distribution.

Table A2. *Cont.*

					Outcome Measure: Willingness to Reduce/Replace		
Author(s), Year	**Design; Year Data Collected**	**Country; Sample**	**Main Research Question**	**Provided Information Prior the Experiment**	**Question or Dependent Variable**	**Response or Finding**	**Effect of Covariates**
Graca, Calheiros, et al., 2015 [37] *	Study 1: Online survey; 2014	Portugal; convenience sample, $N = 1023$, 42% men	Development and validation of a meat attachment questionnaire	In recent times, meat consumption is being increasingly debated on the grounds of environmental sustainability, health and safety concerns, and animal rights/welfare arguments.	Willingness to reduce meat consumption (1 = not willing at all to 5 = very willing). Willingness to follow a plant-based diet (1 = not willing at all to 5 = very willing)	No mean values presented.	Predictors for meat reduction: Meat attachment ($\beta = -0.49$, $p < 0.001$), positive attitudes towards meat ($\beta = -0.11$, $p < 0.05$) Predictors for plant-based diet: Meat attachment ($\beta = -0.54$, $p < 0.001$), positive attitudes towards meat ($\beta = -0.12$, $p < 0.05$), meat consumption frequency ($\beta = -0.12$, $p < 0.01$)
Graca, Calheiros, et al., 2015 *	Study 2: Online survey; 2015	Portugal; Amazon Mechanical Turk, $N = 318$, 58% men	Predictive ability of the meat attachment questionnaire for willingness to reduce meat consumption.	see Study 1	Willingness and intention to reduce meat consumption, avoid eating meat, follow a plan-based diet (items averaged for general measure).	No mean values presented	Predictors for willingness: Meat attachment ($\beta = -0.75$, $p < 0.001$), PBC ($\beta = -0.12$, $p < 0.01$) Predictors for intentions: Attitudes towards meat ($\beta = -0.32$, $p < 0.001$), PBC ($\beta = 0.10$, $p < 0.01$), meat attachment ($\beta = -0.53$, $p < 0.001$).

Table A2. *Cont.*

					Outcome Measure: Willingness to Reduce/Replace		
Author(s), Year	**Design; Year Data Collected**	**Country; Sample**	**Main Research Question**	**Provided Information Prior the Experiment**	**Question or Dependent Variable**	**Response or Finding**	**Effect of Covariates**
Graca, Oliveira, et al., 2015 [24]*	Online survey; 2013	Portugal; convenience sample, $N = 410$, 30% men	Multiple correspondence analysis to identify clusters of meat-related associations	Info was provided related to the negative consequences of meat production and consumption for animals, nature and the environment as well as public health	Intent to change current level of meat consumption Willingness to reduce meat consumption by half Willingness to follow a plant-based diet	60% yes, 27% no, (12% no meat consumers) 49% yes, 38% no, (12% no meat consumers) 44% yes, 53% no	n.a.
Hunter et al., 2016 [34]	Postal survey. Date not specified.	Sweden; stratified simple random sample of single family homes. 55% males. 89.5% of the sample had at least one child. Mean age 55. 219 usable questionnaires were returned by post for a response rate of 22% (95% CI (6.25)).	Understand the factors related to fear or danger that motivate consumers to reduce or alter their meat consumption.	Yes, a cover story stating the negative impact of climate change on the earth and humans and statements about the big impact food has on greenhouse gas emissions as well as statement that reducing meat consumption is the most effective food behavior that can be adopted.	Self-efficacy and response efficacy questions regarding meat curtailment strategies	At the same time, the mean scores for self-efficacy and response efficacy show that the participants in this study on average do not find altered meat consumption to be easy, nor do they believe it to be very effective.	

Table A2. *Cont.*

Author(s), Year	Design; Year Data Collected	Country; Sample	Main Research Question	Provided Information Prior the Experiment	Outcome Measure: Willingness to Reduce/Replace		
					Question or Dependent Variable	Response or Finding	Effect of Covariates
Pohjolainen et al., 2016 [22]	Postal survey; 2010	Finland; representative sample. N = 1890	The level of environmental consciousness among Finnish consumers concerning meat production and consumption	No prior info given.	Support to several actions to curb the meat production impact on the environment	Eating less meat the second less supported, only after techno-optimism; only 25.5% considered meat reduction a possible solution. 39.2% rejected this choice.	Consumers clustered in six groups depending on their awareness of meat-related environmental questions: Those aware (highly conscious and rather conscious), those resistant to the idea (Resistant), those who give neutral answers (highly unsure and rather unsure) and those "careless conscious". Among the highly conscious, 77.2% agree with meat reduction; among the rather conscious, 53% agree with meat reduction.
Schösler et al., 2015 [36] *	Face-to-face interview; 2013	Netherlands; quota samples of second- generation migrants: Turkish/Kurdish N = 350, Chinese/Hong Kongese N = 350, Native Dutch N = 357; 47–49% men	Gender differences in meat consumption and reduction across ethnic group	As an individual, you can make a big difference to nature and climate protection by choosing one (or more) meals without meat every week.	Willingness to reduce meat consumption (including 'yes', 'maybe')	Willingness to reduce: 17% Turks (monoculture), 53% Chinese (monoculture), 40% Native Dutch	Turkish men followed by Turkish women reported lowest willingness to reduce meat consumption; no gender differences for Native Dutch and Chinese.

Table A2. *Cont.*

					Outcome Measure: Willingness to Reduce/Replace		
Author(s), Year	**Design; Year Data Collected**	**Country; Sample**	**Main Research Question**	**Provided Information Prior the Experiment**	**Question or Dependent Variable**	**Response or Finding**	**Effect of Covariates**
Tobler et al., 2011 [33] *	Postal survey; 2010	Switzerland (German- and French-speaking regions); random panel sample, 40% Native Dutch *N* = 6189, 48% men	Consumers' willingness to adopt ecological food consumption	No info.	Intention assessment based on TTM for eating less meat (maximum once or twice per week)	The largest fraction of unwilling consumers was in the domain of reducing meat consumption. 36.3% (not willing) 5.4% (willing but not ready) 11.4 (willing and ready) 46% (doing it already). Those in the change stages (willing...) were influenced by environmental reasons. Those doing it already were influenced by health reasons.	Female gender (OR = 1.76), importance of naturalness (OR = 1.32), less meat is healthier (OR = 1.21) and better for the environment (OR = 0.87) predicted action state for willingness to reduce meat consumption, all $p < 0.001$; age and education were n.s.
Truelove et al., 2012 [26]	Online survey with open ended questions and behavior ratings.; 2008	USA; Undergraduate psychology majors (*N* = 112) (69 women and 43 men)	Students perceptions of the relative impact and effectiveness of certain behaviors on global warming.	No prior info given.	Respondents asked to rate their intention to perform 20 different proenvironmental behaviors. 1 (Strongly unlikely) to 7 (Strongly likely)	Reduce your meat consumption: 2.99/7 SD:2.07	Effectiveness knowledge did not significantly correlate with intention to perform behaviors that mitigate GW. Effectiveness belief did significantly correlate with the intention to reduce meat consumption.

Table A2. *Cont.*

					Outcome Measure: Willingness to Reduce/Replace		
Author(s), Year	**Design; Year Data Collected**	**Country; Sample**	**Main Research Question**	**Provided Information Prior the Experiment**	**Question or Dependent Variable**	**Response or Finding**	**Effect of Covariates**
Vanhonacker et al., 2013 [23] *	Online survey; 2011	Belgium (Flanders); convenience sample, N = 221, 36% men	Attitudes towards more sustainable food choices and consumer segmentation based on their self-evaluated ecological footprint.	Explanation of the concept 'ecological footprint'	Willingness to reduce meat consumption (1 = strongly disagree to 5 = strongly agree)	Meat reduction was rated the most appealing option (approx. M = 3.9, only bar chart shown) out of various options to improve sustainability of food choices (e.g., insects, meat substitutes)	n.a.
EU Report [35]	Telephone survey. 2012	27 EUnion countries; aged 15 and above. In each household, the respondent was drawn at random following the "last birthday rule". 1000 people sample per country. Small countries: 500 people sample.	EU citizens' knowledge of green products and their reasons for buying, or not buying, environmentally-friendly products	The interviewer read out: "Some people say large scale meat production has a negative impact on the environment"	Would you be willing to do the following for environmental reasons? (a) Eat less meat but of certified origin (b) Replace most of the meat you eat by vegetables	(a) 80% EU citizens willing to eat less meat but of certified origin Highest: Portugal (89%) Lowest: Estonia (40%) (b) 50% EU citizens willing to replace most of the meat they eat with vegetables Highest: Romania (69%) Lowest: The Netherlands (29%) (Information by country can be found in the report)	The strongest socio-demographic factor linked to willingness to change one's meat consumption is gender. Female respondents are considerably more willing than male respondents to replace most of the meat they eat with vegetables (59% and 40%, respectively). Women are also more willing to replace beef or pork with poultry or fish (76% versus 67%) and eat less meat but of certified origin (83% versus 76%).

Notes n.a.: not assessed; M = arithmetic mean; SD = standard deviation; SES = socioeconomic status. *: As reported by [55].

Appendix C

Table A3. Vegans, vegetarians, and meat consumption curtailers for environmental reasons.

					Outcome Measure: Reason to Reduce Meat or Become Vegetarian		
Author(s), Year	**Design; Year Data Collected**	**Country; Sample**	**Main Research Question**	**Provided Information Prior the Experiment**	**Question or Dependent Variable**	**Response or Finding**	**Effect of Covariates**
De Backer, Charlotte J.S. Hudders, Liselot; 2014 [49]	Large-scale Online survey; year not specified.	Belgium; *N* = 1566 (76% women) M age = 26.12 SD = 8.92 10.6% = vegetarians; 41.8% semi-vegetarians; 47.6% light-semi-vegetarians.	Motives underlying the different forms of vegetarianism and semi-vegetarianism in a culture where meat continues to play a crucial role in people's diets.	No prior info provided.	Agree or disagree with a 7-point Likert scale with motives for meat reduction/avoidance. Ecological motives: "I don't eat meat every day because it is better for the environment," and "I don't eat meat every day because eating meat increases my ecological footprint".	143/165 vegetarians strongly agreed with ecological motives (6.1 or higher in a Likert scale 1–7). For 28/143 ecological concerns were the main drive (mean of 6.5/7 Liker scale) The rest of the vegetarians (n = 22) disagreed with the ecological concerns (mean of 2.61/7 Likert scale). 323/650 semi-vegetarians: reported ecological concerns as the main motivator for strongly reducing meat. (Mean of 5.57/7 Likert scale) 254/741 light semi-vegetarians reported ecological concerns as the main motivator for avoiding meat one or two days a week. (Mean of 5.12/7 Likert scale)	Ecological concern positively associated with meat reduction, except for light semi-vegetarians.

Table A3. *Cont.*

					Outcome Measure: Reason to Reduce Meat or Become Vegetarian		
Author(s), Year	**Design; Year Data Collected**	**Country; Sample**	**Main Research Question**	**Provided Information Prior the Experiment**	**Question or Dependent Variable**	**Response or Finding**	**Effect of Covariates**
De Boer et al., 2017 [52]	Face-to-face interviews; 2013	Netherlands; two samples of adults (aged 18–35) Native Dutch, n = 357, (Men 48%) Second generation Chinese Dutch, n = 350 (Men 47%) Participants were categorized in four dietary groups (all self-declared) (1) Vegetarians (2) Low meat eaters (2–3 days a week) (3) Medium meat eaters (4–5 days a week) (4) High meat eaters (6 days or more)	Differences between vegetarians and three categories of meat eaters in relation to (1) key characteristics of their hot meal, (2) strength and profile of their food-related motivation, and (3) reasons for and reasons against frequently eating meat?	No prior info provided.	Indicate three reasons for not frequently eating meat. Among them, participants could choose "Because it's better for the environment".	NATIVE DUTCH; Self-declared vegetarians: 21% indicated the environment as a reason for not frequently eating meat. Low meat-eaters: 30% Medium meat-eaters: 44% High meat-eaters: 41% TOTAL: 38% CHINESE DUTCH: Self-declared vegetarians: 42% Low meat-eaters: 38% Medium meat-eaters: 32% High meat-eaters: 15% TOTAL: 26% Environmental and financial reasons were mentioned relatively often, but according to the authors, the fact that they were also mentioned by high meat-eaters indicates that, under the current circumstances, these reasons are not decisive for a reduction in meat consumption.	Native Dutch: the more meat they eat, the more they would give an environmental reason for not eating meat. Chinese Dutch, the less meat they eat, the more report the environment as reason for not eating meat. In both samples, the vegetarians were more often women (about 70%), whereas the high meat-eaters were more often men (about 70%).

Table A3. *Cont.*

					Outcome Measure: Reason to Reduce Meat or Become Vegetarian		
Author(s), Year	**Design; Year Data Collected**	**Country; Sample**	**Main Research Question**	**Provided Information Prior the Experiment**	**Question or Dependent Variable**	**Response or Finding**	**Effect of Covariates**
Dyett, Patricia A., et al., 2013 [40]	Postal survey; (year not reported)	United States; $N = 100$ Population of self-reported vegans for more than 9 months living in different U.S. States. Age: 25–75 yrs old Vegans defined as individuals who used no meat, fish, or poultry, and who used dairy- or egg-containing products less than once per month.	Discover the main reasons for adopting and maintaining a vegan lifestyle and to determine whether participants' diet and lifestyle choices coincided with positive health indices and selected outcome assessment.	No prior info provided.	Reason for being vegan	Because environmental values (2%)	n.a.
Turner-McGrievy, G. et al., 2016 [41]	online quota survey; year not specified;	Majority (90%) from the United States; $N = 422$ ($n = 125$ ULTRA, $n = 152$ FULL, $n = 145$ HALF) More ULTRA participants were men (63%) (vs. FULL (37%) and HALF (23%)	Examine differences in current vegetarian and vegan diets, reasons for it and other dietary behaviors among long distance runners.	No prior info provided.	Participants asked to select all reasons for choosing their current diet that apply to them from a list of 12 reasons (including an option to select no reason or to write in an answer).	More ULTRA participants ($n = 25$, 20%) reported that environmental concerns shaped their diet choice as compared with FULL and HALF participants ($n = 36$, 12%; $\chi 2 = 4.4$, $p = 0.04$).	n.a.

Table A3. *Cont.*

					Outcome Measure: Reason to Reduce Meat or Become Vegetarian		
Author(s), Year	**Design; Year Data Collected**	**Country; Sample**	**Main Research Question**	**Provided Information Prior the Experiment**	**Question or Dependent Variable**	**Response or Finding**	**Effect of Covariates**
Haverstock, Katie, et al., 2012 [48]	Food Choice Questionnaire; year not specified;	International online sample; N = 247 (196 = current animal product limiters and 51 former limiters) 211 = females; Age = 18 to 66 (M = 29.05, SD = 9.39) 222 = Euro-Americans.	Similarities and differences between current and former animal product limiters.	No prior info provided.	Eight items concerning ethical food choice motives were also included [...] These ethical motives include animal welfare, environmental protection, political values, and religion. Likert scale: 1 = not important to 4 = very important.	Importance given to environmental reasons to reduce or avoid meat. CURRENT LIMITERS: Vegans (n = 119) M = 3.10, SD = 0.68 Vegetarian (n = 54) M = 2.71, SD = 0.74 Pescatarian (n = 22) M = 2.79, SD = 0.75 FORMER LIMITERS: Now a regular meat eater (n = 16) M = 2.13, SD = 0.94 Now a occasional meat eater (n = 26) M = 2.67, SD = 0.80 Now a meat avoider (n = 4) M = 2.17, SD = 0.43 Now a pescatarian (n = 5) M = 2.54, SD = 0.88	Few gender differences. Women more strongly endorsed health and the environment motives than did men.

Table A3. *Cont.*

Author(s), Year	Design; Year Data Collected	Country; Sample	Main Research Question	Provided Information Prior the Experiment	Outcome Measure: Reason to Reduce Meat or Become Vegetarian		
					Question or Dependent Variable	Response or Finding	Effect of Covariates
Hoffman, Sarah R. et al., 2013 [39]	Online survey; 2011	USA; People recruited through Facebook, Google, and vegetarian dedicated webpages. $N = 312$ Age: 18–69. (42% = age 20–29) 15.4% men, 84.6% women. 68.3% had some form of Higher Education. 86.5% White-Caucasian 56.7% had an income of <49,000 USD Vegetarian 49.4 (vegetarian) and 50.6 (vegan).	Examine the differences between health and ethical vegetarians by comparing conviction, nutrition knowledge, dietary restriction, and years as vegetarian between the two groups.	No prior info provided.	In order to place subjects into categories (i.e., health, ethical, or other), two multiple choice items were created: "The main reason I became a vegetarian was because of (check only one)," "The main reason I am (still) a vegetarian is because of (check only one)." Fourteen options were given in addition to the option "other"	234 = ethical reasons (animal, ethics, religion, environment) (10 = the environment) as initial reason to become vegetarian.	Not reported.
Izmirli, et al., 2011 [51]	Survey; year not specified;	11 Eurasian countries; $N = 3433$ university students from 103 universities. 47% avoided some meat products. 4% vegetarians 0.4% vegans	Determine the relationship between the consumption of animal products and attitudes towards animals among university students in Eurasia	No prior info provided.	Specify the major reason for meat avoidance like health concerns, religious instruction, concerns for the suffering of animals or for the environment.	479 students (38.1%) gave the environmental reason.	Among "some meat avoidants" (total = 1147) 468 41% because of the environment. (Most chosen reason). Among "vegetarians" (total = 99) 9 (9%) because of the environment. Among "vegans" (total = 7) 2 (29%) because of the environment.

Table A3. *Cont.*

					Outcome Measure: Reason to Reduce Meat or Become Vegetarian		
Author(s), Year	**Design; Year Data Collected**	**Country; Sample**	**Main Research Question**	**Provided Information Prior the Experiment**	**Question or Dependent Variable**	**Response or Finding**	**Effect of Covariates**
Lindeman, Marjaana, et al., 2001 [44]	STUDY 1 Food Choice Questionnaire; year not specified.	Finland; 82 female participants. Age: 17–3 years old. 30.4% semi-vegetarians and 25.3% vegetarians.	The construction of food choice ideologies and the ways dietary groups endorse them.	No prior info provided.	Food Choice Questionnaire. Motives assessed among others: ecological welfare (including animal welfare and protection of nature). Subjects had to rate the statement "It is important to me that the food I eat on a typical day..." on a 4-point scale (1 = not at all important, 4 = very important).	Ecological welfare. Semi and full vegetarians: M = 3, SD = 0.74	Vegetarians regarded ecological food choice reasons as more important than semivegetarians did, $t(45) = -4.12$, $p < 0.001$.
Lindeman, Marjaana, et al., 2001	STUDY 2 Food Choice Questionnaire; year not specified.	Finland; N = 149 women. Age: 19–74 Mean age: 31.5. 44.3 full time students. 41.6% employed women. 16.8% semivegetarians and 10.7% vegetarians.	Idem	No prior info provided.	Idem	Ecological welfare. Semi and full vegetarians: M = 2.94, SD = 0.80	n.a.

Table A3. *Cont.*

					Outcome Measure: Reason to Reduce Meat or Become Vegetarian		
Author(s), Year	**Design; Year Data Collected**	**Country; Sample**	**Main Research Question**	**Provided Information Prior the Experiment**	**Question or Dependent Variable**	**Response or Finding**	**Effect of Covariates**
Péneu, et al., 2017 [50]	Online survey. Ongoing web-based prospective observational cohort study launched in France in May 2009 with a scheduled follow-up of 10 years.	France; N = 22,935 (5688 men)	Investigate the sociodemographic profiles of individuals reporting health and environmental dilemmas when purchasing meat, fish and dairy products, and compare diet quality of individuals with and without dilemma.	No prior info provided.	Respondents have to agree or disagree with the following statement: "I avoid purchasing [meat/fish/dairy products] for environmental issues"	25% strongly agree or agree	
Péneu, et al., 2017					Asked to agree or not with "I am torn between purchasing [meat/fish/dairy products] to follow dietary guidelines or limit purchase for environmental issues".	31.94% said YES	- Women declared more dilemma in the case of meat than men. - In the case of meat, individuals with greater educational level and household including only one adult were more likely to report a dilemma.

Table A3. *Cont.*

					Outcome Measure: Reason to Reduce Meat or Become Vegetarian		
Author(s), Year	**Design; Year Data Collected**	**Country; Sample**	**Main Research Question**	**Provided Information Prior the Experiment**	**Question or Dependent Variable**	**Response or Finding**	**Effect of Covariates**
Povey et al., 2001 [53]	Open ended questionnaires; year not reported.	United Kingdom; Convenience sample; 111 respondents (25 meat eaters, 26 meat avoiders, 34 vegetarians, 26 vegans).	Examine differences between the attitudes and beliefs of four dietary groups (meat eaters, meat avoiders, vegetarians and vegans) and the extent to which attitudes influenced intentions to follow a diet.	No prior info provided.	Record salient thoughts, beliefs and feelings towards these three diets: meat, vegetarian and vegan. A maximum of eight thoughts, beliefs or feelings could be recorded by participants.	MEAT DIET: 6/26 vegans and 6/26 meat avoiders named environmental problems as a salient belief towards eating a meat diet. VEGETARIAN DIET: 4/26 vegans mentioned a vegetarian diet to be environmentally friendly. VEGAN DIET: 12/26 vegans mentioned it to be environmentally friendly.	n.a.
Pribis, et al., 2010 [43]	cross-sectional, observational study; 2007	United States; Andrews University (SDA institution) undergraduate students and their respective families. $N = 609$ participants. (35% male) Mean age = 31 years old. 4% vegans; 25% lacto-ovo vegetarians; 4% pesco-vegetarians; 67% non-vegetarians.	Examine whether reasons to adopt vegetarian lifestyle differ significantly among generations.	No prior info provided.	Using a Likert Scale from 1 to 5 (strongly disagree [1]–agree [2]–no opinion [3]–agree [4]–strongly agree [5]) participants rated reasons why they choose a vegetarian lifestyle. Vegetarian reason: "vegetarian lifestyle is much more protective against the environment"	Responses across generations: 11–20 years old: 3.95 21–40 years old: 3.69 41–60 years old: 3.75 61 or older: 3.79	Younger people (11–20 years) also significantly agreed more with the environmental reason ($p = 0.025$).

Table A3. *Cont.*

					Outcome Measure: Reason to Reduce Meat or Become Vegetarian		
Author(s), Year	**Design; Year Data Collected**	**Country; Sample**	**Main Research Question**	**Provided Information Prior the Experiment**	**Question or Dependent Variable**	**Response or Finding**	**Effect of Covariates**
Rozin, et al., 1997 [45]	Questionnaire; 1987	United States; N = 104 self-identified as at least reluctant to meat. 34 = male Mean age: 26.6 (SD = 8.95)	Describe moralization in the domain of vegetarianism.	No prior info provided.	A list of 20 possible reasons for avoiding meat. Subjects indicated both current agreement (5-point scale ranging from disagree strongly to agree strongly) with each reason and, if relevant, the time of onset of the reason ("this was your first reason for avoiding meat," "this was one of the earliest reasons for avoiding meat," "this was not one of the earliest reasons for avoiding meat," or "this was never a reason for avoiding meat"). Ecological reason: "I resist [avoid] eating "meat" because it is wasteful of resources to eat animal rather than vegetable products, especially in a world where people are starving."	5.8% "initial reason" to avoid meat. 38.2% strongly agreed (22.5% agreed) with ecological reason as current reason.	n.a.

Table A3. *Cont.*

					Outcome Measure: Reason to Reduce Meat or Become Vegetarian		
Author(s), Year	**Design; Year Data Collected**	**Country; Sample**	**Main Research Question**	**Provided Information Prior the Experiment**	**Question or Dependent Variable**	**Response or Finding**	**Effect of Covariates**
Schösler et al., 2015 [36] *	Face-to-face interview; 2013	Netherlands; quota samples of second-generation migrants: Turkish/Kurdish N = 350, Chinese/Hong Kongese N = 350, Native Dutch N = 357; 47–49% men	Gender differences in meat consumption and reduction across ethnic group		Reasons for not frequently eating meat (selection of maximum 3 reasons out of 9 reasons)	It's better for the environment' was selected by 2% Turks, 26% Chinese, 38% Native Dutch	
Study 3: Schösler, de Boer, & Boersema, 2014 [46] *	Online survey; 2010	Netherlands; quota sample, N = 1083, 50% men	Cluster analysis based on type of eating-related motivation and profiling of segments related to meat consumption	No info before questioning	Reasons for not frequently eating meat (selection of maximum 3 reasons out of 9 reasons)	19% selected 'It's better for the environment'	34% of those consumers who internalized the importance of the food-nature relationship agreed that eating less meat is better for the environment.
Vanhonacker et al., 2013 [23] *	Online survey; 2011	Belgium (Flanders); convenience sample, N = 221, 36% men	Attitudes towards more sustainable food choices and consumer segmentation based on their self-evaluated ecological footprint.	Explanation of the concept 'ecological footprint'	Participants had to indicate environmentally-friendly behaviors (what they actually do)	4% consume less meat per meal to reduce their ecological footprint	n.a.

Table A3. *Cont.*

					Outcome Measure: Reason to Reduce Meat or Become Vegetarian		
Author(s), Year	**Design; Year Data Collected**	**Country; Sample**	**Main Research Question**	**Provided Information Prior the Experiment**	**Question or Dependent Variable**	**Response or Finding**	**Effect of Covariates**
Verain et al., 2015 [47] *	Online survey; 2011	Netherlands; quota sample, $N = 942$, 50% men.	Segmentation of consumers based on sustainable food behaviors and profiling of segments	No info	Performance of sustainable food behaviors at least once a month in the previous year ('yes', 'no').	One meat-free day a week (56%) and smaller meat portions (52%) were the most popular sustainable food behaviors compared with other behaviors (e.g., buying organic meat or dairy)	Female gender ($\beta = 0.08$, $p < 0.001$), age ($\beta = 0.21$, $p < 0.05$) and variables on personal/social norms and subjective knowledge about sustainable food choices positively predicted curtailment behavior (average of four items: eating smaller portions of meat, eat less diary, eating less, one meat-free day a week)
White et al. 1999 [42]	Survey; date not specified	United States; Random sample. $N = 2500$ women from each of the past decades' graduating medical school classes 8% Self-described vegetarians	Investigate the prevalence and characteristics of vegetarian subjects in the Women Physician's Health Study and compare them with the omnivore cohort.	No prior info provided.	Self-categorized vegetarians were asked why they were vegetarian.	32.1% cited environmental reasons.	n.a.

Notes n.a.: not assessed; M = arithmetic mean; SD = standard deviation; SES = socioeconomic status. *: As reported by [55].

References

1. Steinfeld, H.; Gerber, P.; Wassenaar, T.D.; Castel, V.; Rosales, M.; Rosales, M.; de Haan, C. *Livestock's Long Shadow: Environmental Issues and Options*; Food & Agriculture Org.: Rome, Italy, 2006; ISBN 92-5-105571-8.
2. Slingo, J.M.; Challinor, A.J.; Hoskins, B.J.; Wheeler, T.R. Introduction: Food crops in a changing climate. *Philos. Trans. R. Soc. B Biol. Sci.* **2005**, *360*, 1983–1989. [CrossRef] [PubMed]
3. Bouwman, L.; Goldewijk, K.K.; Van Der Hoek, K.W.; Beusen, A.H.; Van Vuuren, D.P.; Willems, J.; Rufino, M.C.; Stehfest, E. Exploring global changes in nitrogen and phosphorus cycles in agriculture induced by livestock production over the 1900–2050 period. *Proc. Natl. Acad. Sci. USA* **2013**, *110*, 20882–20887. [PubMed]
4. Godfray, H.C.J.; Beddington, J.R.; Crute, I.R.; Haddad, L.; Lawrence, D.; Muir, J.F.; Pretty, J.; Robinson, S.; Thomas, S.M.; Toulmin, C. Food security: The challenge of feeding 9 billion people. *Science* **2010**, *327*, 812–818. [CrossRef] [PubMed]
5. Thornton, P.K. Livestock production: Recent trends, future prospects. *Philos. Trans. R. Soc. B Biol. Sci.* **2010**, *365*, 2853–2867. [CrossRef] [PubMed]
6. Dauvergne, P. *The Shadows of Consumption: Consequences for the Global Environment*; MIT Press: Cambridge, MA, USA, 2010; ISBN 0-262-26057-3.
7. Willett, W.; Rockström, J.; Loken, B.; Springmann, M.; Lang, T.; Vermeulen, S.; Garnett, T.; Tilman, D.; DeClerck, F.; Wood, A.; et al. Food in the Anthropocene: The EAT–Lancet Commission on healthy diets from sustainable food systems. *Lancet* **2019**, *393*, 447–492. [CrossRef]
8. Sabate, J.; Soret, S. Sustainability of plant-based diets: Back to the future. *Am. J. Clin. Nutr.* **2014**, *100*, 476S–482S. [CrossRef]
9. Glanz, K.; Rimer, B.K.; Viswanath, K. *Health Behavior and Health Education: Theory, Research, and Practice*; John Wiley & Sons: Hoboken, NJ, USA, 2008; ISBN 0-470-43248-9.
10. Segovia-Siapco, G.; Sabaté, J. Health and sustainability outcomes of vegetarian dietary patterns: A revisit of the EPIC-Oxford and the Adventist Health Study-2 cohorts. *Eur. J. Clin. Nutr.* **2018**. [CrossRef]
11. Ruby, M.B. Vegetarianism. A blossoming field of study. *Appetite* **2012**, *58*, 141–150. [CrossRef]
12. Kessler, C.S.; Holler, S.; Joy, S.; Dhruva, A.; Michalsen, A.; Dobos, G.; Cramer, H. Personality Profiles, Values and Empathy: Differences between Lacto-Ovo-Vegetarians and Vegans. *Complement. Med. Res.* **2016**, *23*, 95–102. [CrossRef]
13. Beardsworth, A.D.; Keil, E.T. Vegetarianism, Veganism, and Meat Avoidance: Recent Trends and Findings. *Br. Food J.* **1991**, *93*, 19–24. [CrossRef]
14. Fox, N.; Ward, K. Health, ethics and environment: A qualitative study of vegetarian motivations. *Appetite* **2008**, *50*, 422–429. [CrossRef] [PubMed]
15. Hussar, K.M.; Harris, P.L. Children who choose not to eat meat: A study of early moral decision-making. *Soc. Dev.* **2010**, *19*, 627–641. [CrossRef]
16. Jabs, J.; Devine, C.M.; Sobal, J. Model of the process of adopting vegetarian diets: Health vegetarians and ethical vegetarians. *J. Nutr. Educ.* **1998**, *30*, 196–202. [CrossRef]
17. Neale, R.J.; Tilston, C.H.; Gregson, K.; Stagg, T. Women vegetarians: Lifestyle considerations and attitudes to vegetarianism. *Nutr. Food Sci.* **1993**, *93*, 24–27. [CrossRef]
18. Dagevos, H.; Voordouw, J. Sustainability and meat consumption: Is reduction realistic? *Sustain. Sci. Pract. Policy* **2013**, *9*, 60–69. [CrossRef]
19. Moher, D.; Liberati, A.; Tetzlaff, J.; Altman, D.G.; The PRISMA Group. Preferred Reporting Items for Systematic Reviews and Meta-Analyses: The PRISMA Statement. *PLoS Med.* **2009**, *6*, e1000097. [CrossRef] [PubMed]
20. de Boer, J.; Schosler, H.; Boersema, J.J. Climate change and meat eating: An inconvenient couple? *J. Environ. Psychol.* **2013**, *33*, 1–8. [CrossRef]
21. De Groeve, B.; Bleys, B. Less Meat Initiatives at Ghent University: Assessing the Support among Students and How to Increase It. *Sustainability* **2017**, *9*, 1550. [CrossRef]
22. Pohjolainen, P.; Tapio, P.; Vinnari, M.; Jokinen, P.; Räsänen, P. Consumer consciousness on meat and the environment—Exploring differences. *Appetite* **2016**, *101*, 37–45. [CrossRef] [PubMed]
23. Vanhonacker, F.; Van Loo, E.J.; Gellynck, X.; Verbeke, W. Flemish consumer attitudes towards more sustainable food choices. *Appetite* **2013**, *62*, 7–16. [CrossRef]

24. Graca, J.; Oliveira, A.; Calheiros, M.M. Meat, beyond the plate. Data-driven hypotheses for understanding consumer willingness to adopt a more plant-based diet. *Appetite* **2015**, *90*, 80–90. [CrossRef] [PubMed]
25. Cordts, A.; Nitzko, S.; Spiller, A. Consumer Response to Negative Information on Meat Consumption in Germany. *Int. Food Agribus. Manag. Rev.* **2014**, *17*, 83–106.
26. Truelove, H.B.; Parks, C. Perceptions of behaviors that cause and mitigate global warming and intentions to perform these behaviors. *J. Environ. Psychol.* **2012**, *32*, 246–259. [CrossRef]
27. Lea, E.; Worsley, A. Australian consumers' food-related environmental beliefs and behaviours. *Appetite* **2008**, *50*, 207–214. [CrossRef] [PubMed]
28. Clonan, A.; Wilson, P.; Swift, J.A.; Leibovici, D.G.; Holdsworth, M. Red and processed meat consumption and purchasing behaviours and attitudes: Impacts for human health, animal welfare and environmental sustainability. *Public Health Nutr.* **2015**, *18*, 2446–2456. [CrossRef] [PubMed]
29. Campbell-Arvai, V. Food-related environmental beliefs and behaviours among university undergraduates a mixed-methods study. *Int. J. Sustain. High. Educ.* **2015**, *16*, 279–295. [CrossRef]
30. de Boer, J.; de Witt, A.; Aiking, H. Help the climate, change your diet: A cross-sectional study on how to involve consumers in a transition to a low-carbon society. *Appetite* **2016**, *98*, 19–27. [CrossRef] [PubMed]
31. de Boer, J.; Schosler, H.; Aiking, H. "Meatless days" or "less but better"? Exploring strategies to adapt Western meat consumption to health and sustainability challenges. *Appetite* **2014**, *76*, 120–128. [CrossRef]
32. Siegrist, M.; Visschers, V.H.M.; Hartmann, C. Factors influencing changes in sustainability perception of various food behaviors: Results of a longitudinal study. *Food Qual. Prefer.* **2015**, *46*, 33–39. [CrossRef]
33. Tobler, C.; Visschers, V.H.M.; Siegrist, M. Eating green. Consumers' willingness to adopt ecological food consumption behaviors. *Appetite* **2011**, *57*, 674–682. [CrossRef]
34. Hunter, E.; Roos, E. Fear of climate change consequences and predictors of intentions to alter meat consumption. *Food Policy* **2016**, *62*, 151–160. [CrossRef]
35. European Commission. *Brussels DG Communication COMM A1 'Research and Speechwriting' Flash Eurobarometer 367 (Attitudes of Europeans towards Building the Single Market for Green Products)*; European Commission: Brussels, Belgium, 2013.
36. Schösler, H.; de Boer, J.; Boersema, J.J.; Aiking, H. Meat and masculinity among young Chinese, Turkish and Dutch adults in the Netherlands. *Appetite* **2015**, *89*, 152–159. [CrossRef]
37. Graca, J.; Calheiros, M.M.; Oliveira, A. Attached to meat? (Un)Willingness and intentions to adopt a more plant-based diet. *Appetite* **2015**, *95*, 113–125. [CrossRef] [PubMed]
38. de Boer, J.; Aiking, H. Prospects for pro-environmental protein consumption in Europe: Cultural, culinary, economic and psychological factors. *Appetite* **2018**, *121*, 29–40. [CrossRef] [PubMed]
39. Hoffman, S.R.; Stallings, S.F.; Bessinger, R.C.; Brooks, G.T. Differences between health and ethical vegetarians. Strength of conviction, nutrition knowledge, dietary restriction, and duration of adherence. *Appetite* **2013**, *65*, 139–144. [CrossRef] [PubMed]
40. Dyett, P.A.; Sabate, J.; Haddad, E.; Rajaram, S.; Shavlik, D. Vegan lifestyle behaviors. An exploration of congruence with health-related beliefs and assessed health indices. *Appetite* **2013**, *67*, 119–124. [CrossRef] [PubMed]
41. Turner-McGrievy, G.M.; Moore, W.J.; Barr-Anderson, D. The Interconnectedness of Diet Choice and Distance Running: Results of the Research Understanding the Nutrition of Endurance Runners (RUNNER) Study. *Int. J. Sport Nutr. Exerc. Metab.* **2016**, *26*, 205–211. [CrossRef] [PubMed]
42. White, R.F.; Seymour, J.; Frank, E. Vegetarianism among us Women Physicians. *J. Am. Diet. Assoc.* **1999**, *99*, 595–598. [CrossRef]
43. Pribis, P.; Pencak, R.C.; Grajales, T. Beliefs and Attitudes toward Vegetarian Lifestyle across Generations. *Nutrients* **2010**, *2*, 523–531. [CrossRef] [PubMed]
44. Lindeman, M.; Sirelius, M. Food choice ideologies: The modern manifestations of normative and humanist views of the world. *Appetite* **2001**, *37*, 175–184. [CrossRef]
45. Rozin, P.; Markwith, M.; Stoess, C. Moralization and becoming a vegetarian: The transformation of preferences into values and the recruitment of disgust. *Psychol. Sci.* **1997**, *8*, 67–73. [CrossRef]
46. Schoesler, H.; de Boer, J.; Boersema, J.J. Fostering more sustainable food choices: Can Self-Determination Theory help? *Food Qual. Prefer.* **2014**, *35*, 59–69. [CrossRef]
47. Verain, M.C.; Dagevos, H.; Antonides, G. Sustainable food consumption. Product choice or curtailment? *Appetite* **2015**, *91*, 375–384. [CrossRef] [PubMed]

48. Haverstock, K.; Forgays, D.K. To eat or not to eat. A comparison of current and former animal product limiters. *Appetite* **2012**, *58*, 1030–1036. [CrossRef] [PubMed]
49. De Backer, C.J.S.; Hudders, L. From Meatless Mondays to Meatless Sundays: Motivations for Meat Reduction among Vegetarians and Semi-vegetarians Who Mildly or Significantly Reduce Their Meat Intake. *Ecol. Food Nutr.* **2014**, *53*, 639–657. [CrossRef] [PubMed]
50. Peneau, S.; Fassier, P.; Alles, B.; Kesse-Guyot, E.; Hercberg, S.; Mejean, C. Dilemma between health and environmental motives when purchasing animal food products: Sociodemographic and nutritional characteristics of consumers. *BMC Public Health* **2017**, *17*, 876. [CrossRef] [PubMed]
51. Izmirli, S.; Phillips, C.J.C. The relationship between student consumption of animal products and attitudes to animals in Europe and Asia. *Br. Food J.* **2011**, *113*, 436–450. [CrossRef]
52. de Boer, J.; Schoesler, H.; Aiking, H. Towards a reduced meat diet: Mindset and motivation of young vegetarians, low, medium and high meat-eaters. *Appetite* **2017**, *113*, 387–397. [CrossRef]
53. Povey, R.; Wellens, B.; Conner, M. Attitudes towards following meat, vegetarian and vegan diets: An examination of the role of ambivalence. *Appetite* **2001**, *37*, 15–26. [CrossRef]
54. Beardsworth, A.; Keil, T. Health-related beliefs and dietary practices among vegetarians and vegans: A qualitative study. *Health Educ. J.* **1991**, *50*, 38–42. [CrossRef]
55. Hartmann, C.; Siegrist, M. Consumer perception and behaviour regarding sustainable protein consumption: A systematic review. *Trends Food Sci. Technol.* **2017**, *61*, 11–25. [CrossRef]
56. Lakoff, G. Why it matters how we frame the environment. *Environ. Commun.* **2010**, *4*, 70–81. [CrossRef]
57. Graham, T.; Abrahamse, W. Communicating the climate impacts of meat consumption: The effect of values and message framing. *Glob. Environ. Chang.-Hum. Policy Dimens.* **2017**, *44*, 98–108. [CrossRef]
58. Evans, L.; Milfont, T.L.; Lawrence, J. Considering local adaptation increases willingness to mitigate. *Glob. Environ. Chang.-Hum. Policy Dimens.* **2014**, *25*, 69–75. [CrossRef]
59. Feinberg, M.; Willer, R. Apocalypse soon? Dire messages reduce belief in global warming by contradicting just-world beliefs. *Psychol. Sci.* **2011**, *22*, 34–38. [CrossRef] [PubMed]
60. Howell, R.A. Investigating the long-term impacts of climate change communications on individuals' attitudes and behavior. *Environ. Behav.* **2014**, *46*, 70–101. [CrossRef]
61. Gussow, J.D.; Clancy, K.L. Dietary guidelines for sustainability. *J. Nutr. Educ.* **1986**, *18*, 1–5. [CrossRef]
62. Campbell, J.; Macdiarmid, J.I.; Douglas, F. Young people's perception of the environmental impact of food and their willingness to eat less meat for the sake of the environment: A qualitative study. *Proc. Nutr. Soc.* **2016**, *75*, E224. [CrossRef]
63. Macdiarmid, J.I.; Douglas, F.; Campbell, J. Eating like there's no tomorrow: Public awareness of the environmental impact of food and reluctance to eat less meat as part of a sustainable diet. *Appetite* **2016**, *96*, 487–493. [CrossRef]
64. Poulain, J.-P. *The Sociology of Food: Eating and the Place of Food in Society*; Bloomsbury Academic: London, UK; New York, NY, USA, 2017; ISBN 978-1-4725-8620-9.

International Journal of
Environmental Research and Public Health

MDPI

Article

Consumers' Willingness to Pay for Organic Foods in China: Bibliometric Review for an Emerging Literature

Rui Li [1,2,*], Hsiu-Yu Lee [3], Yu-Ting Lin [4], Chih-Wei Liu [3] and Prony F. Tsai [3,5,6]

1 Department of Food Quality and Safety, FoShan University, XianXi Lakeside, DaLi, NanHai District, Foshan 528231, China
2 South China Food Safety Research Center, Foshan 528225, China
3 Department of Business Administration, Cheng Shiu University, Kaohsiung 83347, Taiwan; leehsiuyu7533967@outlook.com (H.-Y.L.); maniera01@gmail.com (C.-W.L.); tsaifs@gcloud.csu.edu.tw (P.F.T.)
4 Department of Food & Beverage Management, Cheng Shiu University; Kaohsiung 83347, Taiwan; k3912@gcloud.csu.edu.tw
5 Super Micro Mass Research and Technology Center, Cheng Shiu University, Kaohsiung 83347, Taiwan
6 Center for Environmental Toxin and Emerging-Contaminant Research, Cheng Shiu University, Kaohsiung 83347, Taiwan
* Correspondence: ruili@fosu.edu.cn; Tel.: +86-0757-82272751

Received: 11 March 2019; Accepted: 8 May 2019; Published: 16 May 2019

Abstract: We conducted a bibliometric review on a small but promising body of literature on consumers' willingness to pay for organic foods in China. Results found that consumers' health consciousness, individual norms, consumer knowledge, food safety, environmental concerns, animal welfare, and purchasing power are major influencing factors for willingness to pay for organic foods in China. Notably, most research methods utilized are quantitative methods, leading us to call for the adoption of more qualitative, review, or mixed-methods. These findings increase our understanding of the knowledge structure of this emerging context-specific literature.

Keywords: willingness to pay (WTP); organic foods; China; Bibliometrics

---

1. Introduction

Organic food has proposed both opportunities and challenges for market consumers (and other stakeholders) in all of the world, especially in developing economies. There are numerous benefits that are associated with organic food; this includes the fact that it is better for the environment, richer in certain nutrients, healthier, safer, and good for the welfare of animals and future sustainability. Also, organic agriculture is gaining enormous popularity regarding providing food and income.

Willer and Lernoud [1] indicated that the total area in Asia dedicated to organic agriculture was about 4.9 million hectares of organic agricultural land in 2016, whereby the leading countries were China (2.28 million hectares) and India (almost 1.2 million hectares). Therefore, the rest of the global countries are left to share the 20%, which is shocking and raises concern on what influences the willingness of consumers to purchase organic food in emerging economies such as China, Brazil, and India. Nonetheless, some studies have suggested that domestic markets for organic food and products in emerging markets, such as China, have been increasing in the last decade [2]. Currently, there are more people that are willing to eat organic food as well as pay a premium price for it. This is mainly due to health concerns that are linked to inorganic food. Specifically, the demand for food quality is increasing in China, and the quality of food has become an essential component of food quality [3]. The change in attitude has mainly been influenced by recent health concerns in China. For instance, some of the health crises in China associated with food safety include: the baby milk incident, Avian Influenza (Bird Flu), and Spongiform Encephalopathy (Mad Cow disease). See a comprehensive review from the following book [3].

However, as has commonly been accepted by public, one of the major factors that is mainly associated with consumers' willingness to pay for organic food is the price, due to the fact that such foods have commonly higher prices. Based on such thought, the main aim of this study is to critically investigate various research topics that are found in agriculture, economics, nutrition, marketing, and food journals addressing the factors that influence customers' willingness to purchase organic food. The research will analyze the emergent but representative literature about Chinese consumers' willingness to pay for organic foods. This will be achieved through a bibliometric mode of investigation. Therefore, the main aim of this study is to critically explore Chinese consumers' perception of organic foods and their willingness to purchase organic food. The research question is: What are the major factors (addressed in the literature) that influence Chinese consumers' willingness to pay (WTP) for organic food and products?

2. Methodology

Bibliometric studies are used with the theoretical perspective that the examination of citations enhances comprehension of the growth of contributions within a particular scientific field; it can identify when papers were written, and the relevance of such a publication in the currently. If a publication continues to be cited over time, historical value is assumed, making the source considered as a primary reference. Moreover, bibliometric identifies key topics contained in those keywords in the field. Progressive use of a keyword across numerous works and time indicates important areas, concepts, or topics in a field. If there is a significant change in keyword usage over some time, it is an indication that there is a paradigm shift. In such perspective, this study will assess journals and other scholarly works to identify the aspect that influences consumers' willingness to pay for organic food and products in China. To our knowledge, there are not many published papers in the Social Science Citation Index, but it still has great potential because of its practical significance in the Chinese food market.

2.1. Data Selection

The selection of data from an extensive search of prior research in the field of consumer behavior, and specifically on the Chinese consumers' willingness to pay for organic foods was an important primary step in answering the research question, and framing the current study in the context of the wide research in the field. The first step in search of previous research involved checking through various databases such as Blackwell, Proquest, SSRN, and AgEcon Search. These databases were selected as they have a large pool of resources, with both published works and also working papers that are relevant in answering the research question under consideration. Moreover, most of the works published in these journals are original papers whose research has been carried out either quantitatively or qualitatively.

The researcher identified the keyword queries for the different sub-areas of Chinese consumer behavior in reference to their willingness to pay for organic foods. The keywords associated with purchase of organic foods by the Chinese are specific to that area, and are important in ensuring that all factors are explored, and the emergent themes from the analysis are exhaustive. The keywords used to address all the subareas of the research questions included; 'Chinese organic foods', 'willingness to pay', 'organic foods purchase', organic foods purchase intentions', 're-purchase intentions', and 'Chinese consumer behavior.' These keywords were used comprehensively for purposes of covering the multidimensionality of the Chinese consumers' behavior towards the purchase of organic foods. The search was intended to collect a large representative pool of original research that could allow this research to draw informed and fundamental conclusions.

Given that thousands of article results retrieved from the databases did not provide useful estimates and variables related to 'willingness to buy' organic foods among Chinese consumers, the researcher set a criterion that limited the scope to articles that precisely addressed the research problem and the research population. First, articles selected covering the concept of 'willingness to pay' were limited to those that focused on organic foods only. Other articles focusing on other aspects of food safety and sustainability such as pest-reduced foods, pesticide-free foods, and labeled foods were eliminated from this study. However, articles that focused on specific organic foods such as organic rice and organic fruits were included.

The concept of currency was also considered during the selection of the articles. Only articles published within the last five years were included in the study, limiting the articles only for those published between 2013 and 2018. The rationale for the time limit is derived from the fact that five years is a critical time in behavioral fields where adequate changes in customer needs and preferences can undergo significant changes. According to Simões [4], Chinese consumers, especially mature citizens, are very conscious about issues related to sustainability, and their behavior continues to change in alignment with their view of sustainable development. Since organic foods are related to sustainability in the long run, this time period was deemed appropriate in ensuring that the study reflected the current studies about Chinese consumer behavior in reference to the willingness to pay for organic foods.

The selected materials were also limited to Chinese consumers only. Most of the retrieved articles yielded results on willingness to pay for organic foods from consumers in other Asian countries, especially Thailand, India, Indonesia, Taiwan, and Malaysia among others. Since the inclusion of different Asian countries expanded the scope of the current study, only studies that were conducted from the perspective of the Chinese consumer were considered relevant for inclusion in the research.

2.2. Data Collection Procedure

The data collection process was conducted in stages. First, the primary search terms were used in each database one at a time. Then, the keywords were combined to form multiple phrases that refined the search and resulted in a wide range of results. The same procedure was repeated for all the databases until all the relevant articles had been retrieved. Articles were included if they meet the aforementioned inclusion criteria, otherwise the rest were not included. Only quantitative, qualitative, and systematic reviews that were peer-reviewed prior to publication were considered. Others such as newspaper articles and general commentary in economics were excluded. The study only utilized published works.

During the first stage of search, a total of 164 articles were identified as containing relevant information on the field of consumer behaviour, and especially the willingness to pay for organic foods. Due to the large body of previous research available for review and the broad scope of their literature, the inclusion criteria already identified was used to narrow the search. The abstract was used to refine the search and determine whether or not the article contained the required information. As such, all the articles that did not have an abstract were automatically excluded.

After the review of the abstract for relevant information, the sample of articles selected was reduced to 67 articles. Broad categories were used to eliminate policy papers and other articles that did not specifically address the issue of organic foods from the perspective of Chinese consumers. Reviewing the articles to only include those published about Chinese consumers and those written in English, or those that have a translated version, narrowed the search to 43 articles. The final exclusion criterion was the time factor, which limited the sample to 10 articles. Figure 1 below demonstrates the sampling process and the narrowing of the number of articles from 164 to 10.

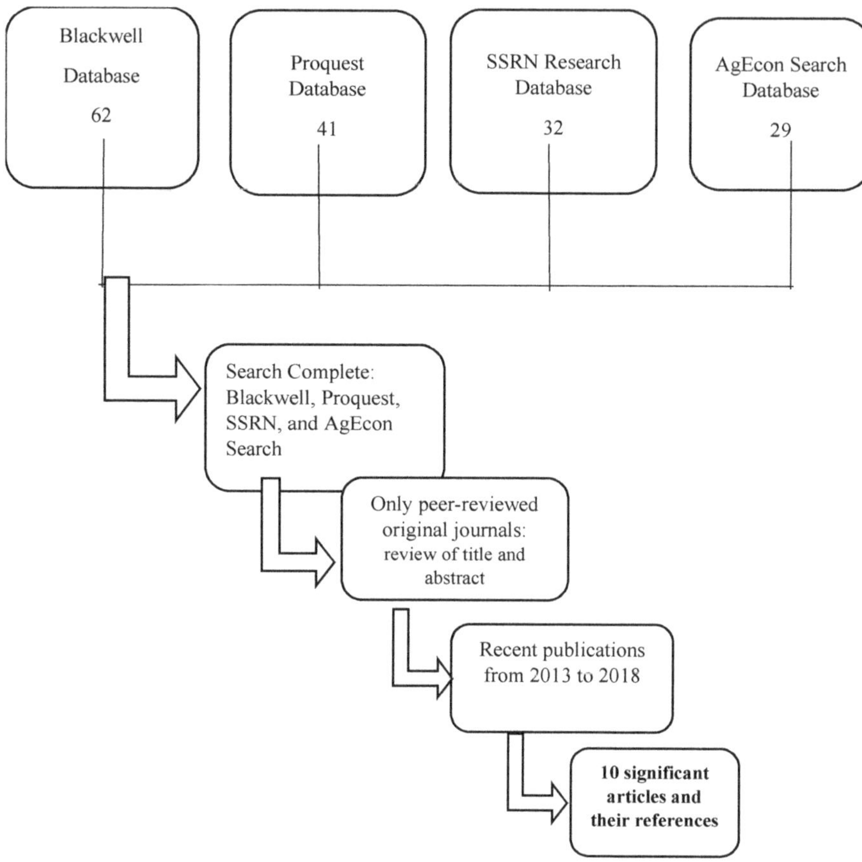

Figure 1. Overview of the sampling procedure.

3. Results and Discussions

3.1. Bibliometric Matrix Synthesis

The results of the analysis of the selected articles based on the author, year of publication, title, methods, and results are presented in the Table 1 below. This analysis was important in performing the bibliometric analysis of the characteristics of selected studies and their contribution to the study.

Table 1. Characteristics of the sampled studies.

No.	Author and Year of Publication	Title	Methods	Results
1	Fang and Levy [5]	'An Analysis of Consumption and Purchasing toward Organic Fruits: Cross-Countries Study between China and France'	A quantitative research design was used to collect data from China and France. An online questionnaire survey utilizing a sample of 261 respondents was used.	The study reported that the Chinese and French consumers have a positive attitude towards the purchase of organic foods, in line with their respective planned behavior. Food-related lifestyles were found to be major drivers that influence purchase decisions.
2	Gan, Zhiyou, Tran, Cohen and Xiangxiang [6]	'Consumer attitudes towards the purchase of organic products in China'	A quantitative research design was adopted where 700 structured questionnaires were used to collect data from a sample in Kunming City. The data sought to measure the attitude of the respondents towards the purchase of organic food products sold in the market.	The findings from this research show household income as the major determinant of the willingness to purchase organic foods. This is because organic foods are expensive compared to conventional foods that are readily available. People with a high social standing are more likely to purchase organic foods as a social marker. Moreover, Chinese households need more information about the benefits of organic foods, especially due to the big scandals in a country of tainted food products.
3	Li and Xin [7]	'Factors influencing organic food purchase of young Chinese consumers.'	This quantitative research used a survey questionnaire to gather data about the factors that influence the Chinese consumers' consumption of organic foods, and their purchase intentions.	The main factor that influences Chinese consumers to purchase organic foods is food safety. Other factors include nutritional values, animal welfare, and environment-friendly. Young Chinese consumers are motivated to pay for organic vegetables and fruits especially due to the recent food scandals in the country.
4	Liu, Pieniak, and Verbeke [8]	'Consumers' attitudes and behaviour towards safe food in China: A review.'	This systematic review includes a sample of 34 studies that focused on the willingness of consumers to pay for safe foods. The article includes organic foods, green foods, and hazard-free foods.	The findings demonstrated that the Chinese consumers have a high level of awareness of safe foods, but low recognition of the concept of safe foods, including labels and identification of safe foods. Generally, there is a positive attitude towards the purchase of safe foods in the Chinese markets.

Table 1. *Cont.*

No.	Author and Year of Publication	Title	Methods	Results
5	McCarthy [9].	'Trends in organic and green food consumption in China: Opportunities and challenges for regional Australian exporters'	A quantitative research design was used to determine the willingness of consumers to pay for food safety, focusing on both organic and green foods. The sample of 250 certified organic and green foods consumers from Beijing, Shanghai, Guangdong, and Chongqing participated in the study.	The willingness of Chinese consumers to purchase organic foods is inspired by health and environmental factors. Moreover, most Chinese consumers who purchase organic foods regularly distrust the Chinese food system; hence their willingness to pay for organic foods is born out of food safety concerns.
6	Xie, Wang, Yang, Wang and Zhang [10]	'Consumer perceptions and attitudes of organic food products in Eastern China.'	The study utilized a mixed methods research design combining both qualitative and quantitative data. The data was collected through a survey administered from consumers in Nanjing and Shanghai, east of China.	The major motivation for willingness to purchase organic foods for consumers in East China is driven by food safety and health concerns. Also, education level and higher purchasing power are attributed to increase the willingness of consumers to purchase organic foods. Lack of awareness and knowledge about the benefits of organic foods is a major barrier to willingness to purchase organic foods in China.
7	Xu, Su and Lone, [11]	'Chinese consumers' willingness to pay for rice.'	A quantitative study conducted using a survey administered in Chongqing and Chengdu, as they are the largest rice consumption cities in China. A conditional logit model was used to analyze the data.	Price is a major contributor of the purchase of organic rice in Chongqing and Chengdu. High demand for organic rice was reported in higher food expense consumers as they are willing to pay a premium for the organic rice.
8	Yang, Al-Shaaban and Nguyen, T. B. [12]	'Consumer attitude and purchase intention towards organic food: A quantitative study of China.'	A quantitative research design was conducted through an online survey of Chinese consumers. The study was based on six hypotheses.	The results of the analysis indicated that the Chinese consumers' willingness to pay for organic foods is influenced by health consciousness, individual norms, and consumer knowledge.

Table 1. *Cont.*

No.	Author and Year of Publication	Title	Methods	Results
9	Yen [13]	Managing self-congruity to influence behavioral intention in organic food contexts in Fujian province, China.	This quantitative study employed 200 questionnaire surveys from organic food consumers in Fujian province.	The results of the analysis demonstrated that self-congruity has a major influence on the willingness to pay for organic foods, including aspects such as self-image, consistence with self-perception, self-reflection, and similar people purchasing organic foods. The ideal social concept should be a bridge between the consumers' self-image and the image of organic foods.
10	Zhu [14].	'Using the theory of planned behavior to investigate what influences Chinese intention to purchase organic food.'	A quantitative research design that utilized data collected from 216 in-depth surveys. The sample population was students from various universities in Hubei Province.	The study found that consumer identity and ecological motive were the biggest drivers of consumers' willingness to purchase organic products. The study concluded that since people's motivation to pay for organic foods is out of ecological concerns, and attitudes, marketers of organic foods should make such variables a priority by connecting organic foods to environmental values.

3.2. Most Cited Keywords

The analysis of the articles revealed intersecting themes including health consciousness, individual norms, consumer knowledge, food safety, environment/animal welfare, and nutritional factors. Some articles reported multiple themes. Figure 2 below presents the major themes emerging from the keywords most cited in the selected articles. Due to space in the figure, please find the caption's full description for each category in the main text.

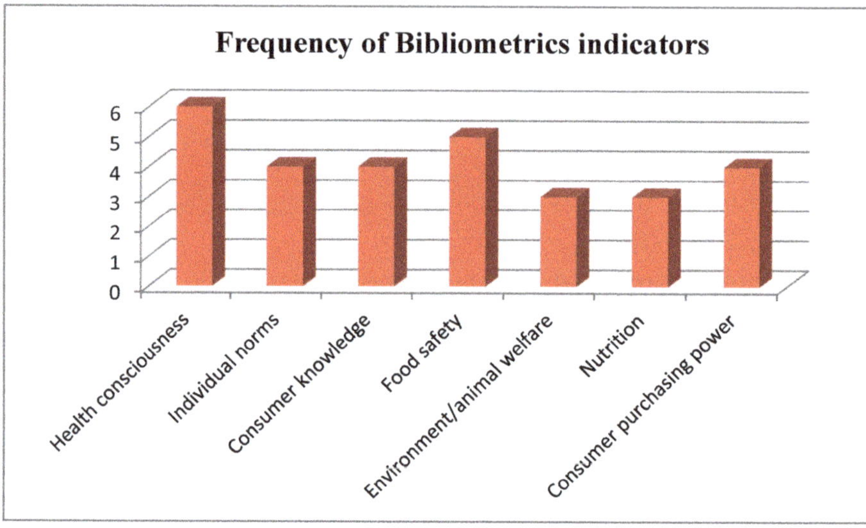

Figure 2. Keywords that were most cited.

This figure has demonstrated that the major concerns for the Chinese consumers' WTP include the following.

3.2.1. Health Consciousness

The result from the materials disclosed in the sampled literature showed that the more conscious the Chinese consumers are about their health, the more their willingness to pay for organic foods. As aforementioned, the Chinese people have a positive attitude towards organic foods due to the understanding that organic foods are good for health, as opposed to the conventional counterparts that are chemical and toxin-laden. Yang, Al-Shaaban, and Nguyen [12] conducted a study in which health consciousness was depicted as the primary factor influencing the purchase intention of consumers towards organic foods. The study found that organic foods are preferred since they are often perceived by consumers to be relatively free from chemicals, and may thus be thought of as safer for consumption. As such, they are more willing to buy the types of foods they are sure to consume without suspicion or worries.

According to Gan et al. [15], Chinese consumers have become increasingly aware of health concerns in the country's food industry in relation to their general wellbeing. This notion was strongly held by the respondents in the study who demonstrated concerns about the quality of the water they are drinking and the foods they are consuming. The respondents also indicated that they constantly worrying about the conventional foods they eat due to the harmful chemicals and pesticides used to grow them. Others reported that were regular readers of health-related articles in books, magazines, and also in the newspaper. They also read the ingredients labels before making any purchase decisions to ensure that they do not pay for foods that have high levels of preservatives and additives. Gan et al. [15] attribute these tendencies to the rise in the number of health-conscious people in China, who demonstrate preferences for organic foods, hence a high level of willingness to pay for organic foods as they are aligned with their lifestyle choices. The study also notes that households with a history of consumption of organic food have fewer cases of chronic illnesses.

3.2.2. Individual Norms

Many of the sampled studies have argued that individuals whose personal norms and attitudes are aligned with a positive attitude towards organic food products are more likely to demonstrate willingness to pay for organic foods compared to those who do not feel that paying for organic foods is the right thing to do. For example, Yang, Al-Shaaban, and Nguyen [12] found that Chinese consumers who feel they make a conscious choice to pay for organic foods, as well as those who believe that choosing organic foods is the right decision, are more willing to buy organic foods because their motivation is intrinsic and based on their attitudes.

In another study, Fang and Levy [5] found that there was a statistically significant relationship between the attitude of Chinese consumers towards organic foods and their purchase intentions. In this case, the study notes that the attitudes of the consumers are aligned with their personal norms, and there is a link between organic foods and the achievement of various values in life. The lifestyle of individuals is important in determining the values they give organic food products. If the value of organic foods is highly aligned with the achievement of life values, then Chinese consumers are highly likely to pay for the organic foods. In most cases, consumers who have a tendency to pay more attention to the foods they buy are more likely to pay for organic foods.

Yen [13] also provided valuable insight into this body of evidence by positing that Chinese consumers' willingness to pay for organic foods is influenced by self-congruity and the need to self-identify with the social self-concept. Respondents indicated that eating organic foods fits well with their image of self, is consistent with the way they view themselves, and has helped them reflect on who they are at a personal level.

3.2.3. Consumer Knowledge

Bibliometric-based evidence from this analysis demonstrates that Chinese consumers are more willing to pay for organic foods when they have sufficient awareness about the knowledge of organic foods, the benefits of organic foods, the nutritional values, and health impact in the long-run. For instance, Gan et al. [6] also reported that the level of knowledge of the consumers regarding organic foods is related to their willingness to buy organic food products. The positive correlation confirms the notion that consumers who have a high level of knowledge regarding the benefits of purchasing and consuming organic foods are highly likely to pay for organic foods. Gan et al. [6] explain that since organic foods are credence goods, information is critical in influencing consumers to buy organic foods as opposed to other conventional food products that are readily available, cheaper, and at the convenience of the consumer. The results of this study are also consistent with Yang, Al-Shaaban, and Nguyen [12] to the extent that limited consumer knowledge is a key impediment to the willingness of consumers to pay for organic food products. As such, it is essential that marketers of organic foods take into account the knowledge level of the market, understanding the concept of organic foods and their many benefits. The more people become knowledgeable about organic foods, the more their willingness to pay for them.

Similarly, Liu, Pieniak, and Verbeke [8] found that among safe foods, green foods, organic foods, and hazard-free foods, organic foods are the less known among the Chinese consumers due to the limited efforts to enhance consumer awareness. The study found that only a quarter of the respondents were aware of organic foods. Knowledge level of a consumer is a primary factor in forming a positive attitude towards the purchase of organic foods.

3.2.4. Food Safety

With the increased cases of scandals in the Chinese food system, food safety has become a major factor to consider when making purchase decisions, especially for consumables. McCarthy [9] noted that most of the food safety issues are attributed to the high levels of additives and pesticides especially in genetically modified foods.

Gan et al. [6] also support the notion that food safety is a major contributor of consumers' willingness to pay for organic food products. Health and food safety factors are important considerations that influence consumers to make purchase decisions in favor of organic foods. Organic foods are more nutritious in terms of vitamins and minerals, and they are also safe from pesticides, toxins, preservatives, and additives. With the increase in cases of food scandals in China, most consumers have become increasingly conscious about the safety of conventionally grown foods. The study also notes that there is limited information about food safety in China, which limits the extent to which consumers can make informed decisions. Increased awareness of the level of safety of foods in the Chinese markets can play a critical role in educating people to make informed health and nutritional choices that are good for their health.

According to Li and Xin [7], young Chinese consumers are motivated to pay for organic vegetables and fruits especially due to the recent food scandals in the country. Li and Xin [7] attribute the increasing demand for organic foods among Chinese consumers to the poor record of the country's food system, which has increased the desire of the people to demand for health and safe foods that are free from contaminants and other disease-causing elements. The study explores various scandals related to conventional food products from counterfeit eggs, gutter cooking oils to milk and baby formula adulterated with melamine. These cases are just an isolated few that raise food safety concerns, giving organic food marketers an edge in terms of the influence to meet consumer needs and expectations of safe foods.

Xie et al. [10] found that most Chinese consumers are highly willing to pay for organic foods because they are assured that the safety of the food is guaranteed. Organic foods are grown without the use of chemicals and pesticides that may have traces in the end products that is taken to the consumers' plates. A respondent was noted saying that she had found worms in organic apples several times, but that was an assurance that the apple was organically grown without any pesticides used to prevent the infection of the fruits by the worms. Another respondent also explained that she had previously worked in an organic certification body and she was aware of animals that were pumped with hormones and medicine, which was not the case in organically grown animals. As such, consumers are highly willing to pay for foods that are organically grown as that assures them of the safety of their foods.

In another study, Yang, Al-Shaaban, and Nguyen [12] also found that food safety was an important consideration in the willingness to pay for organic foods. Organic foods are safe for human consumptions as they are free from chemicals and other harmful substances that would result in ill health. Moreover, organic foods give consumers the confidence that they can consume the foods without fear or suspicion.

3.2.5. Environment/Animal Welfare

The results of this study indicate that environmental and animal welfare concerns are related to a high willingness of Chinese consumers to pay for organic food products. A study conducted by Xie et al. [10] found that the majority of Chinese consumers are ethically minded when making consumption decisions, based on the increased cases of food fraud in the industry. Such consumers have a greater willingness to purchase organic foods as they are aligned with personal objectives of preserving the integrity, beauty, and stability of their households, and also the environment. A respondent in the study stated that the environment in which they live is free from plastic waste, pesticide bottles in the lands, streams and rivers, and it is such a beautiful place that they would wish it for the next generation. The respondents acknowledged the high costs of organic foods, noting that they would rather eat less and preserve their health, the environment, and the welfare of animals than eat more conventional foods.

McCarthy [9] also found that Chinese consumers have a positive attitude towards the ethical principles that guide organic foods farming, including the care for the welfare of the animals. Despite the fact that health concerns outweigh the environmental concerns when making purchase decisions for organic foods, McCarthy [9] notes that consumers are more motivated to pay for organic foods that are linked to moral attitudes, the type of decisions that make people feel that they are doing the right thing for themselves, and for the environment. In this case, consumers are more likely to pay for organic products that are farmed under ethical and moral foundations, and those that ensure environmental conservation and care for the welfare of animals.

According to Zhu [16], the ecological motive has a significant influence on the intention of people to purchase organic foods, which then translates into their willingness to pay for organic foods. The study combines both environmental concerns and animal welfare into the ecological motive, noting that these characteristics are important for organic food consumers, as they are major determinants of whether or not consumers will pay for the organic foods as opposed to the conventionally grown foods. Zhu [14] cites several studies to affirm this conclusion by noting that previous research supports the results in terms of the positive ethical role that influences the purchase decisions of organic food products. The study also explains that organic foods consumers have a high sense of consciousness towards the environmental, and that is their major motivation to buy organic foods. This variable is also related to the self-identity variable, as individuals make the purchase decisions based on their belief systems of doing the right thing.

3.2.6. Nutrition

Nutrition is a critical driver for the purchase of organic foods for Chinese consumers. Xie et al. [10] found that consumers of organic foods are reported to value organic foods because organic foods are laden with nutritional values compared to conventionally grown food products. A total of 65% of respondents in the study reported that they buy organic foods because organic foods are fresher and have a better taste. As such, Chinese consumers are more willing to pay for organic foods if they perceive that the food has more nutritional value compared to the other types of foods that are not fresher, healthier, and rural-like.

The findings in Xie et al. [10] are consistent with the results of the study conducted by Li and Xin [7]. The study found that people reported that organic foods are more nutritious, have a better taste, and are safer and good for human consumption. The labeling of organic food products is also important in providing transparent information that shape the consumers' positive attitudes towards the purchase of organic foods. Li and Xin [7] interpret the finding of the study in the content of a wider body of literature and note that other studies found organic foods to have a high level of minerals such as magnesium, iron, phosphorous, and vitamins such as Vitamin C. These foods are also very nutritious, hence the positive attitude of Chinese consumers towards the purchase of organic foods.

3.2.7. Consumer Purchasing Power

The results of this analysis demonstrate that there is a positive relationship between the consumers' purchasing power and their willingness to pay for organic foods. Gan et al. [6] found that organic foods are expensive compared to conventional foods that are readily available. People with a high social standing are more likely to purchase organic foods as social markers. Education is also attributed to the increase in purchasing power of consumers in China as they have high incomes compared to their uneducated counterparts. Higher income households demonstrate a higher willingness to pay for organic foods because they have the ability to pay for a premium to derive the benefits of organic foods. They believe that organic foods are more preferable, nutritious, healthier, and more aligned with their social values as opposed to the conventional alternatives. On the other hand, price is also discriminative in the sense that consumers who want to consume organic foods cannot do so because of the limited resources. Gan et al. [6] notes that the level of price sensitivity among Chinese consumers is very high, such that people make purchase choices that give them value for money. In this case, they are less willing to pay a premium to get organic foods when conventional foods are cheaper and highly convenient.

The results of Gan et al. [6] are also confirmed by findings reported by Xie et al. [10] in which the respondents reported that the high price for organic foods was a major deterrent to the purchase of organic foods, especially for young Chinese consumers. In Nanjing and Shanghai, the difference in the cost of organic and conventional fruits and vegetables was found to be very high for the average consumers to afford. Some respondents also indicated that while they understood the manual work and the high cost of organic farming compared to conventional farming, the premium price charged for organic farming made it very costly that most people could not buy. Nevertheless, the study found that most people regarded the cost of organic foods as necessary, hence a high willingness to purchase.

Xu, Su, and Lone [11] also found that price is a major determinant of the willingness of Chinese consumers to pay for organic foods. The results of the study support the premise that low food expense consumers are attracted to brands of rice that are affordable, hence low willingness to pay for organic foods. On the other hand, high expense consumers are willing to pay a premium price for organic rice.

3.3. Nature of Evidence

The majority of the studies were made of quantitative studies while the rest of the studies were comprised of mixed methods and systematic reviews.

Such results show an urgent call for utilization of more diverse methodologies, in order to broaden the possible scopes of findings. Quantitative investigation for the issue of consumers' WTP might be more suitable for mature/established research questions, while other methods (especially qualitative methods) might benefit from innovative explorations and thus could facilitate the growth of the literature.

4. Conclusions

The purpose of this bibliometric study was to understand the knowledge structure of scientific literature on Chinese consumers' WTP for organic foods. Due to the fact that there is limited research on the WTP for organic food in China, the bibliometric methodology was vital in gathering information that is likely to influence Chinese consumers' willingness to pay for organic food. The research question was answered with the findings. What are the major factors (addressed in the literature) that influence Chinese consumers' WTP for organic foods? The first relates to health consciousness, which reminds that modern consumers should be vigil regarding foods that can harm their health. The bibliographic analysis also indicated individual norms, consumer knowledge, food safety, environmental concerns, animal welfare, and purchasing power to be very important aspects that influence consumers' WTP in China.

Despite the various perceptions that are related to organic food, the purchasing power remains the strongest fact that should be considered when purchasing organic food [1,10]. As deduced from the study, it is clear that the level of consumption of organic foods in China is positively correlated to the income levels of the consumers. Despite the fact that most consumers may be aware of the health benefits of consuming organic foods, they do not purchase them because they are expensive compared to conventional foods that are cheaper and readily available. Although the various attributes identified can prompt a Chinese consumer to pay a premium price for organic foods, it is very important that advocates and marketers of organic foods consider the price. The price might be a major hindrance for the Chinese consumer to pay for organic foods. Nonetheless, it is evident that despite the premium price associated with organic foods, there are some consumers that are willing to pay more because of issues, such as the perceived health and nutritional benefits that are associated with organic foods. The nutritional value of organic foods is a major motivation that has a positive impact on influencing consumption patterns.

There are some limitations that might stimulate future research ideas. The goal of this paper was not to conduct a thorough check of the literature, but to draw attention and stimulate innovative and timely research thoughts by reviewing a small numbered, but critical and representative, set of papers (for a now arising market for organic foods). In such a sense, it makes sense to review the relevant articles in the past five years. However, the reviewed scope and number of papers could be completed for future studies in order to gain richer information.

Author Contributions: R.L. wrote the original draft of the paper; H.-Y.L. and Y.-T.L. were revisers for the first draft; C.-W.L. was responsible for the analyses; P.F.T. reviewed and validated the manuscript and was in charge of the revision and resubmission.

Funding: This study was partially supported by the Guangdong province characteristic innovation in University Fund's project: "Structure-conduct-performance" Model of Community Supported Agriculture in Guangdong province in China, as well as the Humanities and Social Sciences Research in the Ministry of Education, and Youth Fund Projects in the Border Areas. Funding for the collaborative path research for poverty alleviation (15XJC840002) is acknowledged. Also, this study was subject to the National Social Science Fund's general project "Study on the Synergistic Path of Tourism Accurate Poverty Alleviation and Traditional Village Rejuvenation in Ethnic Areas" (16BSH028).

Conflicts of Interest: The authors declare no conflicts of interest.

References

1. Willer, H.; Lernoud, J. *The World of Organic Agriculture Statistics and Emerging Trends 2018*; Research Institute of Organic Agriculture FiBL and IFOAM: Nuremberg, Germany, 2018.
2. Egelyng, H.; El-Araby, A.; Kledal, P.; Hermansen, J. Certified organic in a North-South and South perspective. In Proceedings of the International Symposium 'Governing through Standards', Copenhagen, Denmark, 24–26 February 2010.
3. Wu, L.; Zhu, D. *Food Safety in China: A Comprehensive Review*; CRC Press: Boca Raton, FL, USA, 2014.
4. Simões, F.D. Consumer Behavior and Sustainable Development in China: The Role of Behavioral Sciences in Environmental Policymaking. *Sustainability* **2016**, *8*, 897. [CrossRef]
5. Fang, Z.; Levy, E. An Analysis of Consumption and Purchasing toward Organic Fruits: Cross-Countries Study between China and France. Available online: http://www.diva-portal.org/smash/record.jsf?pid=diva2%3A821952&dswid=1987 (accessed on 10 May 2019).
6. Gan, C.; Zhiyou, C.; Tran, M.C.; Cohen, D.A.; Xiangxiang, W. Consumer Attitudes towards the Purchase of Organic Products in China. Available online: https://pdfs.semanticscholar.org/15aa/6067059b64c199e522546a70538c668c2565.pdf (accessed on 9 May 2019).
7. Li, X.; Xin, Y. Factors Influencing Organic Food Purchase of Young Chinese Consumers. Available online: http://uu.diva-portal.org/smash/record.jsf?pid=diva2%3A824298&dswid=8377 (accessed on 9 May 2019).
8. Liu, R.; Pieniak, Z.; Verbeke, W. Consumers' attitudes and behaviour towards safe food in China: A review. *Food Control* **2013**, *33*, 93–104. [CrossRef]
9. McCarthy, B.L. Trends in organic and green food consumption in China: Opportunities and challenges for regional Australian exporters. *J. Econ. Soc. Policy* **2015**, *17*, 6.
10. Xie, B.; Wang, L.; Yang, H.; Wang, Y.; Zhang, M. Consumer perceptions and attitudes of organic food products in Eastern China. *Br. Food J.* **2015**, *117*, 1105–1121. [CrossRef]
11. Xu, P.; Su, H.; Lone, T. Chinese consumers' willingness to pay for rice. *J. Agribus. Dev. Emerg. Econ.* **2018**, *8*, 256–269. [CrossRef]
12. Yang, M.; Al-Shaaban, S.; Nguyen, T.B. Consumer Attitude and Purchase Intention towards Organic Food: A Quantitative Study of China. Available online: https://www.semanticscholar.org/paper/Consumer-Attitude-and-Purchase-Intention-towards-A-Yang-Al-Shaaban/5d11b3241ee9d38f668f37e2b848c4810f7fc7d9 (accessed on 9 May 2019).
13. Yen, T.F. Managing self congruity to influence behavioral intention in organic food contexts in Fujian province, China. *MATEC Web Conf.* **2017**, *123*, 32. [CrossRef]
14. Zhu, Y. Using the Theory of Planned Behavior to Investigate What Influences Chinese Intention to Purchase Organic Food. *China-USA Bus. Rev.* **2018**, *17*, 324–333.

15. Yu, X.; Gao, Z.; Zeng, Y. Willingness to pay for the "Green Food" in China. *Food Policy* **2014**, *45*, 80–87. [CrossRef]
16. Chinas Nitrogen Fertilizer Industry 2008 Outlook. Available online: http://www.marketavenue.cn/upload/articles/ARTICLES_1384.htm (accessed on 9 May 2019).

International Journal of
Environmental Research and Public Health

Article

Food System Transformation: Integrating a Political–Economy and Social–Ecological Approach to Regime Shifts

Laura M. Pereira [1,2,3,4,*], Scott Drimie [1], Kristi Maciejewski [1], Patrick Bon Tonissen [5] and Reinette (Oonsie) Biggs [1,3]

1 Centre for Complex Systems in Transition, Stellenbosch University, Stellenbosch 7600, South Africa; scottdrimie@mweb.co.za (S.D.); krismacski@gmail.com (K.M.); oonsie@sun.ac.za (R.O.B.)
2 Centre for Food Policy, City University of London, London EC1V 0HB, UK
3 Stockholm Resilience Centre, Stockholm University, 106 91 Stockholm, Sweden
4 Copernicus Institute of Sustainable Development, Utrecht University, 3584 CB Utrecht, The Netherlands
5 Center for European Integration Research, University of Bonn, 53113 Bonn, Germany; patrick.tonissen@ipbes.net
* Correspondence: pereira.laura18@gmail.com

Received: 12 January 2020; Accepted: 13 February 2020; Published: 18 February 2020

Abstract: Sustainably achieving the goal of global food security is one of the greatest challenges of the 21st century. The current food system is failing to meet the needs of people, and at the same time, is having far-reaching impacts on the environment and undermining human well-being in other important ways. It is increasingly apparent that a deep transformation in the way we produce and consume food is needed in order to ensure a more just and sustainable future. This paper uses the concept of regime shifts to understand key drivers and innovations underlying past disruptions in the food system and to explore how they may help us think about desirable future changes and how we might leverage them. We combine two perspectives on regime shifts—one derived from natural sciences and the other from social sciences—to propose an interpretation of food regimes that draws on innovation theory. We use this conceptualization to discuss three examples of innovations that we argue helped enable critical regime shifts in the global food system in the past: the Haber-Bosch process of nitrogen fixation, the rise of the supermarket, and the call for more transparency in the food system to reconnect consumers with their food. This paper concludes with an exploration of why this combination of conceptual understandings is important across the Global North/ Global South divide, and proposes a new sustainability regime where transformative change is spearheaded by a variety of social–ecological innovations.

Keywords: food systems; global food regimes; innovation; political–economy; social–ecological systems; transformation; regime shifts; resilience

1. Introduction: The Need for Transformation of the Global Food System

Achieving the goal of global food security, together with environmental sustainability and social and economic justice, is one of the greatest challenges of the twenty-first century. The current food system is failing to meet the needs of people, and simultaneously harming the environment and undermining human well-being [1]. With a current world population of more than 7 billion people, expected to reach 9 billion by 2050, global demand for food is undeniably increasing [2]. Food price spikes since 2008 have dramatically affected the affordability of food for much of the world's poor, with hunger on the rise again, largely due to the proliferation of violent conflict, climate-related shocks, and economic downturns [3,4]. As the world moves towards achieving the Sustainable Development

Goals by 2030, goals 2 (*end hunger*), 3 (*improve health*), 8 (*decent work and economic growth*), 12 (*responsible consumption and production*), 13 (*climate action*), 14 (*life below water*), and 15 (*life on land*), are all deeply interlinked with the global food system [1].

Added to the challenge is a burgeoning crisis regarding how we utilize food. The food system is failing to meet the needs of the 820 million undernourished people, 2 billion with micronutrient deficiencies, and more than 600 million diagnosed with obesity [5]. An ailing food system is exacerbating problems in the health system with severe health implications not only arising from under-nutrition and micro-nutrient deficiencies, but with obesity and overweight being linked to 44% of the diabetes burden, 23% of the ischemic heart disease burden, and 7–41% of certain cancer burdens [6,7]. Compounding these trends is the fact that with increasing affluence, diets are shifting dramatically towards more sugar, animal, and fat products, to the exclusion of traditional—and often more sustainable—diets [8,9].

The global food system, especially food production, is a major driver of global environmental change, and has driven massive transformations of terrestrial and marine ecosystems [10]. Human use currently directly affects more than 70% of global ice-free land, and estimates show that up to one third of terrestrial net primary productivity is being used for food, feed, fiber, timber, and energy [11]. A rapidly expanding aquaculture sector is occupying more terrestrial, coastal, and offshore space [12], and projections show that without comprehensive fisheries reforms, over 80% of global fish stocks will be overfished and below their critical biomass by 2050 [13]. Industrialized agriculture is highly reliant on external inputs, contributes to chemical pollution through the use of pesticides and herbicides, changes nitrogen and phosphorous cycles through the addition of synthetic fertilizers, and impacts freshwater stocks through irrigation [14,15]. It is also energy intensive, emitting almost one-third of all greenhouse gases, including methane, thereby contributing to climate change [16].

It is therefore increasingly apparent that a deep structural transformation is necessary in the way we produce and consume food to ensure a more just and sustainable future. The nature of the sustainability challenge means that previously dominant ways of doing things and understanding the world need to be reconsidered in order to make way for knowledge systems that can deal with accelerating change, increasing complexity, contested perspectives, and inevitable uncertainty [17]. This paper uses the concept of regime shifts to understand key drivers and innovations underlying past structural shifts in the food system, and explores how this understanding may help us think about leveraging change to enable a better food system in the future. The paper combines two perspectives on regime shifts: the political–economy perspective derived from the social sciences, and the social–ecological systems perspective derived from the natural sciences. Through an interdisciplinary understanding of food regimes, the paper suggests how connecting different disciplinary understandings of food system change can help to reimagine alternative innovations towards sustainability.

The paper begins with an introduction to regime shifts from the political–economy and social–ecological systems literatures, and then develops an interpretation of food regimes in three case studies drawing from innovation theory. It concludes with an example from a Global South context that highlights how this interdisciplinary conceptualization of regime shifts can be important in highlighting transformative pathways towards a more sustainable future.

2. Two Conceptualizations of Food Regime Shifts

There are a number of frameworks and literature that reference the term "regime" as the dominant way in which processes operate within a system, which is typically associated with distinct system structures. In this paper, we focus on two uses of the regime concept that come from distinct disciplinary backgrounds: (1) the concept of global food regimes introduced by Friedmann and McMichael [18] that emerges from a political–economy perspective, which identifies a regime as a stable period of capital accumulation that is associated with particular configuration of geopolitical power; and (2) a more ecological understanding of regimes that has its foundations in social–ecological systems or resilience thinking, where a regime refers to the combination of factors that constrain the way an (eco) system is structured and functions [19]. By combining the social–ecological and political–economy concepts of regime shifts, we aim to understand past changes in the food system and how future changes might be nudged onto more desirable trajectories.

2.1. A Political–Economy Framing of Food Regimes

In 1989, Friedmann and McMichael published a seminal paper that identified two global food regimes, i.e., the diasporic-colonial food regime of 1870–1914, and the mercantile-industrial food regime of 1947–1973 [18]. The food regime analysis highlighted the important role of food in the global political–economy and provided a structured approach to understanding the role of capital accumulation in agriculture, expressed in the patterns of food circulating in the world economy [20]. Their work also demonstrated a clear shift from one food regime to another during the twentieth century, and linked international modes of food production and consumption to specific periods of capital accumulation.

The diasporic-colonial food regime of 1870–1914 was defined predominantly by food imports to Europe from the colonies; basic grains and livestock from the colonial territories, most notably Australia, Canada, and the United States of America, and tropical imports from the rest of the occupied colonies [20]. By outsourcing its food production, Britain in particular was able to provide its industrial class with cheap food, thereby delineating 'development' in the twentieth century as a dynamic between national agricultural and industrial sectors [20]. In contrast, the mercantile-industrial food regime of 1947–1973 rerouted food from the USA "to its informal empire of postcolonial states on strategic perimeters of the Cold War" [20]: p.141. This shift to a new food regime occurred after the two World Wars, which heralded massive changes in global geo-political power. Framed as a development project that consisted of a suite of interventions such as food aid, Green Revolution technologies, and the extension of markets into the countryside, this regime "universalized a national model of economics as central to the state system following decolonization, whilst simultaneously creating a new international division of labor in agriculture that was centered on transnational commodity complexes" (Raynolds et al. 1993 in [21]: p.141).

In 2005, Friedmann suggested that a third corporate-environmental regime had emerged [22] as the dominance of transnational retailers and agro-food companies—or 'Big Food'—had created a global impact. The premise is that globally powerful food retailers and agro-food companies selectively appropriated demands from environmental and social movements for their own purposes. Friedmann grounds her argument in the restructuring between regimes rather than on the periods of stability, thereby arguing that this food regime emerged out of the contestations between social movements and powerful institutions, which resulted in a new institutional frame [22]. The rise of the third regime is arguably a response to the critique by environmental and social movements of industrial agriculture that began during the pinnacle of the mercantile-industrialist regime in the 1960s. Such critiques put forward alternatives such as 'organic' and 'local', which led to experimentation with agro-ecological practices in the Global North. As these debates unfolded, in the Global South, the Green Revolution and industrial agriculture were transforming agro-ecosystems to increase the yields of staple crops whilst marginalizing rural communities and eroding agro-biodiversity and indigenous knowledge [22]. The rise of this new regime has arguably pivoted on the capture of these environmental critiques and movements by the more powerful corporations to reconfigure the mercantilist regime away from the influence of nation state actors to sit squarely in the hands of the private sector [23]. The appropriation by transnational corporations of the language of social movements has largely captured the rhetoric and façade of sustainability while broadly failing to meet the health and environmental needs of the communities being serviced. Ongoing research on how private food system actors are configuring food environments to the detriment of the health of many communities reinforces this dilemma and, at the same time, shifts the problem space from agricultural production to the dimension of access and consumption [24–26].

Concerns about food consumption patterns within food regimes have been highlighted by authors such as Dixon [27], who traced the history of the 'imperial calorie' through the 'protective' vitamin and more recently the 'empty calorie,' linking these to contemporary concerns around the 'nutrition transition' and rising levels of obesity [28]. The use of the term food deserts—places where it is almost impossible for people to be able to access affordable, healthy food—has sparked contestations in the food system that go beyond food production and engage more substantially with issues of accessibility, nutritional content and cultural appropriateness [29–35]. Such analyses open up questions of how individuals value and interact with their food environment and how understanding these everyday practices can offer non-paternalistic solutions to the crisis of the food system that gives agency to communities to instigate change, rather than relying either on the state or the private sector to save them [34]. However, the focus on consumption remains broadly within the domain of health. Although implicit in many political–economy analyses of food regimes, the environmental implications remain largely side-lined. One exception is Campbell [36], who has used resilience theory to include ecological dynamics into the political–economy analysis of food regimes. His work provides a segue from the political–economy description of food regimes to a social–ecological analysis.

2.2. Social–Ecological Regime Shifts in the Food System

Campbell's use of resilience theory as a way to incorporate ecological feedback into the political–economy perspective of food regimes illustrates that there are advantages to bringing these perspectives closer together because they emphasize different aspects of what is essentially a problem of an inequitable and unsustainable food system [36]. There is an increasing consensus that the food system can be conceptualized as a complex social–ecological system: it consists of cross-scale, multilevel interactions between humans and natural systems, has emergent properties and tipping points, and exhibits nonlinear behavior arising from the interactions and feedbacks in the system [37,38]. Social–ecological regime shifts are defined as long-lasting shifts or changes in the structure and function of social–ecological systems that occur when a tipping point is crossed and a different set of system feedbacks and processes become dominant, restructuring the way the system operates. Such shifts can be precipitated by large shocks as well as slower changes that disrupt or weaken the previously dominant processes and feedbacks [19]. Examples of shocks that can trigger social–ecological regime shifts include climatic fluctuations, large storms, fires, disease, or the outbreak of conflict. Slow changes that may push a social–ecological system towards a tipping point include the gradual consolidation of small farms into large corporate entities, which changes the distribution of benefits and power in the food system, as well as the diversity of food produced.

Regime shifts often result from a combination of slow ongoing gradual changes and an external shock to the system (Figure 1). Once the system is close to the threshold, a regime shift may be triggered by even a small shock to the system, such as a thunderstorm or protest action, that usually would not have any dramatic impacts. For instance, in the case of the food system, gradual changes in livelihood opportunities combined with climate shocks could trigger unexpected larger-scale changes in regional economies and food systems. Social–ecological regime shifts can have major impacts on human economies, security, and health, as they impact the supply of ecosystem services such as crop production and flood regulation [39–41]. Ecosystem services are the benefits that humans gain from the natural environment, which include, for instance, the production of water (provisioning services), the control of climate and disease (regulating services), nitrogen cycles and oxygen production (supporting services), and spiritual and recreational benefits (cultural services) [39]. Changes in ecosystem services directly impact on human wellbeing, including nutrition, livelihoods, social relations and freedom of choice.

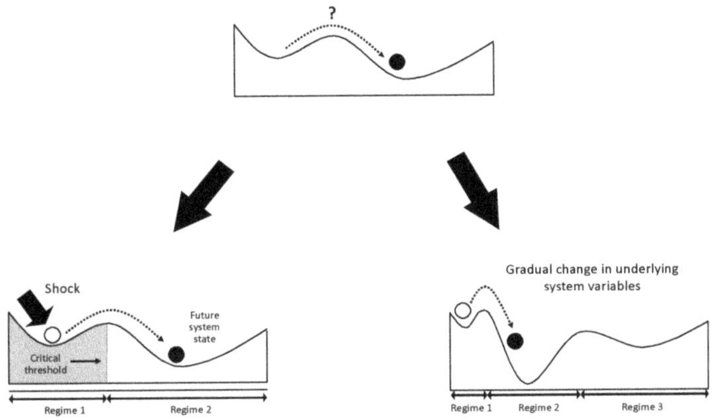

Figure 1. Regime shifts often result from a combination of slow ongoing gradual changes and an external shock to the system that tips the system into an alternative state (Adapted from [19]).

The long-term sustainability of the system therefore relies on a deeper understanding of the mechanisms and drivers that lead to regime shifts, and the capacity to deal with such shifts when they occur. Resilience has been defined as the capacity of a system, subject to ongoing change, to continually self-organize and adapt in a way that retains the same function and structure, i.e., to withstand a regime shift [42]. However, social–ecological resilience is increasingly defined in a more normative way as the capacity of a social–ecological system to sustain human well-being, by adapting or even transforming in the face of change [43,44]. By this definition, regime shifts are seen as being conceptually similar to structural transformations that aim to create more desirable futures. In this conceptualization, the transformation to a sustainable future can be understood as a type of regime shift. Social–ecological regime shift theory highlights the notion that achieving such a transformation involves identifying and changing the dominant feedbacks that currently structure the system dynamics [45,46]. Innovations and innovators can be understood as key actors that help precipitate regime shifts/transformations by changing key systemic feedbacks in ways that change the depth of the basin or create new basins [47–49].

Although the idea of regime shifts as system transformations has started to recognize broader socio-political and economic contexts, these are not yet well integrated in social–ecological regime shifts theory [50]. A combination of perspectives from global food regimes and political–economy allows for a more complete diagnosis of the fundamental power structures and institutional landscapes that maintain systems on particular pathways. Social–ecological systems thinking, on the other hand, allows for an understanding of the complexities of systems, including the ways in which abrupt changes can be precipitated by interacting drivers and gradual shifts that reach a threshold or tipping point. Combining these perspectives offers potential insights for how to shift the global food system onto a new trajectory [47,51]. More generally, it is increasingly acknowledged that any sustainability transformation requires recognition of power and politics, as well as a systemic understanding of how complex social–ecological systems work and change [50,52,53]. In this regard, social and technical innovations can be particularly powerful in disrupting dominant system processes and creating the possibility for larger systemic transformations.

3. The role of Innovation in Regime Shifts

One of the critical areas for food system transformation pivots on behavior change and the aspirations and taste preferences of consumers of food, both in the Global North and South [10]. Current food consumption trends in the developed world—and increasingly in emerging economies—are unsustainable [54]. As people become more affluent, price indicators alone are unlikely to incentivize people to eat more sustainably or healthily. On the food consumption side, companies have invested heavily in shaping the food environment in which consumers choose food, through promoting foods that are aspirational and by creating novel foods that meet demands for tasty, convenient meals that often run counter to cultural traditions around food [55]. The high sugar and fat contents of these foods have been shown to have severe negative health consequences, especially for poorer households that cannot afford healthier alternatives or do not have physical access to fresh fruit and vegetables [56–58].

The food system provides an excellent case study for how innovation has gradually become supply driven rather than demand-driven. Demand-driven innovation is here characterized by companies creating interesting new ideas or products and then creating demand for that good by investing in practices such as excessive marketing: "need" for these goods is therefore manufactured [47]. In addition to those innovations that are predominantly profit-driven, there are other innovations, such as those arising from the public sector, that have also had unforeseen consequences on the food system. The innovations of the Green Revolution emphasized strengthening agricultural production by increasing crop yields through improved inputs: from seeds to machinery and irrigation [59,60]. Although these innovations have undoubtedly improved food security in terms of increasing yields and the total number of calories grown globally, they have ignored other dimensions of the food value chain, particularly concerning access and nutrition. The history of food system regimes presented in Section 4 offers three examples of how innovation has overcome certain challenges, whilst often creating larger problems; a consequence of innovation in a complex social–ecological system. The innovation of social–ecological systems, on the other hand, differs from these conventional innovation strategies because it is founded on notions of complexity, ambiguity, and diversity [47,61]. Examples of these will be the focus of Section 5.

Transformative Social–Ecological Innovation

Disruptive innovations can trigger transitions or transformation out of dominant regimes or stability landscapes into alternate regimes [47]. In social–ecological innovation, this tends to occur as a bottom-up rather than a top-down process, and thus, sits in contrast to corporate power that tends to be top-down [46,62]. At the local level, small, fast variables allow for rapid experimentation to occur; this can be equated with the 'niche' or safe space for innovation [63,64]. Most innovation does not survive, and so there is need for a continual process of design, redesign, and collapse, but every so often, when there is a conducive landscape, an innovation will be successful enough and intermediate processes, such as a more enabling governance arrangement, can start to formalize the novel innovation. At this stage, the innovation can become part of the dominant regime, or it can fail and return to the smaller system where social memory will allow for it to be adapted to the changing landscape environment (Figure 2).

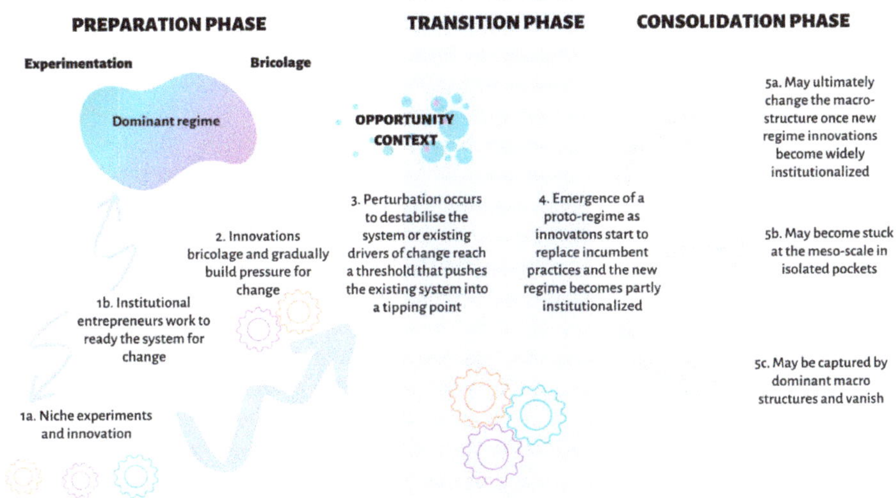

Figure 2. Three phases whereby innovations can cause a system to undergo a regime shift, or transform from one system state to another. The preparation phase requires innovations and experiments to be developed and over time to bricolage, whilst institutional entrepreneurs ready the system for change by bridging between the niche and the regime. A window of opportunity opens up when a perturbation weakens the incumbent system and this allows the innovations that have previously been marginal to become partly institutionalized during the transition phase. The third phase involves three pathways; a- where the resilience of the new regime is built and it becomes the new status quo; b- where the proto-regime becomes stuck and is isolated in pockets; c- where the dominant regime is so resilient that it adapts to the shock and reconfigures itself by capturing the innovations so that they do not result in transformative change. (Adapted from [65]).

There are three phases whereby niche innovations can cause a system to undergo a regime shift, or transform from one system state to another [53]. The preparation phase requires innovations at various levels of the system, often with institutional entrepreneurs operating simultaneously to weaken existing structures and create the possibility for change [53,66]. When a window of opportunity opens up, innovations that have been relatively marginal can suddenly be taken up and shape a new set of system processes, and a new regime. During the transition from one regime to another, small niche innovations can start to have impact at higher levels (See Figure 2). The third phase involves building the resilience of the new regime so that it is not vulnerable to returning either to its previous state or to a less desirable system state. This type of systemic transformation is not a managed process, whereby the final outcome is determined from the outset, but is an emergent one [53]. The system can be nudged (rather than managed) onto a particular trajectory, but the emergence of a new regime is a complex and unpredictable process.

In the next section, we describe three food system regime shifts, linked to three disruptive innovations. Such innovations play an important role in precipitating regime shifts, especially in guiding the direction that these shifts will take, although they in no way act in isolation, and the three we have chosen are examples that serve to illustrate the argument rather than definitive. By understanding the role of disruptive innovation in global food regime shifts, it is possible to speculate on how to nudge future innovations towards a more just and sustainable basin of attraction.

4. Three Examples of Food System Innovations Linked to Regime Shifts

The role that private sector actors play in shaping the global food system has become significant over the last century, especially as large corporations have consolidated the processes across the food system [23,67,68]. The array of actors involved in the food system and their changing roles are fundamental for understanding the regime shifts that the system has undergone over the last two centuries, largely through innovation in response to changing social and institutional contexts. Recognizing that innovations never act in isolation within systems, the following case studies illustrate how combining a political–economy and a social–ecological perspective of understanding regime shifts can help us comprehend some of the complex processes and interactions that, in retrospect, we can identify as regime shifts. By understanding the socio-political, ecological and economic contexts and drivers within which these important innovations occurred, the dimensions of the resulting regime shifts become clearer (See Table 1). These are by no means the only innovations in the food system, but by focusing on one critical intervention in the system, it is easier to conceptualize how change can be triggered, and therefore, how to intervene in a system to nudge it onto a more sustainable pathway. The case studies focus on innovations that occurred in the Global North that have had global impacts on the food system, stressing again where much of the power to shape the global food system has resided. Biophysical as well as social feedbacks necessitated adjustments in the capital allocation of powerful structures. This changing landscape allowed for other innovations to take hold to shift the food system into a new configuration of nature, people, and money.

4.1. Regime Shift 1: From Labor-Intensive Subsistence Agriculture to Commercial-Industrial Agriculture that could Feed Growing Cities: the Haber-Bosch Process to Produce Fertilisers and Subsequent Intensification Technologies

In the early nineteenth century, Europe was still largely an agrarian society, where the majority of the population cultivated food for a living. By the end of that century, as the industrial revolution took hold, peasants left the countryside for the cities and an industrial working class was formed [69]. As people moved from farms into urban areas, there was a need for a steady supply of produce from farms to feed the increasing number of workers in expanding cities. There was, however, less labor available on farms. The industrial revolution not only caused this problem, but also provided the means to a solution: mechanized agriculture. This mechanization coincided with the colonial-diasporic food regime that predicated a need for supplying food to European cities and workers from the colonies [22]. The industrial revolution resulted in increased mechanization of agriculture and a move away from it being labor intensive to being more reliant on other input costs, associated with the scientific discoveries of the age. The 'disruptive innovations' of the Industrial Revolution enabled Western societies to transcend a Malthusian assumption where it was believed that population growth would eventually outstrip natural resources, largely through revolutionizing the instruments of production and the division of labor [70].

A key disruptive innovation in this process was the Haber-Bosch process (of fixing N_2 into NH_4), first observed in 1909, that, after an over-capacity of the manufacture of explosives following World War 2, led to a refocus of the technology into the production of synthetic fertilizers [71]. Fertilizer in turn contributed to vast increases in the amount of food that could be produced on the same amount of land through intensification [72]. As more food was needed to supply burgeoning cities, this intensification of agriculture that increased yields from finite areas of farm land and animal bodies was a welcome development [36]. However, the new technological package of fertilizers, pesticides, heavy machinery, and new stock and crop varieties came at a significant environmental cost, with increasing reliance on industrial inputs, a reduction in internal natural cycles on farms, pressure on landscapes and animals, and other environmental impacts [36]. The combination of these interacting factors led to a reorganization of the system so that it functioned in a very different way, and can be thought of as a regime shift.

This change in the structure and function of the food system also impacted on the availability of livelihood and economic activities for people who had been agricultural laborers, as well as the quality of the food and nutritional content, impacting on human wellbeing. The outcomes of the Green Revolution were used to extend this regime to the Global South. Further, this trend created the space for companies to start specializing in agricultural inputs like farm machines, seeds, fertilizer, and pesticides, and for these to subsequently merge into the agribusiness giants of today, such as Monsanto (an agrochemical company that became a powerful seed company and has now been absorbed by Bayer, which describes itself as a 'multinational pharmaceutical and life sciences company), Agrium (primarily a fertilizer company that is now called Nutrien after a merger with Potash Corp) [21,23], and Bunge, Cargill, Archer-Daniels-Midland, and Louis Dreyfud (the four largest grain trading companies that control up to 90% of the world's cereal trade) [73]. Although cities throughout history have relied on food coming in from rural areas, i.e., the core has always been fed by the periphery, the scale on which industrialization enabled this relationship resulted in a growing disjuncture between food production and its consumption. The "distanciation" of the food system, as argued by Friedmann [74], was reinforced by the second major regime shift, which focuses on the power of global supermarket chains.

4.2. Regime Shift 2: From Local Traders to the Convenience of Global Supply Chains: the Establishment of Supermarkets and Fast Food Restaurants as Conventional Sites of Food Procurement

In response to people moving into cities and needing a central location for their food, the birth of the supermarket can be seen as a disruptive innovation in the food system, whereby one actor was responsible for supplying all the goods that would previously have been supplied by different actors (e.g., the goods from grocers, bakers, butchers etc. were now found under the same roof). The establishment of grocery stores was pioneered in Europe, which grew to dominate the global food retailing business a century later, providing an outlet for the food industry to sell its processed food products in the twentieth century. The first Sainsbury's opened in London in 1869, where it grew into a privately listed company in 1922. The next biggest extant supermarket was founded by Albert Heijn in the Netherlands in 1887, becoming the company Ahold. The French followed with the establishment of the first Casino store in 1898. These retail giants were to become part of the food retail oligopoly that came to dominate the western world over the course of the twentieth century, spreading to all reaches of the globe through the connections established by the colonial system [75–77]. The remainder of this powerful group of food retailers was established in the mid-twentieth century; the United Kingdom's Tesco's in 1952, France's Carrefour in 1958, and the USA's Wal-Mart in 1962.

This second regime shift has at its core a new center of power within the global food system; the global supermarket chain that has profoundly impacted what a large proportion of the world eats, as well as how that food is produced, processed, and marketed [78]. According to Shaw, supermarkets were transformative for four main reasons [79]. Firstly, their patterns of innovation created Schumpeterian 'disruptive competition' (See [80]). More vibrant innovation was created because participants had to compete to offer more varied produce and service. Secondly, by continuously reinventing their format, processes, and channels of distribution, they created novel business models that became impossible to outcompete, as they were constantly adapting [79]. Thirdly, they transformed consumer culture in that customers came to expect a variety of products on offer in one convenient location where they were free to roam the aisles at their own pace [79]. Finally, their competitive edge and adaptive business model made them so successful that it led to the rapid growth of supermarkets, and so their influence spread quickly [79].

This model had universal applicability, and so, by the end of the 20th century, was transferred to a range of economies in a rapid expansion [81]. Having spread the business model throughout the USA and Western Europe, these companies became transnational, establishing themselves in foreign markets either directly or by getting their foothold in the market through partnerships with local companies [76,77].

A similar source of convenience in the food system arose on the other side of the Atlantic in the USA, and made its way across to Europe: the development of fast food restaurants. Schlosser's book on the rise of the fast food chain in the United States gives a detailed account of the innovation that not only developed a model for getting cheap, convenient, and addictively tasty meals to all Americans, but also transformed how agricultural supply chains operated [82]. As the establishment of supermarkets allowed for the convenience of accessing all your food in one place rather than going to the fresh market, the baker, or the butcher, fast food chains provided ready-made food at a cheaper price than cooking it at home [83]. These innovative business models have continued to grow off each other in the twentieth century, and as technologies have provided increased means for efficiency, such as the microwave, the food industry has developed products accordingly, including ready-made meals high in salt, fat, and preservatives like sugar [57,84].

The development of these two models changed the rules of the game of the global food system and made different models of food access acceptable. Slow gradual changes pushed the food system, including both the agri-ecosystems upon which food production is based and the points where consumers access food, towards a threshold. Within a century, this enabled the food industry within the mercantile-industrial food regime to provide consumers with increasingly processed and unhealthy foods [56,57,84], directly impacting on human wellbeing and environmental integrity. The evolution of the agro-industrial complex in response to this second food regime is a defining feature of what constitutes the food that people access through these novel points of procurement.

4.3. Regime Shift 3: From Anonymous Global Supply Chains to Alternative Food Networks: the Rise of Corporate Responses to the Call for Increased Transparency

The phenomenon of the global expansion of supermarket chains into multinational corporations that established themselves in developing country markets since the 1990s is a product of the mercantile-industrial food regime [75,76,81]. These companies now have the network power to source and supply anything from everywhere; it has become commonplace to be able to find Indian mangoes, South African avocadoes, Kenyan green beans, and Chilean grapes in the same aisle in a single UK supermarket. However, the monopolization of the global food system by a few companies brought with it a backlash as consumers demanded transparency: they wanted to know more about what they were eating, how and by whom it was grown or reared or caught, where the seeds came from, and how many inputs were applied. The corporate response has been, by-and-large, the auditing of supply chains through standards from bodies like EUREP-GAP (now GLOBAL-GAP) [22]. This has been capital's attempt to respond to consumer and activist demands about the anonymity of 'Food from Nowhere', i.e., the product of the globalized mercantile-industrial regime, to create 'Food from Somewhere' to sustain market share [85,86].

The rise of 'alternative food networks' is arguably a social movement critique of the increasing disconnect between the majority of consumers from the realities of the production of food. In other words, it reflects the transitioning from the mercantile-industrialist regime into what Friedmann (2005) posits as the corporate-environmental regime. As the environmental and social implications of the dominant food systems in the Global North have been recognized [10], alternatives have started to emerge, resulting in a potential means to rewire the system by altering the social–ecological feedbacks and dynamics [1]. These 'alternatives' include the organic, local, and Slow Food movements, but are increasingly also being mainstreamed by food labelling and certification from bodies like Fair Trade and the Forest Stewardship Council. As Goodman [87]: p. 10 notes:

This shift towards the production of quality local foods, as opposed to the generic 'placeless' commodities of productivist agriculture...is variously conceptualized as the re-embedding, resocializing, and re-localizing of food systems. Slow Food Supply Chains are a major institutional expression of these reconfigured production-consumption relations farmers are encouraged to 'short-circuit' industrial supply chains and to reconstruct the producer-consumer interface by engaging with different conventions and constructions of quality "that evoke locality/region or speciality and nature" (Marsden et al. 2000: p. 425).

Food businesses, as with other private sector entities that need to sell products, have become locked into an unsustainable, growth-oriented regime that has consumerism at its foundation. With a focus on optimizing shareholder value and externalizing social and ecological costs [47], it is unsurprising that innovation in this sector has only been quick to jump on the sustainability bandwagon when there are clear profits in sight (as per Friedmann's third regime argument [22]). Innovations like ethical labelling, in particular, seek to create a sense of trust. 'Fair Trade chocolate', 'Carbon Neutral wine', 'Rainforest

Alliance coffee', and 'Organic cheese' are certified labels that all impact production based on the preferences of 'ethical' consumers, and effectively places them and their certification bodies in charge of how food production is undertaken. This has interesting repercussions for farmers in the developing world, where many of the changes are taking place. Studies on the Roundtable on Sustainable Palm Oil have shown that despite an attempt at full equality and engagement among multiple stakeholders, the major processors and traders still dominate these discussions [88,89]. Fair Trade and other certification schemes similarly reflect good intentions in the North that do not necessarily translate into actual change, especially economically, for producers in the South [90,91]. Even the growth of 'hipster' culture and a reconnection to traditional ecological knowledge has instigated a cultural shift towards better quality, more ethical produce that can be seen in cities around the world, including the Global South [92–94]. This has, however, not yet been able to destabilize the current global food regime that continues to drive unsustainable and unhealthy practices [10]. Rather, it can be seen as an example of capture by the dominant regime of niche innovations that seek to address core social–ecological challenges (pathway 5c in Figure 2).

Table 1. Summary of the regime shifts in the food system linked to key characteristics used to define these shifts.

	Regime Shift 1: Low-Input Labor-Intensive Farming to Commercial-Industrial Agriculture	**Regime Shift 2: Food Procured from a Variety of Local Traders to Food Procured from a Supermarket**	**Regime Shift 3: From Anonymous Global Supply Chains to Alternative Food Networks**
Key innovations	Haber-Bosch Process; Green Revolution technologies	Supermarkets; Fast and convenience foods	Certification and labelling; local and Slow Food movements
Key drivers	Industrial revolution Over-capacity in the manufacture of explosives after World War II Urbanization	Globalization of supply chains; Efficiency of international transport of goods	Consumer demands for transparency in value chain; Rise of 'alternative food networks'
Key feedbacks	Increased agricultural production efficiency; Monoculture farming; Increased dependence on the companies providing inputs	Global expansion of supermarket chains; Disruptive competition; customer demand for variety of products; Cheaper food; Less time and skill spent on cooking	Higher percentage of income being spent on higher quality food; Growth in certification bodies and institutionalized auditing of supply chains; Greater costs to farmers to be enrolled; Support for local and niche food producers to enable their viability
Key ecological impacts	Increased pressure on land and animal bodies; Decreased agro-biodiversity; Decrease in water and soil quality; Increased emission of Greenhouse gases (GHGs); Loss of pollinators; Eutrophication of lakes and seas from agricultural run-off	Agricultural expansion leading to deforestation in the tropics; Increase in carbon emissions from transporting food around the world 'food miles'; Increased meat production on feedlots emitting more GHGs and driving expansion of feed crops like soybean; Increased food waste; Reduced post-harvest losses as food is processed	Organic and less input intensive agriculture decreases impact on soils and water; Reduced food waste as food is more expensive; Improved agro-biodiversity

Table 1. *Cont.*

	Regime Shift 1: Low-Input Labor-Intensive Farming to Commercial-Industrial Agriculture	Regime Shift 2: Food Procured from a Variety of Local Traders to Food Procured from a Supermarket	Regime Shift 3: From Anonymous Global Supply Chains to Alternative Food Networks
Key social and health impacts	Increase in calories available per person; Diversification of livelihood options away from agriculture; Shift from subsistence to commercial agriculture	Decrease in health due to increasingly processed and unhealthy foods being easily accessible and affordable; Increase in the variety of foods available; Exploitation of labor and land to meet international demand; Consolidation of food businesses into multinational corporations MNCs; Processing of food enables women to work and spend less time in the kitchen	Increased inequality in who can access good, healthy food; Improved conditions for those producers who can afford certification; Culinary knowledge valued; Improved nutrition for those who can afford better quality food
Key references	[70–72,95]	[57,75,76,78,79,81,82]	[87–92,94]

5. Quality, Taste, Cuisine, and the Role of Chefs as Social Innovators: Precursors for a Future Regime Shift

An idea or concept can transcend the stark binaries evident in food system debates—binaries that are often underpinned by ideological positions. The "farm to table" or "farm to fork" locavore movement that promotes serving local food at restaurants, preferably through direct acquisition from the producer, is an example of a mobilizing idea that has changed many people's perceptions about where food comes from, reflecting some of the shifts in the third food regime described above ([96]). The idea of "farm to table" has, however, been limited, in that farmers end up serving the table, not vice versa, which makes good agriculture difficult to sustain. An example of a potential solution is Barber's concept of cuisine, which encourages a collection of dishes that reflect a whole system of agriculture [97]. It comes from a similar perspective on food system transformation as espoused by Michael Pollan who emphasizes that change needs to come from a return to the skills of cooking and linking taste and nutrition directly to how food is produced and cooked [83,98].

Barber's book "The Third Plate" doubles as a manifesto for the future of food and highlights the potentially transformative role that engaged chefs can make in the food system [97]. The manifesto argues for a radical shift in what a standard plate of dinner should look like, demonstrating that flavor must start with an understanding of soil, bringing emphasis to the provisioning and supporting of ecosystem services, and the role this plays in the quality of food (https://www.theguardian.com/lifeandstyle/2017/jan/15/dan-barber-mission-to-change-food-and-farming (Accessed 12 January 2020)). A central tenant is that a cook has a duty not only to know the farmers who provide his or her ingredients, but also to be actively involved from seed to table, by selecting for taste. There are increasing numbers of platforms around which chefs are starting to mobilize their role in healthy and sustainable food, including the Slow Food Chefs Alliance, which rallies around protecting biodiversity (https://www.fondazioneslowfood.com/en/what-we-do/slow-food-chefs-alliance/ (Accessed 12 January 2020)) and the Sustainable Development Goal 2 Hub's Chef's Manifesto on how the food industry can deliver a better food system (http://www.sdg2advocacyhub.org/chefmanifesto (Accessed 12 January 2020)). Furthermore, the recognition of traditional cuisine and the knowledge of local cooks, not only chefs, especially in the Global South, offers further evidence of this movement gaining traction [99]. Another clear link can be made to SDG 12 on sustainable consumption and production. SDG 12 reflects the need to develop sustainability practices that bridge production and consumption; however, it is also the SDG most associated with trade-offs against achieving the other goals [100]. Strategic linkages between achieving the targets of SDG 12 through a food system lens could enable key leverage points to be identified that create synergies between the other SDGs, such as life below water (14) and life on land (15), rather than trade-offs.

Critiques of the notion of chefs supporting farmers need to be acknowledged, however. A prominent critic of the small farmer ideal is Julie Guthman, who has argued that agrarianism—which Barber and Pollan espouse—romanticizes the small-scale family farm (which can be an extraordinarily difficult lifestyle), reinforces patriarchy, and perpetuates injustice for farm workers [101]. These important questions bring to bear the issue of food justice and further emphasize why it is important to be able to analyze what is elevating these niches into potential new regimes, and what the pitfalls might be. Drawing these into the argument for strengthening the relationship between food, people, and planet, we can help identify and address many of the social and environmental issues that plague contemporary food systems.

A key question is how enough energy can be brought to a revolution and lead to a regime shift that is premised on sustainable and ethical practices in local contexts. This alternative regime holds great potential, not only from an environmental perspective, but by substantially improving human well-being by enhancing the quality of food and nutrition [93]. Countries in the Global South are often envisaged as passive recipients of regime-defining processes with little to no agency to act in the face of broader power structures. Strategies that build on and enhance local capacities are needed in these contexts rather than a new set of paternalistic solutions that do not address the complexity of the challenge. Emerging economies such as South Africa are confronted by the existence of multiple regimes coexisting within one country: a dual agrarian system consisting of a minority of white, commercial farmers who produce much of the country's produce and a majority of small-scale black farmers [102], the triple burden of malnutrition (hunger, obesity, and micronutrient deficiencies), a high prevalence of diet-related diseases [103–107], and a diverse production landscape, vulnerable to climate change, with only 13% of land considered arable [108]. The South-African context, therefore, brings in a whole array of social–ecological and political economic issues linked to the food system, such as income inequality, malnourishment, obesity, food insecurity, water shortage, and soil degradation, amongst others.

Being able to tackle this complex array of interconnected concerns within the food system requires transformative change towards a more sustainable and just regime. It requires not only understanding the social–ecological implications, but also the political economic context, and how this needs to change. Unlike previous regimes that manifested as a result of the actions of powerful actors exploiting key innovations, we argue that the sustainability regime needs to disrupt from the bottom up, where local innovations and initiatives that meet local needs are able to effect systemic change whilst maintaining their contextual nuance. This theory of bottom up transformation follows the theory of transformative social–ecological transformation presented in Figure 2, and there are increasing examples of how this kind of change might be galvanized [109–111]. A key aspect of this approach lies in acknowledging diverse pathways of scaling impact: scaling up (getting bigger), scaling out (replicating), or scaling deep (changing norms and values) [99,112].

A study from Cape Town, South Africa has shown that niche eco-gastronomic initiatives like Bread Rev—a charity that teaches people to bake bread in wood-fired rocket ovens and sets up local community BREADshops—have the potential to create new interactions between humans and the environment [92]. For her MSc research, Markey (2017) conducted in-depth, semi-structured interviews with 13 niche actors in Cape Town's bread, beer, gin, and pork industries to assess what potential these niche businesses had as seeds of transformation to disrupt and change the broader system [92]. Her findings showed that niche eco-gastronomic initiatives in the Greater Cape Town area have created new configurations of human-environmental relationships, whilst also addressing issues of inequality [92]. Despite being in different sectors, each of the actors was highly connected in response to the low institutional support for these small-scale businesses; however, exchange with the regime of large-scale producers remained limited [92]. Instead, niche actors mainly focused on scaling out and scaling deep to achieve change, fostering the further development of the eco-gastronomic sector, as well as engaging in conversations to shift perceptions and beliefs around what good food means and who can access it [92]. The businesses acted as brokers between producers and consumers to create new social–ecological relations through their eco-gastronomic initiatives, institutionalizing through their network a more holistic approach to creating social–ecological sustainability in their business models [92]. Although the approach to address these aspects varies depending on the specific local conditions and the business model, the study showed that the success of the innovations lay in the sharing of ideas and knowledge, experimentation, and creative usage of available resources, thus bricolaging new constellations of existing system components [92].

These examples are certainly not the only initiatives trying to drive food system transformation in South Africa. Innovations ranging from phone apps to help link small-scale fishers to consumers, organizations bridging indigenous knowledge holders, and marginalized producers with chefs and restaurants, and even social movements around access to land and agro-ecology, are all important in shaping a landscape from which a new food system can emerge [15,113–116]. At the core of these innovations is a disruption of the conventional, paternalistic solutions that are usually offered in response to food insecurity [34]. Rather than being externally-driven, these innovations have emerged from within their unique food environments to address the challenges that specific communities are facing and are looking at scaling in very different ways. Thus, these social–ecological innovations offer genuine alternatives to neoliberal strategies that tend to remove the agency from communities. By understanding how past innovations were captured and shaped the political–economy, it might be possible to guard against similar manifestations of power in the emergence of future regimes.

Similar innovations are underway in Europe, where a return to appreciating traditional ecological knowledge is gaining a following [94]. Links between traditional ecological knowledge and resilience in European landscapes of food production have been identified and called biocultural refugia [117,118]. This work focuses on how knowledge, experience, and practices of managing a local ecosystem and its services are captured, stored, revived, and transmitted through time as social memory [119]. The authors find that the biodiversity of many cultural landscapes has been maintained through local management practices, developed in the context of the relation between local environmental fluctuations and agricultural production. In Europe's agricultural landscapes, loss of traditional ecological knowledge and practices has resulted in an associated erosion of biodiversity and regulating ecosystem services, leading to the conclusion that nurturing biocultural diversity is a fundamental principle for planetary stewardship [120]. Again, this shift acknowledges the agency of local knowledge-holders, thereby disrupting conventional top-down solutions that can be captured by powerful actors. Leveraging traditional knowledges and relationalities could allow for the emergence of an alternative global food system that is embedded in local realities, rather than in the powerful dominance of global actors who, during colonization and globalization, have been able to capture previous innovations to further their own ends.

Although social–ecological innovations currently remain niche, global shifts in discourses emphasizing more sustainable food systems could be reconfiguring the political economic landscape and opening up a window of opportunity for these to scale. By scaling appropriately, these alternatives could alter the feedbacks that maintain the status quo and allow for the global food regime to shift into an alternative state. There are significant numbers of examples of alternative, social–ecological innovations that are emerging in niches around the world [120]. Fostering an environment within which they may be able to flourish and influence the trajectory of the food system is becoming a key concern in creating a more sustainable global food system.

6. Concluding Perspectives

In this paper, we have argued that there is a novel academic contribution to be made by bringing together a political–economy conceptualization of food regimes with a social–ecological systems perspective of regime shifts in order to understand the complex dynamics of past transformations in the food system. We propose that this can best be done by unpacking three historical innovations (and their associated actors and power dynamics) that we believe have helped to enable past regimes shifts. These are meant to serve only as examples of how this approach can be used to understand better how food system regimes have shifted in the past, and are not meant to be seen as definitive. We argue that using a food regimes lens to learn how these innovations helped to create new food system configurations provides us with information on how to navigate future regimes that could address some of the key equity and sustainability challenges of the current global food system.

Analysis of past regime shifts highlights how the concept can be used to identify the main drivers responsible for shifting the system into an alternative state, and the feedbacks and power structures that maintain existing regimes. Reflecting on the nascent potential of future regime shifts, this can help identify leverage points or places to either push a failing system into tipping or intervening to facilitate transformation of the system [94]. However, we are cognizant that using such historical examples to learn about the future requires a level of reflexivity and humility in the claims that are made. Such an approach can benefit from insights from the field of historiography.

Mazlish [121] has argued for a similar conceptualization of measuring change in the past to that of regime shifts, namely by contributing the idea of a 'historical rupture' to mark abrupt change within an otherwise nuanced understanding of history, which describes change as being both continuous and discontinuous. As he contends, the lens through which understanding is gleaned, and in turn, a narrative produced and assertions made, is largely constructed by way of individual perspective, privilege, and purpose:

"Stringing together facts, in what often appears to be a functionally deterministic way, historians draw a smooth line through the past. Both the facts and the continuity are, we now realize, constructed. We construct what is a fact and recognize that we could have emphasized other ones. Equally important, however, is the recognition of ruptures in history ... Against the view of the human past as marked by continuity, ruptures mark abrupt change." (Mazlish 2011:p. 32).

In order to better understand food system regimes and regime shifts within these systems, it is crucial to understand the narratives of past changes, the actors who constructed them, and the agents and factors with the greatest relative impact on facilitating change. With a deconstruction of power and privilege behind past ruptures or shifts, today's actors are better equipped to situate themselves within on-going and future regime shifts in order to bring about more desirable outcomes. With an awareness that the facts emphasized and narratives told are in great part happenstance constructs, a space opens up for the empowerment of actors to take an active role in not only reimagining past shifts, but in shaping new futures.

Author Contributions: Conceptualization, L.M.P. Writing-Original Draft Preparation, L.M.P. and S.D. Writing-Review & Editing, K.M., R.O.B. and P.B.T. All authors have read and agreed to the published version of the manuscript.

Funding: R.O.B. received support from the South African Research Chairs Initiative (SARChI) of the Department of Science and Technology and National Research Foundation of South Africa (grant 98766); the Guidance for Resilience in the Anthropocene: Investments for Development (GRAID) project funded by the Swedish International Development Agency (Sida); and a Young Researchers Grant from the Vetenskapsrådet in Sweden (grant 621-2014-5137). L.M.P. is partly funded by the National Research Foundation of South Africa (grant 115300).

Acknowledgments: The authors would like to thank participants of the International Institute of Social Studies conference in the Hague, 2016, where the ideas behind this manuscript were first presented. The authors would also like to thank the three anonymous reviewers for their insightful comments and suggestions and to Charlotte Ballard for final proof-reading.

Conflicts of Interest: The authors declare no conflict of interest.

References

1. Gordon, L.J.; Bignet, V.; Crona, B.; Hendriksson, P.G.J.; Van Holt, T.; Malin, J.; Lindahl, T. Rewiring food systems to enhance human health and biosphere stewardship. *Envrionmental Res. Lett.* **2017**, *12*, 100201. [CrossRef]

2. Beddington, J.R.; Asaduzzaman, M.; Clark, M.; Fernández, A.; Guillou, M.; Jahn, M.M.; Erda, L.; Mamo, T.; Van Bo, N.; Nobre, C.A.; et al. *Achieving Food Security in the Face of Climate Change: Final Report from the Commission on Sustainable Agriculture and Climate Change*; CGIAR Research Program on Climate Change, Agriculture and Food Security (CCAFS): Copenhagen, Denmark, 2012.
3. FAO; IFAD; UNICEF; WFP. *WHO The State of Food Security and Nutrition in the World: Building Resilience for Peace and Food Security*; FAO: Rome, Italy, 2017.
4. FAO; IFAD; UNICEF; WFP. *WHO The State of Food Security and Nutrition in the World 2019: Safeguarding Against Economic Slowdowns and Downturns*; Development Initiatives: Rome, Italy, 2019.
5. Development Initiatives. *Global Nutrition Report 2017: Nourishing the SDGs*; Development Initiatives: Bristol, UK, 2017.
6. Cordain, L.; Eaton, S.B.; Sebastian, A.; Mann, N.; Lindeberg, S.; Watkins, B.A.; O'Keefe, J.H.; Brand-Miller, J. Origins and evolution of the Western diet: Health implications for the 21st century. *Am. J. Clin. Nutr.* **2005**, *81*, 341–354. [CrossRef] [PubMed]
7. *WHO Obesity and Overweight Fact Sheet 11*; WHO: Geneva, Switzerland, 2015.
8. Hawkes, C. Uneven dietary development: Linking the policies and processes of globalization with the nutrition transition, obesity and diet-related chronic diseases. *Global. Health* **2006**, *2*, 1–18. [CrossRef] [PubMed]
9. Kastner, T.; Rivas, M.J.I.; Koch, W.; Nonhebel, S. Global changes in diets and the consequences for land requirements for food. *Proc. Natl. Acad. Sci. USA* **2012**, *109*, 6868–6872. [CrossRef] [PubMed]
10. Willett, W.; Rockström, J.; Loken, B.; Springmann, M.; Lang, T.; Vermeulen, S.; Garnett, T.; Tilman, D.; DeClerck, F.; Wood, A.; et al. Food in the Anthropocene: The EAT–Lancet Commission on healthy diets from sustainable food systems. *Lancet* **2019**, *6736*, 3–49. [CrossRef]
11. IPCC Summary for Policymakers. In *Climate Change and Land: An IPCC Special Report on Climate Change, Desertification, Land Degradation, Sustainable land Management, Food Security, and Greenhouse Gas Fluxes in Terrestrial Ecosystems*; Shukla, P.R.; Skea, J.; Buendia, E.C.; Masson-Delmotte, V.; Pörtner, H.-O.; Roberts, D.C.; Zhai, P.; Slade, R.; Connors, S.; Diemen, R.; et al. (Eds.) Cambridge University Press: Cambridge, UK, 2019.
12. Lester, S.E.; Stevens, J.M.; Gentry, R.R.; Kappel, C.V.; Bell, T.W.; Costello, C.J.; Gaines, S.D.; Kiefer, D.A.; Maue, C.C.; Rensel, J.E.; et al. Marine spatial planning makes room for offshore aquaculture in crowded coastal waters. *Nat. Commun.* **2018**, *9*, 945. [CrossRef]
13. Worm, B. Averting a global fisheries disaster. *Proc. Natl. Acad. Sci. USA* **2016**, *113*, 4895–4897. [CrossRef]
14. IPES-Food. *Food from Uniformity to Diversity: A Paradigm Shift from Industrial Agriculture to Diversified Agroecological Systems*; IPES-Food: Brussels, Belgium, 2016.
15. Pereira, L.M.; Wynberg, R.; Reis, Y. Agroecology: The Future of Sustainable Farming? *Environ. Sci. Policy Sustain. Dev.* **2018**, *60*, 4–17. [CrossRef]
16. Vermeulen, S.J.; Campbell, B.M.; Ingram, J.S.I. Climate Change and Food Systems. *Annu. Rev. Environ. Resour.* **2012**, *37*, 195–222. [CrossRef]
17. Lotz-Sisitka, H.; Wals, A.E.J.; Kronlid, D.; McGarry, D. Transformative, transgressive social learning: Rethinking higher education pedagogy in times of systemic global dysfunction. *Curr. Opin. Environ. Sustain.* **2015**, *16*, 73–80. [CrossRef]
18. Friedmann, H.; McMichael, P. Agriculture and the State System: The rise and decline of national agricultures, 1870 to the present. *Sociol. Ruralis* **1989**, *29*, 93–117. [CrossRef]
19. Biggs, R.; Blenckner, T.; Folke, C.; Gordon, L.; Peterson, G.D. Regime Shifts. In *Encyclopedia of Theoretical Ecology*; University of California Press: Berkeley/Los Angeles, CA, USA, 2012; pp. 609–617.
20. McMichael, P. A food regime genealogy. *J. Peasant Stud.* **2009**, *36*, 139–169. [CrossRef]
21. McMichael, P. A food regime analysis of the 'world food crisis'. *Agric. Human Values* **2009**, *26*, 281–295. [CrossRef]
22. Friedmann, H. From Colonialism to Green Capitalism: Social Movements and Emergence of Food Regimes. *Res. Rural Sociol. Dev.* **2005**, *11*, 227–264.
23. Patel, R. *Stuffed and Starved: From Farm to fork, the Hidden Battle for the Wold Food System*; Portobello Books Ltd.: London, UK, 2007.
24. Haysom, G. Urban-scale food system governance: An alternative response to the dominant paradigm? In *Untamed Urbanisms*; Routledge: London, UK, 2015; pp. 76–87. ISBN 9781317599098.

25. Peyton, S.; Moseley, W.; Battersby, J. Implications of supermarket expansion on urban food security in Cape Town, South Africa. *Afr. Geogr. Rev.* **2015**, *34*, 36–54. [CrossRef]
26. Battersby, J.; Peyton, S. The Spatial Logic of Supermarket Expansion and Food Access. In *Rapid Urbanisation, Urban Food Deserts and Food Security in Africa*; Springer International Publishing: Cham, Germany, 2016; pp. 33–46.
27. Dixon, J. From the imperial to the empty calorie: How nutrition relations underpin food regime transitions. *Agric. Human Values* **2009**, *26*, 321–333. [CrossRef]
28. Popkin, B.M.; Adair, L.S.; Ng, S.W. Global nutrition transition and the pandemic of obesity in developing countries. *Nutr. Rev.* **2012**, *70*, 3–21. [CrossRef] [PubMed]
29. Wrigley, N. "Food deserts" in British cities: Policy context and research priorities. *Urban Stud.* **2002**, *39*, 2029–2040. [CrossRef]
30. Morton, L.W.; Blanchard, T.C. Starved for Access: Life in Rural America's Food Deserts. *Rural Realities* **2007**, *1*, 1–10.
31. Walker, R.E.; Keane, C.R.; Burke, J.G. Disparities and access to healthy food in the United States: A review of food deserts literature. *Health Place* **2010**, *16*, 876–884. [CrossRef]
32. Battersby, J. Beyond the Food Desert: Finding ways to speak about urban food security in South Africa. *Geogr. Ann. Ser. B Hum. Geogr.* **2012**, *94*, 141–159. [CrossRef]
33. Battersby, J.; Crush, J. Africa's Urban Food Deserts. *Urban Forum* **2014**, *25*, 143–151. [CrossRef]
34. Shannon, J. Food deserts: Governing obesity in the neoliberal city. *Prog. Hum. Geogr.* **2014**, *38*, 248–266. [CrossRef]
35. Widener, M.J.; Shannon, J. When are food deserts? Integrating time into research on food accessibility. *Health Place* **2014**, *30*, 1–3. [CrossRef] [PubMed]
36. Campbell, H. Breaking new ground in food regime theory: Corporate environmentalism, ecological feedbacks and the 'food from somewhere' regime? *Agric. Human Values* **2009**, *26*, 309–319. [CrossRef]
37. Folke, C. Resilience: The emergence of a perspective for social–ecological systems analyses. *Glob. Environ. Chang.* **2006**, *16*, 253–267. [CrossRef]
38. Ericksen, P.J. Conceptualizing food systems for global environmental change research. *Glob. Environ. Chang.* **2008**, *18*, 234–245. [CrossRef]
39. Millennium Ecosystem Assessment. *Ecosystems and Human Well-being: Synthesis*; Island Press: Washington, DC, USA, 2005.
40. Díaz, S.; Settele, J.; Brondizio, E.S.; Ngo, H.T.; Agard, J.; Arneth, A.; Balvanera, P.; Brauman, K.A.; Butchart, S.H.M.; Chan, K.M.A.; et al. Pervasive human-driven decline of life on Earth points to the need for transformative change. *Science* **2019**, *13*, 366. [CrossRef]
41. IPBES. *The Global Assessment Report on Biodiversity and Ecosystem Services of the Intergovernmental Science-Policy Platform on Biodiversity and Ecosystem Services*; IPBES: Bonn, Germany, 2019.
42. Holling, C.S. Resilience and Stability of Ecological Systems. *Annu. Rev. Ecol. Syst.* **1973**, *4*, 1–23. [CrossRef]
43. Biggs, R.; Schlüter, M.; Schoon, M.L. *Principles for Building Resilience: Sustaining Ecosystem Services in Social-Ecological Systems*; Biggs, R., Schlüter, M., Schoon, M.L., Eds.; Cambridge University Press: Cambridge, UK, 2015; ISBN 1316299929.
44. Folke, C.; Biggs, R.; Norström, A.V.; Reyers, B.; Rockström, J. Social-ecological resilience and biosphere-based sustainability science. *Ecol. Soc.* **2016**, *21*. [CrossRef]
45. Moore, M.-L.; Olsson, P.; Nilsson, W.; Rose, L.; Westley, F.R. Navigating emergence and system reflexivity as key transformative capacities: Experiences from a Global Fellowship program. *Ecol. Soc.* **2018**, *23*. [CrossRef]
46. Olsson, P. Synthesis: Agency and opportunity. In *The Evolution of Social Innovation. Building Resilience through Transitions*; Westley, F.R., McGowan, K., Tjornbo, O., Eds.; Edward Elgar: Cheltenham, UK, 2017.
47. Westley, F.; Olsson, P.; Folke, C.; Homer-Dixon, T.; Vredenburg, H.; Loorbach, D.; Thompson, J.; Nilsson, M.; Lambin, E.; Sendzimir, J.; et al. Tipping Toward Sustainability: Emerging Pathways of Transformation. *Ambio* **2011**, *40*, 762–780. [CrossRef] [PubMed]
48. Westley, F. Social Innovation and Resilience: How One Enhances the Other. *Stanford Soc. Innov. Rev.* **2013**, *11*, 6–8.
49. Westley, F.R.; Tjornbo, O.; Schultz, L.; Olsson, P.; Folke, C.; Crona, B.; Bodin, Ö. A theory of transformative agency in linked social-ecological systems. *Ecol. Soc.* **2013**, *18*, 27. [CrossRef]

50. Stirling, A.C.; Scoones, I.; Abrol, D.; Atela, J.; Charli-Joseph, L.; Eakin, H.; Ely, A.; Olsson, P.; Pereira, L.M.; Priya, R.; et al. Transformations to Sustainability: Combining structural, systemic and enabling approaches. *Curr. Opin. Environ. Sustain.* **2020**. [CrossRef]
51. Olsson, P.; Galaz, V. Social-ecological innovation and transformation. In *Social Innovation: Blurring Boundaries to Reconfigure Markets*; Nicholls, A., Murdock, A., Eds.; Palgrave Macmillan: Basingstoke, UK, 2012.
52. Leach, M.; Stirling, A.C.; Scoones, I. *Dynamic Sustainabilities*; Routledge: London, UK, 2010; ISBN 9781849775069.
53. Olsson, P.; Galaz, V.; Boonstra, W.J. Sustainability transformations: A resilience perspective. *Ecol. Soc.* **2014**, *19*. [CrossRef]
54. Godfray, H.C.J.; Beddington, J.R.; Crute, I.R.; Haddad, L.; Lawrence, D.; Muir, J.F.; Pretty, J.; Robinson, S.; Thomas, S.M.; Toulmin, C. Food Security: The Challenge of Feeding 9 Billion People. *Science* **2010**, *327*, 812–818. [CrossRef]
55. Nestle, M. *Unsavoury Truth: How Food Companies Skew the Science of What We Eat*; Basic Books: New York, NY, USA, 2018.
56. Nestle, M. *Food Politics*; Unviersity of California Press: Berkeley/Los Angeles, CA, USA, 2007.
57. Nestle, M. Big Food, Food Systems, and Global Health. *PLoS Med.* **2012**, *9*, e1001242.
58. Battersby, J.; Peyton, S. The Geography of Supermarkets in Cape Town: Supermarket Expansion and Food Access. *Urban Forum* **2014**, *25*, 153–164. [CrossRef]
59. Ingram, J. A food systems approach to researching food security and its interactions with global environmental change. *Food Secur.* **2011**, *3*, 417–431. [CrossRef]
60. Pingali, P.L. Green Revolution: Impacts, limits, and the path ahead. *Proc. Natl. Acad. Sci. USA* **2012**, *109*, 12302–12308. [CrossRef] [PubMed]
61. Olsson, P.; Galaz, V. Social-Ecological Innovation and Transformation. In *Social Innovation*; Nicholls, A., Murdock, A., Eds.; Palgrave Macmillan: Basingstoke, UK, 2011; p. 25.
62. Schäpke, N.; Stelzer, F.; Caniglia, G.; Bergmann, M.; Wanner, M.; Singer-Brodowski, M.; Loorbach, D.; Olsson, P.; Baedeker, C.; Lang, D.J. Jointly Experimenting for Transformation? Shaping Real-World Laboratories by Comparing Them. *GAIA Ecol. Perspect. Sci. Soc.* **2018**, *27*, 85–96. [CrossRef]
63. Geels, F.W.; Schot, J. Typology of sociotechnical transition pathways. *Res. Policy* **2007**, *36*, 399–417. [CrossRef]
64. Geels, F.W. Ontologies, socio-technical transitions (to sustainability), and the multi-level perspective. *Res. Policy* **2010**, *39*, 495–510. [CrossRef]
65. Pereira, L.M.; Bennett, E.M.; Biggs, R.O.; Peterson, G.D.; McPhearson, T.; Norström, A.V.; Olsson, P.; Preiser, R.; Raudsepp-Hearne, C.; Vervoort, J.M. Seeds of the future in the present: Exploring pathways for navigating towards "Good Anthropocenes". In *Urban Planet*; Elmqvist, T., Bai, X., Frantzeskaki, N., Griffith, C., Maddox, D., McPhearson, T., Parnell, S., Roberts, D., Romero Lankao, P., Simon, D., et al., Eds.; Cambridge University Press: Cambridge, UK, 2018; ISBN 9781107196933.
66. Gelcich, S.; Hughes, T.P.; Olsson, P.; Folke, C.; Defeo, O.; Fernández, M.; Foale, S.; Gunderson, L.H.; Rodríguez-Sickert, C.; Scheffer, M.; et al. Navigating transformations in governance of Chilean marine coastal resources. *Proc. Natl. Acad. Sci. USA* **2010**, *107*, 16794–16799. [CrossRef] [PubMed]
67. Von Braun, J.; Diaz-Bonilla, E. *Globalisation of Food and Agriculture and the Poor*; Oxford University Press: New Delhi, India, 2008.
68. IPES-Food. *Too Big to Feed*; IPES-Food: Brussels, Belgium, 2016.
69. Rosenberg, E. *A World Connecting: 1870–1945*; Harvard University Press: Cambridge, MA, USA, 2012.
70. Fukuyama, F. *Political Order and Political Decay: From the Industrial Revolution to the Globalisation of Democracy*; Profile Books: London, UK, 2015.
71. Galloway, J.N.; Cowling, E.B. Reactive Nitrogen and the World: 200 Years of Change. *Ambio* **2002**, *31*, 64–71. [CrossRef] [PubMed]
72. Mosier, A.; Syers, J.; Freney, J. *Agriculture and the Nitrogen Cycle. Assessing the Impacts of Fertilizer Use on Food Production and the Environment*; Island Press: London, UK, 2004.
73. Murphy, S.; Burch, D.; Clapp, J. *Cereal Secrets: The World's Largest Grain Traders and Global Agriculture*; Oxfam Research Reports; Oxfam: Oxford, UK, 2012.
74. Friedmann, H. After Midas's feast. In *Food for the Future: Conditions, Contradictions of Sustainability*; Allen, P., Ed.; Wiley: New York, NY, USA, 1993.

75. Weatherspoon, D.R.D.; Reardon, T. Supermarkets in Africa: Implications for Agrifood systems and the rural poor. *Dev. Policy Rev.* **2003**, *2*, 1–17.
76. Reardon, T.; Timmer, C.P.; Berdegue, J.; Barrett, C.B. The rise of supermarkets in Africa, Asia and Latin America. *Am. J. Agric. Econ.* **2003**, *83*, 1140–1146. [CrossRef]
77. Reardon, T.; Hopkins, R. The Supermarket Revolution in Developing Countries: Policies to Address Emerging Tensions among Supermarkets, Suppliers and Traditional Retailers. *Eur. J. Dev. Res.* **2006**, *18*, 522–545. [CrossRef]
78. Burch, D.; Lawrence, G. *Supermarkets and Agri-food Supply Chains: Transformations in the Production and Consumption of Foods*; Burch, D., Lawrence, G., Eds.; Edward Elgar: London, UK, 2007.
79. Shaw, G. Transferring a Retail Innovation: The early stages of Supermarket Development in Post-war Britain. In *Economic History Yearbook*; Verlag De Gruyter Oldenbourg: Berlin, Germany, 2005; pp. 57–70.
80. Schumpeter, J. *Capitalism, Socialism and Democracy*; Allen and Unwin: London, UK, 1943.
81. Reardon, T.; Barrett, C.B.; Berdegue, J.A.; Berdegué, J.A.; Swinnen, J.F.M. Agrifood Industry Transformation and Small Farmers in Developing Countries. *World Dev.* **2009**, *37*, 1717–1727. [CrossRef]
82. Schlosser, E. *Fast Food Nation: The Dark Side of the All-American Meal*; Houghton Mifflin Harcourt: New York, NY, USA, 2001.
83. Pollan, M. *The Omnivore's Dilemma: The Search for a Perfect Meal in a Fast-Food World*; Bloomsbury: London, UK, 2007.
84. Lang, T.; Heasman, M. *Food Wars: The Global Battle for Mouths, Minds and Markets*; Earthscan: Oxford, UK, 2004; ISBN 1853837016.
85. Campbell, H. Consultation, commerce and contemporary agri-food systems: Ethical engagement of new systems of governance under reflexive modernity. *Integr. Assess. J.* **2006**, *6*, 117–136.
86. Campbell, H. The rise and rise of EurepGAP: The European (re)invention of colonial food relations? *Int. J. Sociol. Agric. Food* **2005**, *13*, 6–19.
87. Goodman, D. *Place and Space in Alternative Food Networks: Connecting Production and Consumption*; King's College: London, UK, 2009.
88. Paoli, G.D.; Yaap, B.; Wells, P.L.; Sileuw, A. Opinion Article CSR, Oil Palm and the RSPO: Translating boardroom philosophy into conservation action on the ground. *Trop. Conserv. Sci.* **2010**, *3*, 438–446. [CrossRef]
89. Ruysschaert, D.; Salles, D. The strategies and effectiveness of conservation ngos in the global voluntary standards: The case of the roundtable on sustainable palm-oil. *Conserv. Soc.* **2016**, *14*, 73. [CrossRef]
90. Valkila, J.; Nygren, A. Impacts of Fair Trade certification on coffee farmers, cooperatives, and laborers in Nicaragua. *Agric. Hum. Values* **2010**, *27*, 321–333. [CrossRef]
91. Ibanez, M.; Blackman, A. Is Eco-Certification a Win–Win for Developing Country Agriculture? Organic Coffee Certification in Colombia. *World Dev.* **2016**, *82*, 14–27. [CrossRef]
92. Markey, E. Bread and Beer for a Better Biosphere: The Transformative Potential of the Eco-gastronomic niche in the Greater Cape Town Area. Master's Thesis, Stockholm University, Stockholm, Sweden, 2017.
93. Stringer, L.C.; Fraser, E.D.; Harris, D.; Lyon, C.; Pereira, L.; Caroline, W.; Simelton, E. Adaptation and Development Pathways for Different Types of Farmers. *Environ. Sci. Pol.* **2020**, *104*, 174–190.
94. Guerrero Lara, L.; Pereira, L.M.; Ravera, F.; Jiménez-Aceituno, A. Flipping the Tortilla: Social-ecological innovations and traditional ecological knowledge for more sustainable agri-food systems in Spain. *Sustainability* **2019**, *11*, 1222. [CrossRef]
95. Vitousek, P.M.; Mooney, H.A.; Lubchenco, J.; Melillo, J.M. Human domination of Earth's ecosystems. In *Urban Ecology: An International Perspective on the Interaction Between Humans and Nature*; Springer: New York, NY, USA, 2008; pp. 3–13.
96. Weiss, B. Configuring the authentic value of real food: Farm-to-fork, snout-to-tail, and local food movements. *Am. Ethnol.* **2012**, *39*, 614–626. [CrossRef]
97. Barber, D. *The Third Plate: Field Notes on the Future of Food*; Little, Brown: London, UK, 2014.
98. Pollan, M. *Defense of Food*; Penguin: New York, NY, USA, 2008.
99. Pereira, L.M.; Contreras, R.C.; Norström, A.V.; Espinosa, D.; Willis, J.; Lara, L.G.; Khan, Z.; Rusch, L.; Palacios, E.C.; Amaya, O.P. Chefs as change-makers from the kitchen: Indigenous knowledge and traditional food as sustainability innovations. *Glob. Sustain.* **2019**, *2*, 1–9. [CrossRef]
100. Pradhan, P.; Costa, L.; Rybski, D.; Lucht, W.; Kropp, J.P. A Systematic Study of Sustainable Development Goal (SDG) Interactions. *Earth's Futur.* **2017**, *5*, 1169–1179. [CrossRef]

101. Guthman, J. *Agrarian Dreams: The Paradox of Organic Farming in California*; University of California Press: Berkeley, CA, USA, 2004.
102. Aliber, M.; Cousins, B. Livelihoods after Land Reform in South Africa. *J. Agrar. Chang.* **2013**, *13*, 140–165. [CrossRef]
103. Shisana, O.; Labadarios, D.; Simbayi, L.; Zuma, K.; Dhansay, A.; Reddy, P.; Parker, W.; Hoosain, E.; Naidoo, P.; Hongoro, C.; et al. *South African National Health and Nutrition Examination Survey (SANHANES-1)*, 2014 ed.; HSRC Press: Cape Town, South Africa, 2014; ISBN 9780796924766.
104. Igumbor, E.U.; Sanders, D.; Puoane, T.R.; Tsolekile, L.; Schwarz, C. '"Big Food,"' the Consumer Food Environment, Health, and the Policy Response in South Africa. *PLoS Med.* **2012**, *9*, 1–7. [CrossRef] [PubMed]
105. Mayosi, B.M.; Flisher, A.J.; Lalloo, U.G.; Sitas, F.; Tollman, S.M.; Bradshaw, D. The burden of non-communicable diseases in South Africa. *Lancet* **2009**, *374*, 934–947. [CrossRef]
106. Hofman, K.J.; Tollman, S.M. Population health in South Africa: A view from the salt mines. *Lancet Glob. Health* **2013**, *1*, e66–e67. [CrossRef]
107. Kimani-Murage, E.W. Exploring the paradox: Double burden of malnutrition in rural South Africa. *Glob. Health Action* **2013**, *6*, 19249. [CrossRef]
108. Carter, S.; Gulati, M. *Understanding the Food Energy Water Nexus Climate change, the Food Energy Water Nexus and food security in South Africa*; WWF-SA: Cape Town, South Africa, 2014.
109. Sellberg, M.M.; Norström, A.V.; Peterson, G.D.; Gordon, L.J. Using local initiatives to envision sustainable and resilient food systems in the Stockholm city-region. *Glob. Food Sec.* **2020**, *24*, 100334. [CrossRef]
110. Pereira, L.M.; Bennett, E.; Biggs, R.O.; Mangnus, A.; Norström, A.V.; Peterson, G.; Raudsepp-Hearne, C.; Sellberg, M.; Vervoort, J. Seeding Change by Visioning Good Anthropocenes. *Solutions* **2019**, *10*.
111. Bennett, E.M.; Solan, M.; Biggs, R.; McPhearson, T.; Norström, A.V.; Olsson, P.; Pereira, L.; Peterson, G.D.; Raudsepp-Hearne, C.; Biermann, F.; et al. Bright spots: Seeds of a good Anthropocene. *Front. Ecol. Environ.* **2016**, *14*, 441–448. [CrossRef]
112. Moore, M.-L.; Riddell, D.; Vocisano, D. Scaling Out, Scaling Up, Scaling Deep: Strategies of Non-profits in Advancing Systemic Social Innovation. *J. Corp. Citizsh.* **2015**, *58*, 67–84. [CrossRef]
113. Drimie, S.; Pereira, L. Advances in Food Security and Sustainability in South Africa. In *Advances in Food Security and Sustainability*; Barling, D., Ed.; Academic Press: Burlington, NJ, USA, 2016; pp. 1–31. ISBN 9780128098639.
114. Zgambo, O.; Pereira, L.; Boatemaa, S.; Drimie, S. *T-labs for Alternative Food Systems in the Western Cape*; Food Jams, Salt River: Stellenbosch, South Africa, 2018.
115. Cramer, C.; Pereira, L.M.; Drimie, S.; Willis, J.; Kushitor, S. *T-lab on Coastal Wild Foods in the Western Cape Report on T-Lab Held 2–3 May 2019*; Food Jams, Salt River: Stellenbosch, South Africa, 2019.
116. Mabhaudhi, T.; Chibarabada, T.P.; Chimonyo, V.G.P.; Murugani, V.G.; Pereira, L.M.; Sobratee, N.; Govender, L.; Slotow, R.; Modi, A.T. Mainstreaming underutilized indigenous and traditional crops into food systems: A South African perspective. *Sustainability* **2018**, *11*, 172. [CrossRef]
117. Barthel, S.; Crumley, C.; Svedin, U. Bio-cultural refugia-Safeguarding diversity of practices for food security and biodiversity. *Glob. Environ. Chang.* **2013**, *23*, 1142–1152. [CrossRef]
118. Barthel, S.; Crumley, C.L.; Svedin, U. Biocultural refugia: Combating the erosion of diversity in landscapes of food production. *Ecol. Soc.* **2013**, *18*. [CrossRef]
119. Barthel, S.; Folke, C.; Colding, J. Social-ecological memory in urban gardens-Retaining the capacity for management of ecosystem services. *Glob. Environ. Chang.* **2010**, *20*, 255–265. [CrossRef]
120. Gómez-Baggethun, E.; Reyes-García, V.; Olsson, P.; Montes, C. Traditional ecological knowledge and community resilience to environmental extremes: A case study in Doñana, SW Spain. *Glob. Environ. Chang.* **2012**, *22*, 640–650. [CrossRef]
121. Mazlish, B. Ruptures in History. *Hist. Speak.* **2011**, *12*, 32–33. [CrossRef]

International Journal of
Environmental Research and Public Health

MDPI

Article

Rome, a Policy without Politics: The Participatory Process for a Metropolitan Scale Food Policy

Giampiero Mazzocchi * and Davide Marino

Department of Biosciences and Territory, University of Molise, 86100 Campobasso, Italy; dmarino@unimol.it
* Correspondence: gia.mazzocchi@gmail.com

Received: 15 October 2019; Accepted: 8 January 2020; Published: 11 January 2020

Abstract: In light of the challenges that all cities face today, food is offered as a prism through which to read and intervene on various areas that affect the quality of life of the population: circular economy, urban metabolism, social relations, economies, and food quality. In the Roman context, in recent years, numerous initiatives have revitalized the debate on food and brought the discussion to the center of the interest of an ever-increasing number of citizens. However, these experiences appear unrelated and there is a lack of coordination and political coherence. Faced with this evidence, starting from a territorial analysis, this contribution analyzes the process that led a local group of stakeholders to formulate a proposal for a food policy for the city of Rome. The proposal contains a series of possible actions that aim, on the one hand, to recompose the relations between the city and its territory, with a view to re-localization and re-territorialization of agro-food productions and, on the other hand, to reconnect the economic and social relations that the industrialization of food chains has compromised. The network analysis of the bottom-up process, which mainly investigates networking and negotiation skills between various interests, is carried out and related to a careful analysis of the food system in the Roman context. Furthermore, an overview of the state of the art of urban food policies in Italy has been provided to better contextualize the study case. The findings show actors and topics involved in the process, identifying further development towards a more comprehensive participatory process for a systemic food strategy at the metropolitan level.

Keywords: food policy; food system sustainability; urban agriculture; local development; food governance; stakeholder engagement

---

1. Introduction: Why Cities Need Food Policies

The recent report of the Lancet Commission "Food in the Anthropocene" [1] confirms some dramatic evidence concerning today's food systems: (1) The current composition of diets does not allow adequately feed the entire world population preserving, at the same time, ecosystems and natural resources. (2) Food consumption patterns of richer economies contribute substantially to global warming, producing huge amounts of greenhouse gases, while those that are emerging in developing countries are based on increasing consumption of red meat [2] and cereals, exacerbating pressures on water resources, soil, and biodiversity [2–4] (In the period 1961–2016, the greenhouse gas emissions due to agricultural production increased from 3.1 Gt CO_2-eqyr-1 to 5.8 Gt 17 CO_2-eqyr-1, mainly due to breeding, the use of synthetic fertilizers, and the cultivation of rice [2]). (3) Food production, processing, and distribution are among the most impacting factors in terms of climate change, loss of biodiversity, use of water resources, impairment of nutrient cycles, and changes in land use. These issues clearly highlight the paradox that afflicts food systems: despite the challenge for the coming decades to feed a growing world population that is increasingly concentrated in cities [5,6], the current global food model, dominated by diets strongly impacting on ecosystems and human health, is not able to guarantee this function in a sustainable manner. Furthermore, while 820 million people, equal to 10.9% of the world population, do not have access to sufficient quantities of food has achieved (in 2014, the number of undernourished people was 783 million, equal to 10.7% of the world population), the number of non-transmissible diseases linked to obesity and overweight is growing rapidly [5]. All this happens at a time when, as attested by the last Intergovernmental Panel on Climate Change (IPCC) Special Report on climate change and soil, the availability of food per capita is higher than the average requirement [7,8]. Furthermore, as the 2007/2008 agricultural prices downturn has shown, globalized food systems are fragile, especially in cities, also in light of estimates of urban population growth rates in the coming decades, and put the issue of food security at the center of political agendas. These changes have been so sudden and of such remarkable impact that a new paradigm has been proposed to describe them, the "New Food Equation" [9,10]. The New Food Equationì is based on the evidence that the dynamics concerning the co-evolution between food systems and demographic development are rapidly reshaping the way in which food is produced, processed, distributed, and consumed; impacts the metabolism of a city and its economy; affects public and private spaces [11]; and impacts food consumption patterns and foodscapes.

Numerous factors require a theoretical revision of the relationships that the city weaves with the surrounding territory for the purposes of food production, processing, and distribution. The de-territorialization of food systems [12] has manifested itself in a very decisive manner in recent decades, with evident and dramatic consequences on the ability to manage and govern material (materials raw, processed products, and food waste) and immaterial (knowledge, traditions, and consumer-producer relations) flows related to food [13]. Recently, simultaneously with a growing interest of research and institutions for urban–rural relations, cities have begun to think about how to integrate, connect, and protect green agricultural areas and the social, economic, and environmental functions they provide to the well being of the population [10,14]. Among the approaches to respond to the challenges that involve the transformation of urban food systems, the most widespread today is the City-Region Food System, supported by the Food for the Cities program of FAO in collaboration with the RUAF foundation (Resource Centers on Urban Agriculture and Food Security). It has been defined as the complex network of actors, processes, and relationships affecting the production, transformation, marketing, and consumption of food that exists in a given geographical region that includes a more or less concentrated urban center and its surrounding peri-urban and rural hinterland [15]. The underlying assumption of this approach is that, if policies are developed on a city-region scale, the recognition of the agro-ecosystems' specificities makes it possible to deal simultaneously with urban issues (food safety and health), agricultural issues (opportunities for local farmers), and environmental issues (risk management). In this context, cities—through diversified approaches and various administrative and functional solutions—are taking on an increasingly central role in trying to manage the food system for better governance of many of the challenges they are facing today. In other words, cities are becoming strategic geographical transition nodes that can exploit the political vacuum generated by the absence of global, coherent, and integrated food policies—national or supranational—to develop more sustainable food systems [16,17]. (When referring to processes towards sustainability of food systems, the authors mean holistic approaches that integrate all aspects of the food system. This includes urban and peri-urban agriculture; strengthening the rural–urban interface to ensure connections between rural supplies and urban contexts; taking into account street food, retailers, food processing and distribution, nutrition, and health; and linkages to water, waste, transport, and energy systems [18].) In this context, urban food policies find their ways and are increasing popularity. Urban food policies have been defined [19] as a concerted action on the part of city government to address food-related challenges. Urban food policies often emerge through significant involvement of civil society and other actors. Most urban food policies consist of targeted actions with specific goals, such as addressing a specific public health or environmental concern (e.g., obesity and food waste). Such actions can pave the way for—and be incorporated into—integrated food policies at a later stage and may also have benefits in other policy areas. In most urban food policies, the reasoning on local food systems is directed towards a physical (and therefore ecological), symbolic, and economic reconnection between the city and the countryside through a series of actions and tools that involve all phases of food chains [19,20]. This implies the recognition of agro-ecosystems and agricultural production not as antithetical activities compared to urban ones, but as integrated phenomena, capable of playing a key role in the development of urban systems. Cities are where the social movements of food re-signification find their place, a privileged lens for observing the transformations of food systems and governing many of the challenges of the coming decades. Indeed, the urban scale has been considered the most appropriate one for a couple of decades [21] to intercept and govern the economic and social dynamics and, above all, the policies, tools, and actions that make food systems more sustainable and compatible with the challenges of the new millennium that the Sustainable Development Goals of the United Nations 2030 Agenda have highlighted. Acknowledging that cities have a strategic role to play in developing sustainable food systems and promoting healthy diets, the Milan Urban Food Policy Pact (MUFPP) was established in 2015. MUFPP represents a global commitment among more two hundred mayors from around the world who consider food as an entry point for the sustainable development of growing cities. It represents the main framework for cities and international stakeholders active in innovative urban food policies for the management and governance of local food systems, and includes more recent experiences as well as cities with relevant good track records in institutionalized food strategies such as Toronto, New York, Bristol, Quito, Belo Horizonte, Montpellier, and many others. This research contributes to the momentum that is creating a fertile ground for the establishment of a food policy for the metropolitan city of Rome. Indeed, as highlighted in the next paragraphs, several political statements and declarations, along with growing pressure for a more sustainable food system from civil society, are strong factors that could lead to an institutionalization of a systemic food strategy for Rome. In this context, analyzing the local food policy group—and its relationships with the administration—contributes to the understanding of its role in the future and its ability to survive political cycles, reflecting upon the pros and cons of the space in which the food policy group is to operate. To better position the case of Rome, an overview of the state of the art of urban food policies in Italy has been provided.

2. Research Design

The relationships between agricultural landscape and city have always determined consumption choices and dietary patterns [22]. The forms of food production give a transformation—physical, but not only—to the natural landscape, through a conscious and systematic anthropic action [23]. Agricultural landscape, diets, and food consumption models are therefore interconnected and mutually dependent. In the Mediterranean diet, these relationships have shaped territories and compositions of foodstuffs that are exchanged and consumed, orienting them towards products such as fruit, vegetables, legumes, seeds, and olive oil, widely recognized as "healthy" for human health [24,25]. However, their effects in terms of environmental sustainability, agricultural systems, and rural landscapes are not unique [26]. This clarification seems necessary to contextualize the agri-food system of Rome, in which the associative movements and the bottom-up participatory path for a food policy proposal have arisen in response to the tendency towards the "westernization" of diets, which, besides being considered risky for human health [27], risk dispersing cultural capital, traditions, and connections with the rural landscape embedded in typical products of the Mediterranean diet. (The Western Pattern Diet (WPD) or Standard American Diet (SAD) is a dietary model generally characterized by high intakes of red meat, processed meat, pre-packaged foods, butter, fried foods, fat-rich dairy products, refined cereal eggs, potatoes, corn (and high fructose corn syrup), and high-sugar drinks.) In the following paragraphs, the process that led a group of local stakeholders (associations, researchers, groups of farmers, and civil society) to formulate a Food Policy proposal for the city is analyzed. The analysis of the process, which mainly investigates the networking and negotiation capacities between various interests, is carried out and related to an overview of the food system in the Roman context, which highlights its territory, agro-food productions, the food distribution model, consumption patterns, and already active experiences, which can play a fundamental role in directing the transition of the local food system towards greater economic, social, and environmental self-sustainability. In Figure 1, the research design has been shown, in order to better orient the reading of the paper.

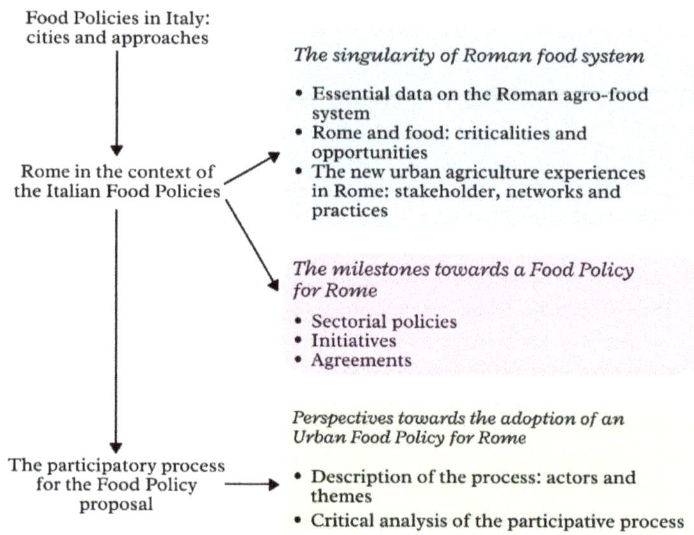

Figure 1. Research design (source: our elaborations).

3. Food Policies in Italy: Cities and Approaches

In Italy, the models proposed by the so-called Territorialist School [28,29], ascribable to the approaches attributable to the urban bioregion [30,31], were based on cultural and scientific models such as ecological agriculture, the territorial approach, and the analysis of territorial metabolism and supply chains [32]. In the national territory, the initiatives linked to food policies, in the absence of a national strategy dedicated to food systems, have developed according to specific paths at the local level, around some practices (school canteens, quality food education programs, and short food chains), often based on private corporate and civil society initiatives. Today, 26 Italian cities, out of a total of 207 cities, have signed the MUFPP (as of 2 October 2019), but not all of them are equally committed to more sustainable food systems. The following experiences have been reported to highlight the different entry points towards formalized urban food policies or bottom-up processes in Italy. Our aim is not to compare them, as we are referring to different scales, from metropolis (Milan and Turin) to medium-sized cities (Livorno and Pisa), to small towns (Castel del Giudice and Tollo). Instead, it is useful to provide a synthetic overview of the approaches adopted by Italian municipalities, local groups, or mixed partnerships, in order to understand what role Rome can play in a national view. In Tuscany, the university and researchers have played a fundamental role in the promotion and cultural dissemination of issues related to food planning: the Food Plan of Pisa, adopted by the Province in 2010 [33], represented for some years the national reference point for both participatory processes and relations between the various sectors of the administrations and among administration, citizens and other stakeholders [34]. After Expo 2015 "Feeding the planet, energy for life", the municipality of Milan implemented its own Food Policy, a strategy based on five priorities: (1) healthy food for everyone; (2) sustainability of food system; (3) food education; (4) fight against food waste; and (5) scientific research on agro-food systems. The Polytechnic University of Turin, in collaboration with the Metropolitan City, recently drafted the Turin Metro Food Atlas, an analysis, representation, and communication initiative of the urban metropolitan food system [35]. A similar initiative was launched in Matera, where, however, the experience of the Food Atlas does not yet involve the local administration but, as stated on the project website, "could become support for the construction of a Food Chart in Matera". Recently implemented is the Pact for Urban Policies for Food signed by the municipalities of Lucca and Capannori, born following the experiences of the European project ROBUST (Risk and Opportunity management of huge-scale BUSiness communiTy cooperation), a project that analyzes the interrelations between urban and peri-urban areas with respect to territorial planning, culture, production enhancement, and services. Even small municipal entities have realized the potential of local food policies: this is the case, for example, of Castel del Giudice (Molise Region), which through the Food Plan and a virtuous collaboration with the University of Molise has set up a strategy to counter the depopulation of the area and provide opportunities for socio-economic development by leveraging the landscape, agricultural, and environmental characteristics of the place [36]. The Municipality of Tollo (Abruzzo) has instead approved its own Food Policy, with the intention of making the strong wine prevalence a catalyst for triggering integrated processes between local economies, increased biodiversity through crop diversification, improved quality of diet, and increased territorial development. In Bergamo, the Bergamo Green project produced a mapping of the Alternative Food Networks of the province, with the aim of giving visibility to the realities of production, distribution, and consumption of sustainable, local, and organic products, and to favor the path towards the Food Policy of Bergamo. The project Feeding Trento also relied on a mapping of the virtuous experiences of its territory, also making a platform available and establishing a working table managed by a collaboration between the municipality and the University of Trento. Finally, Venice is carrying out an ex-ante analysis, trying to understand how the remarkable tourist flow affects the lagoon food system, in particular the foodscape and the food narratives that are proposed in the city. From a preliminary analysis carried out through an online survey among the members of the Italian Network of Local Food Policies, there are several critical issues related, in the first instance, to the difficulties inherent in the translation of plans, strategies and programs into actual actions on the territory and on the food system. First, there is a strong concentration of policies in the central and northern regions, in which the Food Policy of Milan emerges as the only major city in which the administration has supported the policy thanks to institutional support and a dedicated and participated budget by a foundation banking. An important area of innovation is represented by Tuscany, in which the experiences of Lucca, Pisa, and Livorno suggest the role of the university, and sometimes of individual departments, represents a propelling factor of considerable importance. The themes around which plans and programs are developed mainly include Green Public Procurement in collective catering, often accompanied by attention to the flows—and the consequent agricultural and territorial dynamics—that should fuel the food supply according to the sustainability criteria promoted. Another key theme is food waste, around which the debate is on today but which is often addressed without a systemic view of the problem. In fact, as highlighted in the ISPRA report on food waste [37], the objective of the systemic approach is the protection of joint socio-ecological systems, and not only the efficient use of resources or food security. Finally, among the critical points observed so far in the Italian panorama of food policies, there is a certain propensity of administrations to devote themselves to programs that can be achieved through a modest—or sometimes zero—use of economic and human resources, with the result of a failure or watered-down effectiveness of the measures. This could represent the effect of the lack of a medium- to long-term political vision that could reverberate its own measures on the territorial and land-based structures of the city-region. In this sense, the administrators are probably more worried about communicating and highlighting the efforts within an electoral mandate that is certainly too short compared to a possible transformation and transition of the urban food system. In this context, we are interested in observing how Rome is moving in the panorama of Italian food policies. To this end, an analysis of the Roman foodscape (productions, agricultural models, critical issues, challenges, new economies, and networks) is provided and then the participatory process that led a group of stakeholders, which have been clustered and presented in the next paragraphs, to present some food policy proposals to the administrations is critically investigated. In this context, we refer to the definition of foodscape offered by Mikkelsen [38], which identifies it both as a tool for the study of "food environments" and for the evaluation of potential impacts on food choices and behavior, both as a framework for analyzing how food, places, and people are interconnected and how they interact with each other.

4. The Rome Foodscape

4.1. Essential Data on the Roman Agro-Food System

The metropolitan area of Rome has a population of about 4.34 million inhabitants for an area of 5352 km^2. At the municipal level, the Total Agricultural Area (SAT) of Rome is approximately 58,000 ha, or 45.1% of the territory. In the countryside and on the Roman hills, many quality food products are produced and processed (8 Protected Designation of Origin and 7 Protected Geographical Indication), among which stand out products that strongly characterize the territory, the supply chains, and the local cuisine such as the Abbacchio Romano, the Pecorino Romano, and the Ricotta Romana. Historically, sheep and goat farms have represented a fundamental economy for the Roman countryside, substantially determining the landscape, customs, and traditions of the Roman countryside. In Lazio Region, and in particular in the Municipality of Rome, there is a significant presence of livestock holdings in the sheep and goat sector (in 2014, 6500 units in Lazio and 500 units in the Municipality of Rome, according to data of the Livestock Registry established by the Ministry of Health), which, in terms of size, are concentrated in small and medium size classes (number of heads less than 1000) [39]. Sheep milk production includes some P.D.O (Pecorino Romano, Pecorino Toscano, Pecorino di Picinisco, and Ricotta Romana) and numerous Traditional Agri-food Products, into which tangible and intangible values are embedded from the high interaction between natural and cultural capital [40], being the Roman agricultural landscape immersed in a network of pre-existing archaeological sites, monuments, villas, and farmhouses. This framework is enriched by the role that the "Agro Romano" has historically held, in particular for the relevance that some productions have in the local productive fabric. This is the case, for example, of the dairy livestock practiced in the northwest quadrant of the city towards the Via Aurelia, of the horticulture linked to the coastal reclamation plain and the ovine livestock and the agro pastures [41]. In the Municipality of Rome, farmers are mainly dedicated to the production of arable crops (cereals and forage), followed by permanent meadows, pastures, and woody and agricultural crops such as olives and vines.

In structural terms, the latest data available on the Roman primary sector show a growth in the number of farms, which mainly concerns small- and very small-scale realities (areas of less than 2 ha). At the provincial level, there has been, over the last decade, a 58% decrease in farms and 8.2% in agricultural areas, while, at the municipal level the figures show opposite signs, and for significant values: +43.8% of farms and +12.1% of Total Agricultural Area, for a total of 57,959 ha (45.1% of the 128,540 ha occupied by the municipality). A fundamental role for the ecosystem services offered to the population is played by the protected natural areas, which constitute an important system of green infrastructures (see Figure 2). These areas are located mainly in the peri-urban area, but also extend to the most central areas and, overall, reach an area of 41,500 ha, equal to 32% of the entire municipal area. According to the current General Master Plan, two-thirds of the municipal territory constitutes the city's current Ecological Network, an articulated and functional system of areas of naturalistic, agricultural, and recreational importance. It is composed, in fact, by a complex fabric of protected natural areas, urban green areas (historic villas, gardens, tree-lined streets, etc.), blue infrastructures, and agricultural areas. These green areas cover approximately 86,000 ha, equal to 67% of the entire surface of the municipality, offering spaces for the conservation of habitats of particular naturalistic value and a variety of natural environments and ecological niches. (Most of the protected natural areas (14 out of a total of 19 plus the protected marine area) are managed by the regional body "RomaNatura", as they all fall entirely within the municipal territory.) In this context, agricultural parks play a central role for the green connectivity of Rome and for their high potential in terms of reconstructing the relationship between agriculture and cities, citizens and farmers, and public and private spaces, actively contributing to improving the resilience of the city [42]. (One of the most significant cases is certainly represented by the Casal del Marmo Agricultural Park, in the northwest quadrant of Rome. The area is completely immersed in the urban mosaic and therefore plays a role—real and potential—of providing ecosystem services that are very relevant to the surrounding area. In particular, the analysis of cultural ecosystem services provided by the park was carried out by Davide Pellegrino in a doctoral thesis on landscape and environment entitled "The green infrastructure for landscape governance: the contribution of ecosystem services" [43].) Still referring to the Roman foodscape, it is necessary to mention the abundance of initiatives linked to urban community gardens in the city, historically present and recently reported at the center of the discourses on urban metabolism and social justice. From a survey by the Zappata Romana Urban Architecture Project, which in 2011 compiled a map of shared gardens in Rome, there are 218 recognized associative experiences. A study [44] found 3200 plots of residential gardens (85% of the total), shared gardens, institutional gardens, and informal gardens within the municipality borders. An important project on urban agriculture in Rome is Ru:rban, a European project started in 2019 that aims to exchange good practices between cities and European experiences in the context of urban gardens and agriculture, taking an example from the paradigmatic experience of Rome. Simultaneously with the return of urban agriculture, a phenomenon that has resumed in recent years following a long period of housing development that slowed its development [45,46], is the remarkable development of alternative food networks in Rome: all forms of short supply chains—from farmers' markets to Solidarity Purchasing Groups (SPGs) and box scheme experiences—have achieved significant success. The numbers of short supply chains in the Roman area present proportions of primary importance in the national context: the municipality of Rome has 33 farmers' markets and 55 SPGs, while 744 out of 2,656 farms of the area sell direct (with a 40% increase in the last census interval 2000–2010).

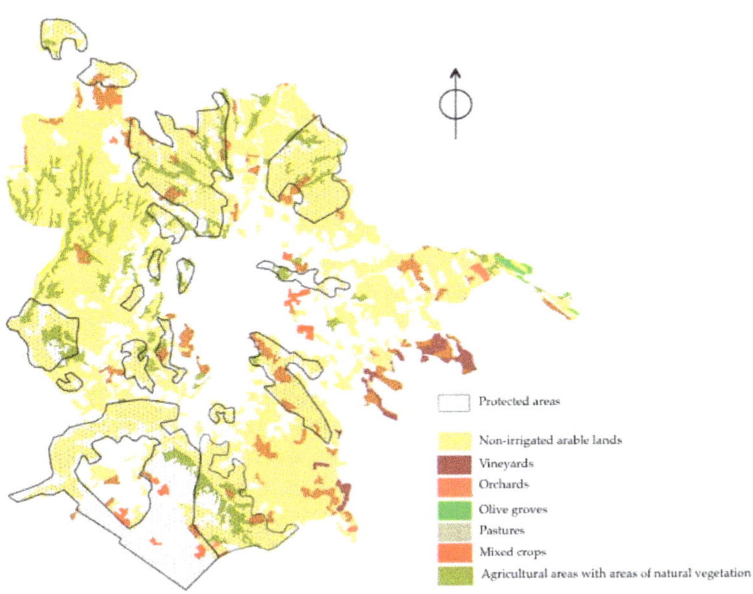

Figure 2. Land uses and protected areas in the Municipality of Rome, scale 1: 100,000 (Source: [42] using data from Corine Land Cover, 2006).

4.2. Rome and Food: Criticalities and Opportunities

It has been highlighted that the foodscape of the Roman area is particularly rich in experiences, productions, and practices related to innovation in the agro-food sector. Rome and the surrounding countryside represent a considerable basin for the production of food and agricultural services, a characteristic that is not common if one looks at the territorial composition of many western metropolises. However, practices and experiences appear disconnected, poorly interconnected and, above all, exposed to the risks of an agro-industrial market that reduces the space for small traders. Among the critical issues that undermine the value of the Roman agro-food system, it is necessary to mention the land use changes in place in the Municipality of Rome. The 2018 report on the Land use in Rome [47] shows that 23.5% (30,241 ha) of the territory is covered by artificial surfaces. It can be noted that, in the face of an almost zero increase in the population, the consumption of soil is growing. In fact, even in sparsely populated neighborhoods and in demographic decline areas, the impermeable surfaces are stably increasing over time. The same report indicates that an area of 6332 ha is characterized by the maximum hydraulic danger and 14,588 residents are at risk. To the extent that land use change is a theme that is shared and worries most of the metropolis, it is to be noted that, in the Roman context, the disruption of the relationship between farmers and consumers due to the processes of standardization and globalization of the supply chains in recent years has produced a series of critical issues that reverberate on the markets and, therefore, on the purchasing methods and consumption patterns of Romans. Indeed, although the local markets represent a source of food supply for Romans from the post-war period onwards, they are going through a moment of crisis due to confused and inefficient regulation and the competition of the Organized Large-Scale Distribution system, where about 70% of purchases take place at the national level. According to the latest census of local markets, carried out in 2015 by the Economic Development Department, there are 127 food markets in the capital, with a total of over 2500 sellers and 5000 stations. Nevertheless, merchants often mediate access to markets, with only one hundred farmers selling their products directly. Over the years, these markets have experienced an almost constant decline, resulting in many semi-abandoned and customer-reduced facilities [48]. Local markets have the fundamental role as a hub for the distribution of products grown in Rome and Lazio. However, the relationship between city and countryside is influenced substantially by the land composition of the territory, the presence of abandoned lands—whether they are private property or state property—and the contractual relations between farm managers and owners. However, given the wide availability of agricultural land, the supply of regulatory instruments for access to them could be developed and integrated into a food policy in which the settlement of new—possibly young—farmers is one of the cornerstones for the revitalisation of a re-localized agro-food sector. In 2014, the Municipality of Rome promoted the program "Rome, city to be cultivated". The tender for the assignment of land and rural buildings owned by Rome was aimed at protecting and restoring the Roman agro-production, by developing multi-functional farms, rewarding the orientation of competitors towards organic farming and multi-functionality. Nevertheless, the works for the renovation of rural buildings are currently at a standstill and the effective incidence of the assignable lands on the total of the available lands is minimal: only a quarter of the roughly 26,000 ha of public lands is the object of assignment and multi-functional agricultural projects have been approved on only 95 ha. In terms of re-territorialization of production and the triggering of virtuous processes in support of local agriculture and nutritious and sustainable diets, the Green Public Procurement also intervenes substantially. Indeed, in Italy, all contracting stations—including municipalities—are obliged to respect the rules established by the Minimum Environmental Criteria, in accordance with a law issued by the Ministry of the Environment. Rome offers school catering services daily to around 144,000 pupils. In the Municipality of Rome, between September 2007 and June 2012, approximately 67.2 million meals were distributed in school canteens, and between January 2013 and June 2017 approximately 71.4 million meals were distributed, for a total of about 138.6 million meals in ten years. These data tell us that, by working on the tender specifications (the value of the last call for tenders is more than 350 million euros over three years) to follow short supply chain criteria, sustainability, and seasonality, important effects could be obtained on the fabric urban, peri-urban, and rural economic and educational levels. Broadening the analysis of the Roman foodscape towards the world of food distribution, in Rome, supermarkets show growing numbers. The 2016 data of the Ministry of Economic Development identify 30 hypermarkets (+61% from 2013 to 2017) and 520 supermarkets in the province of Rome (329 in the municipality), growing in the last five years. With reference to the municipality, there are over 500 supermarkets, with a total sales area of over 290,000 square meters and 7000 employees, which are associated with 194 minimarkets, which in turn cover an area of approximately 36,000 square meters and have almost 900 employees. Rome is first in Italy for shops specializing in organic food (118 out of 1437). At the same time, Rome is one of the cities in which the phenomenon of minimarkets managed by Egyptian or Bengali personnel has strongly grown in recent years. The ease with which a license can be obtained, combined with the fact that retail trade and self-entrepreneurship are almost obligatory choices for social groups forgotten by employment policies, has led these minorities to taking over hundreds of small shops from Italian merchants. Usually, they remain open until late at night, including Sunday, and offer various kinds of products, including low-cost fruit and vegetables. Finally, it is necessary to mention the distortions linked to food waste, which in Italy are worth over 15 billion euros (almost 1% of GDP). Considering that the estimates on the quantity of food thrown out in Italy by wholesale and organized distribution markets amount to about 400,000 tons of food—40% of which is made up of fruit and vegetables—in the metropolitan area of Rome, there could potentially be 29,000 tons of recoverable and redistributable food a year. In Rome, numerous experiences were born with the aim of guaranteeing a second life for food that is not consumed or purchased. The law of 19 August 2016, No. 166 (the so-called Gadda law) encourages and facilitates the distribution and recovery of foodstuffs to poor people and allows a reduction in the waste tax. In 2017, Rome approved a plan for the reduction and management of post-consumer materials in Rome Capital 2017–2021, which aims to bring Rome closer to a circular economy with zero waste. The goal is to reduce the annual waste production by 200,000 tons by 2021, increase separate waste collection from 44% to 70%, and create new recycling and composting plants. The Food Sharing project is part of the plan, which provides for the systematic withdrawal of fresh food from shops, supermarkets, and local markets to subsequently redistribute them to non-profit organizations dealing with people in difficulty and to the Bioparco Foundation, which allocates unsold foodstuffs to the hosted animals. Other virtuous experiences, such as ReFoodGees, assume principles of circular and solidarity economy to activate food recovery and distribution actions, with the aim of creating spaces for socialization, combating waste, and social exclusion and discrimination.

4.3. The New Urban Agriculture Experiences in Rome: Stakeholder, Networks and Practices

The panorama described thus far reveals, on the one hand, the centrality of agriculture for the Roman territorial system, the liveliness of some movements (urban gardens and alternative food networks), and the relevance of the interaction between cultural and natural capital for Rome, and, on the other hand, the critical issues related to urban development that create significant pressures on agricultural land and the associated economies and a lack of integrated and long-term regulation of spaces, uses, and markets. More than forty years after the movements that led to the creation of numerous agricultural cooperatives linked to the recovery of uncultivated land owned by the state, there are still today several experiences that invest in innovation, diversification, and the relationship with the local community. However, the problems linked to the development of low-density housing settlements, the abandonment of large areas that were previously semi-natural, and the consequent loss of environmental functions are putting at risk the ability to provide citizens of urban areas with those fundamental services for the good quality of life in the city. As mentioned above, Rome boasts one of the largest non-urbanized areas in Europe. This aspect makes it particularly suitable for hosting urban and peri-urban forms of agriculture. These experiences are mostly of a voluntary nature and pursue social and community aims. However, there are an increasing number of multifunctional urban agriculture initiatives for commercial and entrepreneurial purposes, which produce private incomes but reverberate the positive externalities on the community. These farms adopt strategies to adapt the management and commercial organization to seize the opportunities of Rome's large and dynamic urban market. The particular vivacity of the bottom-up movements that distinguish Rome from the post-war period onwards led "Agro Romano" to be the fertile ground for the development of pioneering experiences of social agriculture at the national level starting from the end of the 1970s, concomitantly with the bottom-up recovery movements of abandoned land. This aspect has made the Roman territory a laboratory of social innovation practices in which social agriculture has always played a central role. In the municipality of Rome, there are today 32 farms conducting social agriculture (Capodarco cooperative is among the most famous one, also at national and European level), to which another 20 must be added in the metropolitan area. There are more than 1740 end users at the regional level. With regard to collective processes, an important process of legitimizing and regulating urban gardens in the city of Rome is now carried out by the OrtiInComune network. Animated by passionate horticulturists and citizens, through a constructive and constant dialogue with the Rome administration and representatives of the various municipalities, the network cooperates in facilitating the political processes that govern the management of urban green spaces, including urban gardens. In addition, the network is active in awareness-raising campaigns, training days, and cultural activities, substantially contributing to the public debate in the city linked to the themes of public green areas and food. In this context, the experiences of solidarity-based economy, which in Rome are channeled into the Network of Social and Solidarity Economics (Rete di Economia Sociale e Solidale (RESS)) are growing their relevance. (The Social and Solidarity Economics Networks are concrete and practical alternatives to create an economy and a society oriented to the well-being of all. The idea behind social and solidarity economy networks is the concept of reciprocity, or the exchange of favors directly or indirectly, between different subjects of society, for "duties" of solidarity.) It involves the different Solidarity Purchasing Groups (SPGs) of the territory, consumers, producers, and suppliers; organizations and associations that deal with solidarity economy; and individual citizens. RESS periodically organizes meetings and activities, both to talk about different models of small organized solidarity distribution and to exchange information and knowledge on the topics of food sustainability and distribution of added value along the food chains. Another experience that involves different SPGs is RESS Ciociaria, which brings together different purchasing groups in Lazio to promote sustainable and conscious consumption and production.

5. The Participatory Process for the Food Policy Proposal

5.1. Description of the Process: Actors and Themes

In light of the social liveliness of the Roman foodscape, in October 2018, a bottom-up participatory process started, aiming at formulating a proposal for an urban food policy for the city of Rome to be presented to the city and local institutions. (It has to be specified that the authors participated as moderators and facilitators of the meetings of the local food policy group and that the following analysis is based on the minutes of the meetings, the comments reported by email via the dedicated account, and the feedback provided by the participants.) The work started from the awareness of the lack of integration between the experiences seen above both horizontally—that is among them—and vertically—that is with respect to the various governance levels that intervene in food systems. Rome is one of the 26 Italian cities that has signed the MUFPP Pact; however, the administration has not yet committed itself to defining a food strategy, except for the sectorial initiatives mentioned in the present work and the adhesion to international (100 Resilient Cities) or European (Urbact and Ru:rban) projects. The presence of important international institutions operating in the food and rural development sector (FAO, World Food Program, IFAD etc.), make fertile ground for the opening of a debate on a Food Policy in Rome. Furthermore, Rome has already adopted guidelines in favor of urban resilience by joining the international 100 Resilient Cities program, which provides indications for supporting a more sustainable agro-food system. Finally, with the approval of the climate emergency motion, the municipality has undertaken to "pursue a Food Policy aimed at improving the interconnection between production and consumption, with a view to environmental and economic sustainability". Finally, in the Metropolitan Strategic Plan, among the 10 programmatic macro-objectives, there are the enhancement of the links between city and countryside and the promotion of Natural and Cultural Capital [49] (see Figure 3).

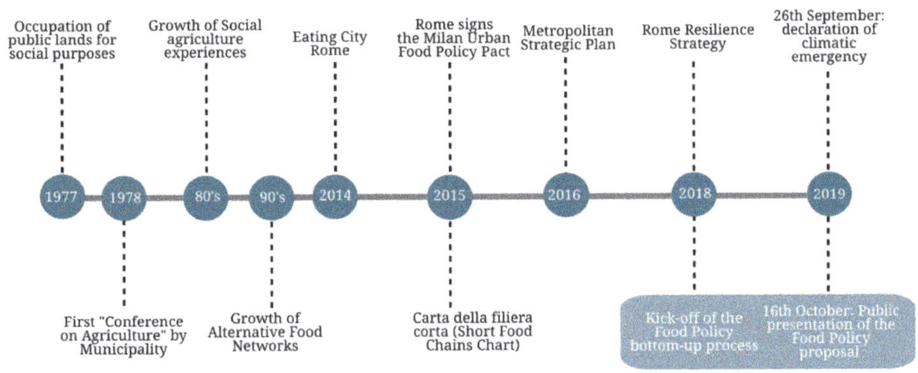

Figure 3. Milestones towards a Food Policy for Rome. Source: our elaborations.

In specifying that the process is still in progress, we can recognize that the objective of the work group was threefold: (i) to show the municipal and metropolitan administration that the Roman agro-food system has enormous potential and many threats and critical points that should be considered by policy-makers to face the different challenges that the city faces; (ii) to stimulate the participation of civil society in the debate on the food system, increasing awareness of the potential impact of a food policy, and increasing the social capital and trust between operators and economic actors; and (iii) to provide and suggest tools, actions, and concrete measures for the implementation of a systemic food strategy for the city of Rome.

Thus far, the participatory process has involved many stakeholders, stimulated by a group of academics from the University of Molise and Roma Tre University, independent researchers, journalists, and activists in the world of associations. Today, the group counts 116 participants, including representatives of: associative movements linked to urban gardens, multi-functional farms, foundations and organizations active on environmental and food issues, professional agricultural organizations, research institutes, associations and social and solidarity economy networks, private operators active in the various phases of the food chain, and interested citizens. The first outcome produced by the promoters has been a report showing why Rome needs a food policy, starting from an analysis of the agro-food system and then formulating a list of ten proposals (the report (in Italian) can be downloaded at this link: https://www.politichelocalicibo.it/wp-content/uploads/2019/10/Una-Food-Policy-per-Roma.pdf). The participatory process has continued through several meetings and the request for amendments and comments to the document by the stakeholders involved, a path that led to the signing of a list of ten policy objectives to be presented to the administration. In the final part of the document, in fact, in light of the evidence that emerged comparing the state of the art, the mapping of innovative practices, and the policies currently in place, it is stated that it is necessary to implement actions aimed at strengthening and supporting small- and medium-sized farms and companies that populate the Roman primary sector along all the steps of the supply chain, from production to marketing, up to the post-consumer phases. It is also stated that, to ensure healthy nutrition and access to quality food for all citizens, while at the same time taking care of the protection of natural resources, it is necessary to support and enhance the good practices of food chains, "having in mind to strengthen the economic and social ties with the rural areas close to the metropolis". Finally, it is reported that a Food Policy should also be aimed at encouraging generational turnover in agriculture, as well as food education and waste reduction, through both prevention and recovery and redistribution of surpluses. (The ten food policy objectives that the document "offers" to the administration can be traced back to the following topics: (1) access to resources (land, water and agro-biodiversity); (2) sustainable agriculture and biodiversity (support for organic farming and agro-ecology); (3) short supply chains and local markets (including local markets); (4) urban–rural relations (integration between supply chain phases; Green Public Procurement); (5) food and territory (territorial labeling, traceability of the supply chain); (6) waste and redistribution (support for recovery and redistribution of surpluses); (7) promotion of multi-functionality; (8) awareness of citizens (food and environmental education plan); (9) landscape (curbing land consumption and other phenomena of land degradation); and (10) planning of resilience (green infrastructures and quantification of services provided by the agro-silvo-pastoral system to the community).) As mentioned above, the participatory process was fueled above all by a series of meetings in which the priorities for the Roman agro-food system were discussed. This last phase led to the organization of five thematic working groups (access to resources; collective/school catering and Green Public Procurement (GPP); agriculture and workers' rights; distribution and consumption; and solidarity economy, food poverty, waste, and redistribution) and a cross-sectional working group on communication.

5.2. Critical Analysis of the Participative Process

To analyze the participatory process from a conceptual point of view, the stakeholders have been grouped into six clusters: research/university; cooperatives and farmers; urban gardeners; Associations (agriculture/food/environment); civil society; networks for local sustainable development. The clusters have been created to detect the topics and themes that have been raised by them during the meeting of the local food policy group. They have been built through a qualitative and discretional work of allocation performed by the authors. Their composition is described in Table 1.

Table 1. Description of the clusters composing the local food policy group in Rome (Source: own elaboration).

Cluster	Number	Percent	Description
Research/University	31	26.7%	Representatives of Universities, center of research on agricultural and food economics, independent researchers, representatives of research departments within foundations, companies and associations which link academic research with local development strategies or international exchange programs.
Cooperatives and farmers	12	10.3%	Farmers (mostly running multifunctional farms), representatives of social agriculture experiences, representatives of agricultural cooperatives and one agronomist.
Urban gardeners	7	6.0%	Representatives of urban gardens associations, public local agencies for the promotion of urban gardens in Rome, representatives of European projects on urban gardens, a private company that organize events for participatory processes aimed at improving urban gardens and that performed a mapping of urban gardens in Rome
Associations (agriculture/food/environment)	36	31.0%	Local associations and Roman seats of national and international associations. The associations are active in the fields of the protection of the environment, on the farmers' rights, on the fair access to agricultural resources, on the promotion of high-quality and fair agro-food products, on food rescue and waste prevention initiatives, on the promotion of agroecology and organic agriculture, on the defense of specific aspects (bees and pollination, organic agriculture in mountain areas), on the support to the achievement of Sustainable Development Goals.
Civil society	20	17.2%	Citizens interested in participating the local group, networks of students, associations active in topics marginally related to food systems (human rights), independent journalist.
Networks for local sustainable development	10	8.6%	Networks for Solidarity and Social Economies, public and private local agencies involved in local development programs, foundations working of sustainable development.
	116	100%	

The direct participation of the authors in the meetings, the minutes of the meetings and the exchanges of views, and the debate that took place in the specifically created mailing list made it possible to outline the main topics of debate and the positions of the participants. The conceptual graph below (Figure 4) shows which themes triggered the interactions between the six clusters.

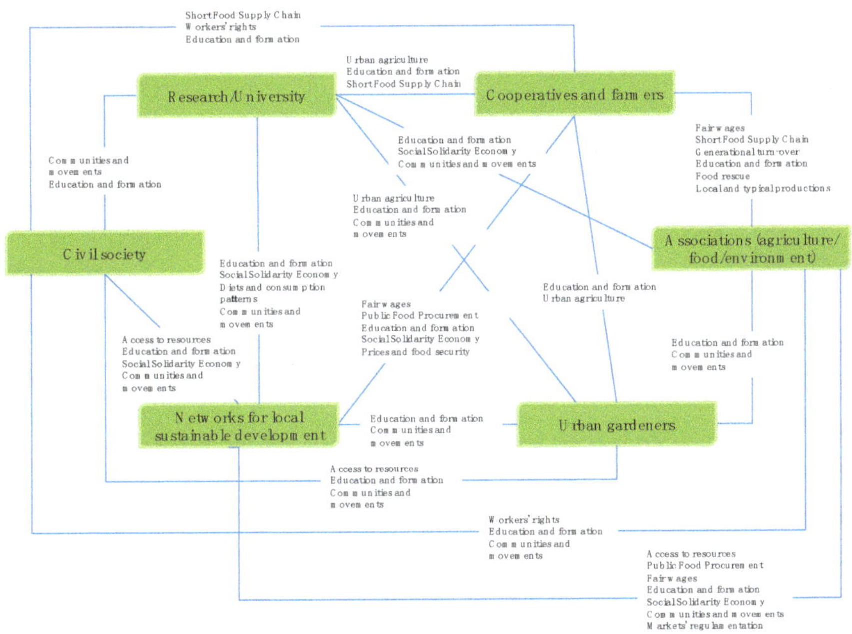

Figure 4. Conceptual map of the actors and themes. The themes common to each pair of clusters are in order of relevance from top to bottom (Source: own elaboration).

It can be observed how some themes have been dealt with in a more widespread and in-depth manner during the meetings. These are issues related to education and training in the agro-food sector, access to resources (mainly the lands) and social and solidarity-based economies, and the associated movements and communities. It is evident that the very composition of the stakeholder group, in which the presence of associations and networks involved in food democracy issues is very strong, has influenced and addressed the debate. Some topics, such as distribution and logistics, have been dealt with to a much lesser extent, due to the absence of operators able to direct the dialogue towards understanding the difficulties and opportunities relating to the modalities with which the food arrives in the city. It can be affirmed that requests and positions relating to alternative forms of marketing of agro-food products are more represented, as many of the actors are directly or indirectly involved in supporting short supply chains. Despite the sustainability of these practices compared to the large-scale retail trade has been demonstrated by the analysis of cases of Italian Alternative Food Networks [50], the participation of representatives of the conventional supply chains in the participatory process would allow grasping the diversity of approaches and needs of a sector that still conveys most of the food that arrives in the city [51] and that must be taken into account in the definition of a systemic food strategy on an urban scale.

The analysis made it possible to identify areas of common interest among the various stakeholder clusters. The results can be viewed in Table 2, which for each pair of clusters identifies projects, possible actions, and collaborations that have arisen during the meetings. At the same time, they represent possible fields of practical application of a food policy for the city of Rome, which could leverage these elements as catalysts for a systemic strategy capable of orienting the entire agro-food system towards an ecological agro-food transition.

Table 2. Thematic areas and common projects developed during the participatory process (Source: own elaboration).

	Research/University	Ag Cooperatives and Farmers	Urban Gardeners	Associations (Agriculture/Food/Environment)	Civil Society	Networks for Local Sustainable Development
Research/University						
Agricultural Cooperatives and Farmers	Search for urban and peri-urban agriculture solutions as a response to pressures (access to resources) and opportunities in the city (markets): multi-functional farms short supply chains, etc.					
Urban gardeners	Collaboration in the quantification of ecosystem services provided by urban gardens (particularly, the cultural ES). Training organization and dissemination activities	Exchange of practices and agronomic technical information. Collaboration in the organization of events and training days.				
Associations (agriculture; food; environment)	Collaboration in structuring the forum on food policy and in identifying and dealing with work areas.	Cooperation in the search for solutions to favor generational turnover and promote sustainable and multi-functional forms of agriculture.	Organization of training days. Studies on the replicability and trans-scalarity of urban gardens. Participation in financing projects.			
Civil Society	Identification of stakeholders from areas such as rights, social inequalities, and civic networks. Collaboration in the promotion of the forum.	Search for short supply chain solutions to meet the demand for quality food and products while respecting the environment and workers' rights.	Exchange of visions about the goals of urban gardens and their future development, as well as in relation to urban social issues such as urban expansion and social inequalities.	Development of proposals and possible solutions regarding the issues of access to quality food (food security) and the development of a food community at the urban level.		
Networks for local sustainable development	Public food procurement as a set of tools for the re-territorialization of agriculture and the shift towards healthy and sustainable diets	Activation of networks of producers and consumers through innovative economic forms (CSA) and recognition of fair wages	Collaboration on the issues of access to resources (for example water) and how to increase citizen participation (stakeholder engagement).	Development of thematic projects such as the Public Food Procurement, the social and solidarity economy, access to resources (primarily land), and training in agriculture.	Development of reasonings and proposals concerning the need to train the population to adopt healthier and more balanced diets, working through the Public Food Procurement	

6. Conclusions

In this contribution, the authors offer a critical analysis of the ongoing process, highlighting the salient passages, strengths, critical points, and above all the future prospects in terms of taking charge of the food question by a city such as Rome, which shows ample room for improvement on the front of integrated and coordinated public policies in support of a more sustainable food system. The variety of practices that animates the Roman context affirms that the city has already started reconsidering food towards more innovative production, distribution, and consumption models, in line with the principles of sustainability and resilience. In fact, three factors intervene to support this hypothesis: (i) Rome is characterized by many experiences related to sustainable food. However, at the political level there is as yet no defined strategic vision and direction, with the risk that such initiatives lose the ability to accompany the transition to sustainable food systems. (ii) There is an interesting mix of bottom-up initiatives and sector institutional tools/incentives/actions. However, these two worlds are often unrelated and lack connections, spaces for debate, and political coordination. (iii) There is an agricultural mosaic of considerable value, but it is not adequately supported. The most pressing challenges include the fragmentation of the agricultural landscape and the fragility of urban markets. Some considerations emerge from the critical analysis of the ongoing participatory process, which also reflect some of the opinions expressed by the stakeholder group during the meetings that began in October 2018. Today, an effort is needed to try to hold together two models, that of alternative production and marketing forms [52] and that of conventional practices related to large-scale distribution, and work towards rationalization—which does not mean trivialization or simplification, but rather the recognition of diversity—of the Roman food system. The system must necessarily mediate—in a rational and critical way—between different models, having in mind that each of them brings different benefits to the community and the territory. This seems even more necessary to avoid the risk that sustainable food production and consumption practices do not remain as niches, but can be compared with the market and with more conventional models. The institutionalization of an urban food policy is a fundamental step to harmonize the multiple instances that come from a fragmented agro-food system and at the mercy of increasingly fluid and financial markets. As recognized in research on the Local Food Policy Groups in Bristol and London [53], local administration support helps to integrate sustainable food-related objectives within policy areas that are under the mayor's direct control. At the same time, mayoral support provides status and legitimacy to food issues. Nevertheless, institutionalizing a Food Council can also come with some downsides: the main risk is being too closely associated with one political personality. Political differences within local government structures can also be an issue. In London, for instance, boroughs (local governments within the city) that were led by a different political party to that of the mayor may have been less willing to cooperate over food. The management of the food issue should be a duty on the part of the Roman administration, which, having signed the MUFPP, has thus far only put in place some sporadic institutional initiatives for the sustainability of the food system, in most cases guided and financed by participation in European projects. As stated in a note from the FAO Working Group "Strengthening Urban Rural Linkages Through City Region Food Systems", "[...] initiatives rooted in multi-stakeholder participation will be more effective in reaching their objectives, more transparent/accountable in the use of resources, and to ensure longer-term sustainability beyond temporary government administrations. The processes of dialogue and discovery of common interests across urban and rural landscapes will have to be institutionalized in order to last, and to institutionalize governance structures and mechanism will need policy support, resources, and capacity building". The local group of actors who are determining their overall strategy must reflect upon the pros and cons of the space in which the food policy group is to operate. What are the benefits of their institutional home? What are the drawbacks? How can they work around these? How can food policy councils take their ability to survive political cycles very seriously? Being aware that the process for a food policy in Rome is at its early stage, the reported analysis suggests some reflections. On the one hand stands the importance of the role of movements from the bottom—above all for their ability to take action and work regardless of the duration of the electoral mandates of the administrations, too short compared to an effective transition of the food systems. On the other hand is the need to institutionalize food policies to provide adequate support in terms of resources, political support, and the spread of consent.

Author Contributions: G.M. drafted the first version of the paper, designed the analysis, created the tables and figures and acquired the data. D.M. conceptualized the research, designed the tables and figures, reviewed the paper and supervised the work. Both the authors have approved the submitted version and agree to be personally accountable for the author's own contributions and for ensuring that questions related to the accuracy or integrity of any part of the work, even ones in which the author was not personally involved, are appropriately investigated, resolved, and documented in the literature. All authors have read and agreed to the published version of the manuscript.

Funding: The authors declare no conflict of interest.

Conflicts of Interest: The research presented in the paper did not receive any funding.

References

1. Willett, W.; Rockström, J.; Loken, B.; Springmann, M.; Lang, T.; Vermeulen, S.; Garnett, T.; Tilman, D.; DeClerck, F.; Wood, A.; et al. Food in the Anthropocene: The EAT–Lancet Commission on healthy diets from sustainable food systems. *Lancet Comm.* **2019**, *393*, 447–492. [CrossRef]
2. IPCC. *Special Report on Climate Change, Desertification, Land Degradation, Sustainable Land Management, Food Security, and Greenhouse Gas Fluxes in Terrestrial Ecosystems (SRCCL)*; Retrieved August 9; IPCC: Geneva, Switzerland, 2019.
3. Allen, T.; Prosperi, P.; Cogill, B.; Flichman, G. Agricultural biodiversity, social-ecological systems and sustainable diets. *Proc. Nutr. Soc.* **2014**, *73*, 498–508. [CrossRef]
4. FAO; WFP; IFAD. The State of Food Insecurity in the World. How does international price volatility affect domestic economies and food security? In *Food Additives and Contaminants–Part A Chemistry, Analysis, Control, Exposure and Risk Assessment*; FAO: Rome, Italy, 2011; Volume 33.
5. FAO. *The Future of Food and Agriculture–Alternative Pathways to 2050*; Food and Agriculture Organization of the United Nations: Rome, Italy, 2018; p. 224.
6. United Nations, Department of Economic and Social Affairs, Population Division. *World Urbanization Prospects: The 2018 Revision*; Online Edition; United Nations: New York, NY, USA, 2019.
7. FAOSTAT. Food and Agriculture Organization Corporate Statistical Database. Available online: http://www.fao.org/faostat/en/#home (accessed on 10 October 2019).
8. Hiç, C.; Pradhan, P.; Rybski, D.; Kropp, J.P. Food Surplus and Its Climate Burdens. *Environ. Sci. Technol.* **2016**, *50*, 4269–4277. [CrossRef]
9. Morgan, K.; Sonnino, R. The urban foodscape; world cities and the new food equation. *Camb. J. Reg. Econ. Soc.* **2010**, *3*, 209–224. [CrossRef]
10. Sonnino, R. The new geography of food security: Exploring the potential of urban food strategies. *J. Geoaphical* **2009**, *182*, 190–200. [CrossRef]
11. Steel, C. *Hungry City. How Food Shapes Our Lives*; Chatto & Windus: Random, UK, 2009.
12. Marino, D.; Mastronardi, L.; Giannelli, A.; Giaccio, V.; Mazzocchi, G. Territorialisation dynamics for Italian farms adhering to Alternative Food Networks. *Bull. Geogr. Socio Econ. Ser.* **2018**, *40*, 113–131. [CrossRef]
13. Mastronardi, L.; Romagnoli, L.; Mazzocchi, G.; Giaccio, V.; Marino, D. Understanding consumer's motivations and behaviour in alternative food networks. *Br. Food J.* **2019**, *121*, 2102–2115. [CrossRef]
14. Viljoen, A.; Wiskerke, H. (Eds.) *Sustainable Food Planning: Evolving Theory and Practice*; Wageningen University Press: Wageningen, The Netherlands, 2012.
15. Forster, T.; Hussein, K.; Mattheisen, E. City Region Food Systems: An inclusive and integrated approach to improving food systems and urban-rural linkages. *RUAF Urban Agric. Mag.* **2015**, *29*, 8–11.
16. Blay-Palmer, A. The Canadian pioneer: The genesis of urban food policy in Toronto. *Int. Plan. Stud.* **2009**, *14*, 401–416. [CrossRef]
17. Sonnino, R.; Tegoni, C.L.S.; De Cunto, A. The challenge of systemic food change: Insights from cities. *Cities* **2019**, *85*, 110–116. [CrossRef]
18. Dubbeling, M. *Linking Cities on Urban Agriculture and Urban Food Systems*; CityFood Project: Bonn, Germany, 2013.

19. IPES-Food. *What Makes Urban Food Policy Happen? Insights from Five Case Studies*; International Panel of Experts on Sustainable Food Systems: Brussels, Belgium, 2017.
20. Marino, D.; Mazzocchi, G. *La Pianificazione Alimentare: Concetti e Modelli*; Rete Rurale Nazionale 2014–2020 Working Paper; Fondo europeo agricolo per lo sviluppo rurale: Roma, Italy, 2019.
21. Pothukuchi, K.; Kaufman, J.L. Placing the food system on the urban agenda: The role of municipal institutions in food systems planning. *Agric. Hum. Values* **1999**, *16*, 213–224. [CrossRef]
22. Marino, D.; Cavallo, A. Agricoltura, cibo e città: Verso sistemi socioecologici resilienti. *CURSA pas (SAGGI)* **2014**, 14. [CrossRef]
23. Sereni, E. *Storia del Paesaggio Agrario Italiano*; Laterza: Roma/Bari, Italy, 1961.
24. WHO. *Global Nutrition Policy Review 2016-2017: Country Progress in Creating Enabling Policy Environments for Promoting Healthy Diets and Nutrition*; World Health Organization: Geneva, Switzerland, 2018.
25. Ruini, L.F.; Ciati, R.; Pratesi, C.A.; Marino, M.; Principato, L.; Vannuzzi, E. Working toward Healthy and Sustainable Diets: The "Double Pyramid Model" Developed by the Barilla Center for Food and Nutrition to Raise Awareness about the Environmental and Nutritional Impact of Foods. *Front. Nutr.* **2015**, *2*. [CrossRef]
26. Campos, S.; Madureira, L. Can Healthier Food Demand be Linked to Farming Systems' Sustainability? The Case of the Mediterranean. *Diet. Int. J. Food Syst. Dyn.* **2019**, *10*, 262–277.
27. Carrera-Bastos, P.; Fontes-Villalba, M.; O'Keefe, J.; Lindeberg, S.; Cordain, L. The western diet and lifestyle and diseases of civilization. *Res. Rep. Clin. Cardiol.* **2011**, *2*, 15–35. [CrossRef]
28. Ferraresi, G. Scenari nel territorio postfordista: Da consapevolezza a responsabilità di territorio per l'attivazione della società civile. In *Scenari Strategici: Visioni Identitarie per il Progetto di Territorio*; Magnaghi, A., Ed.; Alinea: Firenze, Italy, 2007.
29. Magnaghi, A. *Il Progetto Locale*; Bollati Boringhieri: Torino, Italy, 2000.
30. Fanfani, D. La bioregione urbana come forma e progetto della coevoluzione fra dominio urbano e rurale. In Proceedings of the XVII Conference of Italian Society of Urban Planners, Milan, Italy, 15–16 May 2014.
31. Magnaghi, A. *La Bioregion Urbaine. Petit Traité sur le Territoir Bien Commun*; Eterotopia: Paris, France, 2014.
32. Bocchi, S.; Bellingeri, D.; Galli, A. Classification and land evolution in the South Milan Agricultural Park. In Proceedings of the Analysis of Multi-Temporal Remote Sensing Images—1st International Workshop on Multitemp, Trieste, Italy, 14–16 September 2001; pp. 14–16.
33. Provincia di Pisa. *Piano del Cibo–Atto Politico di Indirizzo*; Provincia di Pisa: Pisa, Italy, 2010.
34. Di Iacovo, F.; Brunori, G.; Innocenti, S. Le Strategie Urbane: Il Piano del Cibo. *Agriregionieuropa* **2013**, *9*, 9.
35. Dansero, E.; Pettenati, G.; Toldo, A. *The Atlas of Food. A Space of Representation, a Place for Policies, a Methodology of Territorial Analysis*; FAO: Roma, Italy, 2015.
36. Mazzocchi, G.; Marino, D. Food Policy Councils as Levers for Local Development: The case of Castel del Giudice, Italy. *Urban Agric. Mag.* **2019**, *36*, 40–41.
37. Vulcano, G.; Ciccarese, L. *Spreco Alimentare: Un Approccio Sistemico per la Prevenzione e la Riduzione Strutturali*; ISPRA–Istituto Superiore per la Protezione e la Ricerca Ambientale, Rapporti 279/2018; ISPRA: Roma, Italy, 2018.
38. Mikkelsen, B.E. Images of foodscapes: Introduction to foodscape studies and their application in the study of healthy eating out-of-home environments. *Perspect. Public Health* **2011**, *131*, 209–216. [CrossRef]
39. Messina, S.; Marino, D. La pastorizia urbana, un ossimoro? *Agriregionieuropa* **2016**, *12*, 44.
40. Barbera, G.; Biasi, R.; Marino, D. *I Paesaggi Agrari Tradizionali: Un Percorso per la Conoscenza*; Franco Angeli Editore: Milano, Italy, 2014.
41. Marino, D. (Ed.) *Agricoltura Urbana e Filiere Corte. Un Quadro Della Realtà Italiana*; Franco Angeli: Milano, Italy, 2016.
42. Cavallo, A.; Di Donato, B.; Guadagno, R.; Marino, D. The agriculture in Mediterranean urban phenomenon: Rome foodscapes as an infrastructure. In Proceedings of the 6th Aesop Sustainable Food Planning Conference, Leeuwarden, The Netherlands, 5–7 November 2014.
43. Pellegrino, D.; Marino, D. Il Parco agricolo di Casal del Marmo. Un sistema ecologico, agricolo e alimentare resiliente per Roma. Available online: https://agriregionieuropa.univpm.it/it/content/article/31/44/il-parco-agricolo-di-casal-del-marmo-un-sistema-ecologico-agricolo-e (accessed on 10 October 2019).
44. Lupia, F. *Mappatura Spaziale Dell'agricoltura Urbana. Analisi di Alcune Esperienze Realizzate con Strumenti di Web-Mapping*; Quaderni INEA: Rome, Italy, 2014.
45. Insolera, I. *Roma moderna, da Napoleone al XXI Secolo*; Einaudi: Torino, Italy, 2011.

46. Bianchi, D.; Zanchini, E. (Eds.) *Ambiente Italia 2011, Il Consumo di Suolo in Italia*; Edizioni Ambiente: Milano, Italy, 2011.
47. Roma Capitale—Dipartimento Trasformazione Digitale, ISPRA. *Il consumo di suolo di Roma Capitale. Analisi della copertura di suolo e delle aree di pericolosità idraulica nel territorio di Roma Capitale*; ISPRA: Roma, Italy, 2018.
48. Terra! Onlus. *Magna Roma: Perché nel Comune Agricolo più Grande d'Italia i Mercati Rionali Stanno Morendo*; Terra! Onlus: Roma, Italy, 2018.
49. Città Metropolitana di Roma Capitale. *Approvazione del Documento di indirizzo del Piano Strategico della Città Metropolitana di Roma Capitale*; Law No. 111; Città Metropolitana di Roma Capitale: Roma, Italy, 1 June 2016.
50. Mazzocchi, G.; Marino, D. The value of Farmers' Markets for the territory and the community: The case of Campi Aperti Alternative Food Network (Italy). In Proceedings of the 3rd International Conference on Agriculture and Food in an Urbanizing Society, Porto Alegre, Brasil, 17–21 September 2018.
51. Baritaux, V.; Billion, C. Rôle et place des détaillants et grossistes indépendants dans la relocalisation des systèmes alimentaires: Perspectives de recherche. *Revue de L'organisation Responsible* **2018**, *13*, 17–28. [CrossRef]
52. Torquati, B.; Giacchè, G.; Marino, D.; Pastore, R.; Mazzocchi, G.; Ninño, L.; Arnaiz, C.; Daga, A. Urban farming opportunities: A comparative analysis between Italy and Argentina. Int. Symp. On Greener Cities for More Efficient Ecosystem Services in a Climate Changing World. *Acta Hortic.* **2018**. [CrossRef]
53. Halliday, J.; Barling, D. The Role and Engagement of Mayors in Local Food Policy Groups: Comparing the Cases of London and Bristol. *Adv. Food Secur. Sustain.* **2018**, *3*, 177–209.

International Journal of
Environmental Research and Public Health

MDPI

Article

Toward Livestock Supply Chain Sustainability: A Case Study on Supply Chain Coordination and Sustainable Development in the Pig Sector in China

Ni Zhuo [1,2] and Chen Ji [1,2,*]

1 Department of Agricultural Economics and Management, School of Public Administration, Zhejiang University, Hangzhou 310058, China
2 China Academy for Rural Development (CARD), Zhejiang University, Hangzhou 310058, China
* Correspondence: jichen@zju.edu.cn

Received: 30 June 2019; Accepted: 31 August 2019; Published: 4 September 2019

check for updates

Abstract: Stricter environmental regulations on livestock production pollution have changed the sustainable practices of livestock supply chain stakeholders. By adopting three cases in China's livestock supply chain, this study explores how supply chain coordination facilitates sustainable development of livestock production in China. It is found that close supply chain coordination and the capabilities of the core companies jointly contribute to supply chain sustainability. Thus, this research has theoretical significance in explaining the roles of supply chain coordination and core company capabilities in driving supply chain sustainability, which is not completely understood thus far. This study also has practical implications for livestock supply chain stakeholders and the government in terms of improving supply chain sustainability via closer supply chain coordination and enhancing the capabilities of the core companies involved.

Keywords: supply chain coordination; supply chain sustainability; livestock industry

1. Introduction

In recent years, the Chinese government has enacted and implemented a series of environmental regulations to improve the country's environmental conditions. In October 2013, the state council issued a regulation on environmental pollution produced by the livestock sector (hereafter called the "Regulation"). This is considered the strictest environmental control regulation in the livestock sector. Therefore, there is an emerging need for supply chain members to cope with the Regulation to achieve sustainable development in the livestock sector.

A sustainable supply chain performance refers to traditional measures of profit and loss, as well as an expanded conceptualization of performance that includes social and natural dimensions [1]. It is commonly agreed that the sustainable outcome of a supply chain should include sustainable economic performance [2], social performance [3], and environmental performance [4]. This is known as the triple bottom line in supply chain sustainability literature.

Current studies have discussed the factors that may drive sustainability such as climatic change [5], uncertainties [6], institutional changes [7], governance structures [8], and transparency in the food supply chain [9]. However, discussion on potential contributions from supply chain coordination remains limited. Supply chain coordination is a theme that receives wide attention in supply chain management studies, and its outcome includes relationship satisfaction [10], food safety [11], and enhancement of supply chain performance [12]. Supply chain sustainability as an outcome of supply chain coordination has not captured sufficient attention [13]. Through a literature review, it was found that supply chain members may work together to obtain capabilities at the supply chain level, and the capabilities may contribute to the sustainable performance of the supply chain [3].

Therefore, by adopting three cases in the pork supply chain in China, this study first tries to understand how supply chain members in the pig sector work together to build a more sustainable pork supply chain. Then, the study explores the relationship between supply chain coordination and supply chain sustainability by providing empirical evidence from the pig sector in China. At the same time, this study tries to reveal the underlying mechanism using dynamic capabilities theory.

2. Literature Review

2.1. Agrifood Supply Chain Sustainability

An increasing population, deficiency of natural resources, and climate change require organizations in the agrifood sector to redesign their current supply chains to involve social, economic, and environmental perspectives in their performance [14]. Crowder and Reganold pointed out that sustainability is particularly relevant for the agrifood industry because sustainability contributes to agrifood companies in gaining competitive advantages in the market [15]. This is thought to be closely associated with quality improvement. Firms that adopt self-regulation in sustainability practices are able to better anticipate regulation changes from the government [16]. Thus, these firms can voluntarily improve traceability and accountability across the supply chain, and ensure consistent quality and a reliable supply [17].

However, the issues of supply chain sustainability in agrifood industries have not been adequately empirically studied [18]. Current studies have primarily attempted to understand the environmental sustainability practices in developed countries using life cycle analysis (LCA) methods [19], while empirical evidence is lacking in emerging economies [20]. From the literature review, it was found that on the one hand, scholars have indicated that environmental practices should be performed by different stakeholders of the supply chain so that sustainability of the agrifood chain will be achieved [21]. On the other hand, scholars have agreed on some dimensions to evaluate the sustainable performance of supply chains. For example, environmental sustainability is usually measured by on-farm and off-farm environmental protection practices [22], lower emissions of pollutants [23], adoption of organic fertilizer [24], use of third-party certification [25], and other instruments [26]. The social dimension of sustainable development involves human rights [25], an increase in social welfare [27], the development of local communities [28], and customer health and safety [29]. Economic performance includes productivity, financial outcomes, and customer satisfaction [30].

2.2. Agrifood Supply Chain Coordination

Supply chain coordination is not a new concept in supply chain management. It refers to a collection of formal or informal institutional relationships among supply chain members [31]. The coordinated supply chain relationships in the food sector that have been most widely reported in the literature include market relations, short-term contracts, long-term contracts, joint ventures, strategic alliances, and vertical integration [32]. There are several outcomes of supply chain coordination mentioned in the literature, including the relationships between supply chain coordination and supply chain performance [33], supply chain coordination and supply chain relationship satisfaction [34], and supply chain coordination and food safety [11].

The relationship between supply chain coordination and sustainability has been increasingly discussed in the recent literature. Evidence has shown that business-to-business collaboration in the food industries leads to improvements in economic, environmental, and social standards through effective communication between farmers and traders [35]. It is proposed that the interplay of collaboration behaviors will influence agrifood supply chain sustainability, while empirical evidence is not provided [13]. The mechanism of collaboration will influence supply chain sustainability and needs more investigation. Theories such as the dependency theory and transaction cost economy theories can be adopted [36].

2.3. Dynamic Capabilities of Organizations

Helfat and Winter defined dynamic capacity as the capacity of an organization to purposefully create, extend, and modify its resource base [37]. Dynamic capability theory is rooted in the resource-based view (RBV) [38], which posits that firms combine bundles of valuable, rare, inimitable, and non-substitutable resources in an effort to gain or maintain a competitive advantage [39]. Tsai et al. stated that dynamic capability is a factor that influences organizational sustainability [40]. Dynamic capability theory tackles this challenge by reasoning that organizations consistently operating in dynamic environments create and recombine their resources in new ways [2].

In the agrifood supply chain, focal companies usually drive the main changes in the supply chain, such as supply chain integration and food safety certification adoption. These companies were asked to consider the environmental and social problems present in their entire supply chain [41].

2.4. Relationships Among Supply Chain Coordination, Dynamic Capabilities of Organization, and Supply Chain Sustainability

Supply chain members jointly learn how to build capabilities through learning loops for innovations that will help the entire supply chain become more sustainable [13]. Focal companies are supply chain members that provide leadership for the supply chain, specify supply chain policies to other members, and exercise control over the supply chain's decisions and activities [42].

Soylu et al. pointed out that supply chain collaboration is a common way for companies throughout the supply chain to share information, make strategic alliances, and reduce overall costs, as well as to improve supply chain sustainability [43]. Azevedo et al. pointed out that collaboration is seen as necessary to develop, apply, and establish new innovative ideas and practices, and is linked to the socioeconomic and ecological sustainability of the agrifood system [44]. Chen et al. stated that one of the reasons that companies perform supply chain collaboration is to achieve sustainability [13]. This includes collaboration with suppliers, customers, competitors, and other organizations. This will enrich the companies' resources and enhance their capability for achieving an improved sustainable performance.

Therefore, the research question of this study is:

RQ: How is supply chain sustainability jointly achieved through building the supply chain coordination and dynamic capabilities of the core companies?

3. Data and Methodology

3.1. Case Study Method

According to Yin [45], a case study design should be considered when (a) the focus of the study is to answer "how" and "why" questions, (b) you cannot manipulate the behavior of those involved in the study, (c) you want to cover contextual conditions because you believe they are relevant to the phenomenon under study, or (d) the boundaries are not clear between the phenomenon and context. A multiple-case study enables the researcher to explore differences within and between cases. The goal is to replicate findings across cases. The researcher can predict similar results across cases or predict contrasting results based on a theory [45]. Therefore, a multiple-case method is appropriate to be adopted in this study because there is a "how" research question that requires exploration of the mechanism of livestock supply chain sustainability.

Three different supply chain modes in the pig sector in China were selected for study: (1) "Vertical integration" mode (Mu'yuan Food Co., Ltd.), (2) "company + local family-owned commercial farms" mode (Wen'shi Food Co., Ltd. (Huai'an)), and (3) "cooperative + farmers + company" mode (Chun'ran Agri-Food Tech Co., Ltd.). Table 1 provides the basic information of the three cases. These three cases all undertake supply chain coordination to achieve sustainable development. To do this research, we obtained firsthand data from field studies and secondhand data from the official websites of the companies.

Table 1. Case description.

Name	Location	Year Established	Number of Interviewees
Mu'yuan	Nan'yang, Henan	1992	4
Wen'shi (Huaiyin)	Huai'an, Jiangsu	2006	6
Chun'ran	Long'you, Zhejiang	2010	5

3.2. Data Collection

We conducted 15 interviews (four interviewees from Mu'yuan, six interviewees from Wen'shi (Huai'an), and five interviewees from Chun'ran) in total from May 2018 to April 2019. To ensure a comprehensive view was captured, the interviewees included all managers of the three companies, the presidents of cooperatives, some core members of the local family-owned commercial farms, and some farmers. All of them were interviewed twice or more.

Research instruments included face-to-face semi-structured interviews lasting 60–120 mins per interview and archival data from the official websites of the companies or from direct observation. Our interview protocol addressed the following key issues: (1) The companies' profiles, (2) their supply chain coordination, and (3) their sustainable development.

We carried out at least two rounds of data collection/field visits for each company (Table 1). Generally, in the first round of data collection, we tried to understand the supply chain modes of the case companies. In the second round of field visits, we collected data with regard to supply chain coordination and sustainable development. The field researchers (coauthors) visited each of the three companies. For each visit, the field researchers recorded field notes based on direct observations of the case company's operations.

3.3. Data Analysis

The study adopted case analysis and cross-case analysis to analyze the data. It carried out within-case analysis first for each of the three companies. Each field researcher individually coded the data, and then the coded data between the coauthors were compared to ensure consistency and inter-coder reliability. Disagreements were resolved along the way.

Within-case analysis gained a broad understanding of the business models of each case company and the supply chain coordination and sustainable development of the cases. Cross-case analysis was performed and the findings tabulated to identify the mechanism of how companies in the pork supply chain in China achieve sustainable development with supply chain coordination.

4. Descriptions of Three Case Companies

4.1. Mu'yuan Co., Ltd.

Mu'yuan Co., Ltd., was founded in 1992. It is a top-listed agricultural company in China. Its businesses include feed production, pig production, and pig slaughtering. The feed production volume of Mu'yuan reaches 5 million tonnes annually, and pig production reaches 10 million heads annually. The number of pigs slaughtered reaches 1 million heads. The supply chain of Mu'yuan is integrated: The company manages the feed production, pig production, and slaughtering itself. There are branches of Mu'yuan across nine provinces in the country. Thus, Mu'yuan effectively handles pig production safety. Mu'yuan puts emphasis on the recycling economy. The company has developed a way to use production wastes as fertilizers for planting, thus driving the surrounding farmers to vigorously develop an ecological agriculture.

4.2. Wen'shi Co., Ltd.

Wen'shi Food Co., Ltd., started pig production in Jiangsu Province in 2006. Relying on its "company + local family-owned + commercial farms" mode, Wen'shi provides standardized pigs to local family-owned commercial farms, and shares profits with them so that farmers can ensure the safety and quality of pig production. Thus, Wen'shi has successfully rooted this innovative mode into the Southeast area. Taking Jiangsu province as an example, there are eight regional subcompanies in the province with more than 3000 staff and nearly 200 local family-owned commercial farms. Wen'shi puts great effort into practicing sustainable pig production, and it leads the local farms in changing traditional ways of pig farming.

4.3. Chun'ran Agri-Food Tech Co., Ltd.

Long'zhu cooperative (hereafter Long'zhu) was founded in 2010 with a group of 36 farmer members in Qu'zhou city, Zhejiang province. The initial purpose of establishing the cooperative was to purchase feed together from Ke'sheng, which is a local feed company. With the development of Long'zhu, the president (Mr. Zhao Chungen, hereafter Zhao) found that market price fluctuations made the profits of farmers uncertain, so he decided to create downstream integration to generate a profit premium for farmer members. Zhao started Chun'ran Agro-Tech Co., Ltd., and registered the pork brand Jiu'hao'mu'chang to sell premium quality pork to the marketplace. (Long'zhu collaborated with the Animal Science School of Zhejiang University to create a type of feed that includes tea leaves, which makes the pork taste better; Zhao made the pork a premium product in the marketplace.) Zhao also started a pig production waste company (called Kai'qi Energy Co., Ltd.) to help member farmers avoid risks from the changing environmental policies of the government.

5. Cross-Case Analyses

5.1. Supply Chain Coordination of Case Companies

In this section, the supply chain coordination of the three companies is described (see Table 2).

Table 2. Supply chain coordination.

Name	Form of Supply Chain Coordination Mode	Main Activities
Mu'yuan	"Vertical integration"	Employ farmers to produce pigs for them on farms of Mu'yuan
Wen'shi	"Company + local family-owned commercial farms"	Sign contracts to share profits and risk with locally contracted farmers
Chun'ran	"Cooperative + farmers + company"	Purchase feed products together from Ke'sheng feed company

5.1.1. Supply Chain Coordination of Mu'yuan

Mu'yuan is a traditional pig production company, but it integrates both upstream and downstream stakeholders. It has a very stable relationship with producers because farmers produce pigs as required by Mu'yuan, and they abide by all safety standards dictated by Mu'yuan. In fact, the upstream farmers are "workers" for Mu'yuan, they do not actually own their farms or produce pigs for themselves. Mu'yuan employs the farmers to produce pigs for them on farms owned by Mu'yuan, and the farmers' income is not associated with the number of pigs sold.

5.1.2. Supply Chain Coordination of Wen'shi

Wen'shi has a close relationship with local family-owned commercial farms. To be more specific, Wen'shi chooses to partner with local family-owned commercial farms, which produce between 1000 to 170,000 heads annually. Wen'shi and its partnering farms sign contracts to share profits and share risk.

Contracted local farms receive breed, feed, and other inputs from Wen'shi for free, and they establish production farms according to the instructions of Wen'shi. Wen'shi monitors all production processes of its contracted farms in order to ensure food safety.

Wen'shi shields its farmers from market risks, and it pays for the farmers to fatten pigs over 155 days based on three categories of quality: I, II, and III. The price tiers are 14 RMB/kg for quality class I, 12 RMB/kg for quality class II, and price to be determined for quality class III. In this way, Wen'shi encourages the farmers to produce pigs of high quality. Wen'shi ensures that the farmers receive profits by adjusting the purchasing price to the market price. The costs of breaching a contract for farmers and for Wen'shi are both high. If farmers breach a contract, they will need to pay all production costs plus 20% of the value of a class-I pig. If Wen'shi breaches a contract, it pays the farmers at least US $335 plus a value adjustment for the market.

5.1.3. Supply Chain Coordination of Chun'ran

Chun'ran manages the supply chain coordination with farms through the Long'zhu cooperative. There are 55 cooperative member farmers in the Long'zhu cooperative, and the primary link between the cooperative and Chun'ran is the collaborative purchase of feed products in order to achieve economies of scale. Mr. Zhao Chungen aims to sell premium pork to the market, and thus 10% of the cooperative farms have a market relationship with Chun'ran.

5.2. *Dynamic Capabilities of Organizations*

In this section, the dynamic capabilities of the three companies is described (see Table 3).

Table 3. Dynamic capabilities of focal case companies.

Name	Capability in Disposing Wastes	Capability in Establishing Sustainable Production Base	Capability in Prevention of Epidemic Diseases
Mu'yuan	Organic fertilizer and methane	Ventilation system; Humidity device; Pig house heating and insulation device	A combined use of domestic and imported vaccination; Dead pig bodies used as organic fertilizer and biodiesel
Wen'shi	Urban sewage treatment system is applied; Centralized liquid spray system; vertical fermentation tank	Leaking plate; Urinary septic tank; Fermentation bed	Free training classes to local partner farmers; Strict sterilization system
Chun'ran	Kai'qi Energy to recycle wastes into fertilizer and electricity	Tea leaf feeds for pigs; Organic fertilizer for tea trees, which absorbs the smell of feces and urine	Technological support from universities; A specialized veteran team; Withdrawal time of drugs

5.2.1. Capability of Disposing Wastes of Three Companies

Regarding the capability of disposing of production wastes, Mu'yuan enjoys natural advantages because the lands in Henan province are almost plain and the weather is suitable to producing crops. The Research and Development group of Mu'yuan converts the production wastes into organic fertilizer, which is used for crop production, and methane, which is used for some farm management utilities. Thanks to the dry weather in Henan province, the organic fertilizer is not easily washed away.

Wen'shi pays attention to the treatment of liquid wastes, gas wastes, and residue wastes. The solution to liquid wastes is to use an urban sewage treatment system that can separate open ditches and underground pipes so that rainwater does not mix with liquid waste. The solution to gas waste is to use a centralized liquid spray system to cover the gas waste so that the transmission distance of gas waste is largely reduced. The waste residue is directly returned to the field by using a vertical fermentation tank for fermentation treatment.

Chun'ran invested US $2.9 million to start a pig production waste recycling company (Kai'qi Energy), which helps cooperative member farmers to dispose of pig production waste. All member farmers transport their production waste to Kai'qi Energy Co., Ltd. This disposal service is free, so member farmers need not worry about the cost of waste disposal. In addition, the energy produced after the waste process can be used by the farmers because Chun'ran inputs the electricity of Kai'qi Energy Co., Ltd., into the national electricity network.

> *"We feel proud of our recycling system, which received awards from local government and was set as an example in livestock waste recycling. With this plant, our member farms were successfully shielded from stricter environmental policy risks, and our cooperative became even more prosperous,"* said Zhao.

5.2.2. Capability of Establishing Sustainable Production Base

Mu'yuan has created its own production environment management system, which consists of three sections. The first is the ventilation system. A siphoning system in the production house helps with air circulation, and the air inlet window together with the exhaust window strengthens the ventilation. The second section is a device to ensure humidity, including wire rope and a wet sack on the wire rope. These help to add humidity to the pig house. The third section is a pig house heating and insulation device, which includes a main outlet pipe and main return pipe. A water heater is connected to the main outlet pipe and main return pipe. The main outlet pipe is connected to multiple thin heat pipes, the heat pipes are laid parallel to each other on the back of the pig net bed, and the heat is cooled. The other end of the tube is connected to the main return pipe.

Wen'shi has built its own production house. The pig house is designed with a diffused ventilation system and is kept at a constant temperature of 25 °C. It is equipped with a leaking plate. The urine and feces can be leaked through the plates. A urinary septic tank is placed under the plates. The feces is softened by the urine and then transferred to a fermentation bed. The fermentation bed is rolled continuously so that the heat produced by fermentation is disseminated. Strains are added simultaneously so that the urine and feces are converted to organic fertilizer.

Chun'ran has a special way of establishing a sustainable production base. Researchers from universities found that adding tea leaf extracts to feed can help reduce the odor of pork, which is deemed unpleasant by some Chinese consumers. Thus, Chun'ran collaborates with researchers from universities in Zhejiang and the Ke'sheng feed company to add tea leaf extracts to feed for pigs. The tea trees are planted outside the pigpens; the practice has two advantages in terms of sustainability. First, the production waste can be used as fertilizer for those tea trees after proper disposal. Second, tea trees can absorb the smell of feces and urine excreted by the pigs to a certain extent.

5.2.3. Capability in Prevention of Epidemic Diseases

Mu'yuan is good at examining vaccination, and they are strict with vaccination quality when making purchasing decisions. The domestic vaccination market is mixed with high-quality and low-quality products. Mu'yuan uses a combination of domestic and imported vaccinations in order to ensure that the pigs are well protected. Dead pig bodies are crushed completely, and after high-temperature sterilization, they are made into organic fertilizer and biodiesel.

Wen'shi established Wen'shi College, which provides free training classes to local partner farmers. In addition, Wen'shi assigns professional veterans to live near the family-owned farms in case of emergency. There is a strict sterilization system on the pig farm: Whenever people enter a farm, everything should be sterilized. In order to make farmers use vaccinations in the correct way, Wen'shi removes the drugs from farmers when pigs reach 140 days because the pigs need a period of withdrawal from the drugs.

> *"Every time when people exit and enter a farm, we require them to sterilize, wash hands, and put on pasteurized clothes. The veteran personnel of Wen'shi provides us with constant guidance on how to protect the farm from being influenced by epidemic disease, and we have been disease-free for at least six months,"* said Jin Yunhao, a core family-owned commercial farmer of Wen'shi.

Long'zhu cooperative members are large-scale producers, so they have gained many skills and experience in raising pigs. Chun'ran emphasizes collaboration with universities in Zhejiang province. Based on the technological support from universities, Chun'ran formed a specialized veteran team as consultants for cooperative member farmers so that the farmers constantly receive professional assistance.

5.3. Supply Chain Sustainability

Through close coordination with—and education from—the core companies, farmers also achieve sustainable performance. Based on the literature review, this study adopts three dimensions to describe supply chain sustainability: The economic performance, social performance, and environmental performance of both the farmers and the company (see Table 4).

Table 4. Supply chain sustainability.

Name	Economic Performance (Productivity and Profitability)	Social Performance (Social Welfare Improvement and Local Community Development)	Environmental Performance (Wastes Reduction and Dead Pig Treatment)
Mu'yuan	Farmers: More pigs produced in safer ways; Less exposure to market price risks; Farmers' incomes are enhanced Company: Ensure the safe production of pigs stabilize pig supply	Farmers: Gain good practice in producing high-quality pigs Company: Take great social responsibility in paying great attention to environmental protection and quality enhancing	Farmers: Collectively follow the environmental protection practices Company: The dead pigs are ground down in a machine and reproduced into organic fertilizer and biodiesel
Wen'shi	Farmers: 10% more pigs than in the same place before; Costs of utility are almost reduced by 15%; Never lose profits after contacting with Wen'shi Company: To stabilize the pig supply	Farmers: Female farmers actively participate in the management and operation of family-owned commercial farms Company: Address the food safety and quality in pig production; Pay attention to social responsibility	Farmers: Collectively follow the Wen'shi instructions Company: Use septic chemicals to dispose of dead pig bodies
Chun'ran	Farmers: More productive through knowing more good practices after joining Long'zhu; Reduced 15% of feed costs; 20% more price premium through tea-leaf pork Company: Premium pork market leader	Farmers: Know-hows in safe production Company: Long'you area is now a community that provincial livestock bureaus in Hunan and Hubei came to learn the wastes disposal management experience	Farmers: Deliver the dead pig bodies to the collection team from the local government to get them treated Company: Recycling the wastes through Kai'qi

In terms of economic performance, this study uses productivity and profitability as two dimensions. Regarding productivity, Mu'yuan farmers have enhanced their capability to produce more pigs in safer ways through coordination with Mu'yuan, and Mu'yuan has succeeded in ensuring the safe production of pigs by having closer integration with its farmers. Meanwhile, Mu'yuan has stabilized its pig supply. Wen'shi farmers have substantially reduced their production costs. By constructing the production base in a more effective way, the farmers are able to raise 10% more pigs than before, and the utility costs are reduced by almost 15%. For Chun'ran, farmers are more productive because they are familiar with more good practices (i.e., farm management and treatment of pigs that died of illness) after joining the Long'zhu cooperative.

With regard to profitability, farmers from Mu'yuan benefit through coordination, are less exposed to market price risks, and their income from producing pigs was enhanced. Mu'yuan also stabilizes its profits from having good relationships with the farmers. The Wen'shi farmers have largely increased and stabilized their income in raising pigs, and it is reported that farmers never lose revenue after collaborating with Wen'shi. Chun'ran members have reduced 15% of their feed costs, and they can charge a 20% higher price premium through the tea leaf pork they produce. Thus, their profits are increased.

In terms of social performance, this study adopts two dimensions: Social welfare improvement and local community development. As a leading pig producer in China, Mu'yuan has taken great social responsibility in paying attention to environmental protection and quality enhancements. Wen'shi also addresses food safety and quality in pig production, and encourages female farmers to participate in the management and operation of family-owned commercial farms. The recycling system of production wastes innovated by Chun'ran has wide social impacts, and staff from provincial livestock bureaus in Hunan and Hubei have come to learn the wastes disposal management experience.

Mu'yuan has contributed to the development of the local community by involving local farmers in pig production and helping them to increase their income. Wen'shi has nurtured 20 family-owned commercial farms in the Huai'an area. Chun'ran has positively contributed to local development by establishing the Long'zhu cooperative, which is a national leading cooperative in China.

In terms of environmental performance, we used waste reduction and the disposal of dead pigs from illness as two dimensions. For waste reduction, all three companies have successfully found ways to reduce production wastes, and they have achieved almost zero pollution, meaning that all wastes are recycled.

Regarding the treatment of pigs that died from illness, Mu'yuan has a very advanced method of response: The dead pigs are ground down in a machine to produce organic fertilizer and biodiesel. Mu'yuan have indicated that this treatment of dead pigs achieves zero transmission of epidemic diseases. Wen'shi has determined its own method to dispose of pigs that die from illness. The CEO (Mr. Wang' Sifeng) has reported that Wen'shi uses septic chemicals to dispose of dead pig bodies, which is environmentally sustainable. Chun'ran delivers the dead pig bodies to the collection team from the local government to have them treated.

6. Development of Pathways Toward Sustainable Livestock Supply Chain

Based on the analysis, it was found that there are several important steps that contribute to achieving livestock supply chain sustainability.

The first two important steps include building the dynamic capabilities of focal companies and forming close and stable supply chain coordination. In our study, the companies have strived to establish dynamic capabilities in sustainable development; they have also created a close and stable coordinative relationship with their upstream farmers. The third step is to enhance the dynamic capabilities at the supply chain level. In our study, the upstream farmers have enjoyed the learning process and gained benefits from coordinating with the focal companies. When farmers' sustainable practices improve, the dynamic capabilities of the entire supply chain are enhanced. Finally, the entire supply chain achieves sustainable performance in improving economic interests, social welfare, and environmental protection and recycling. The path to achieving a sustainable livestock supply chain is shown in Figure 1.

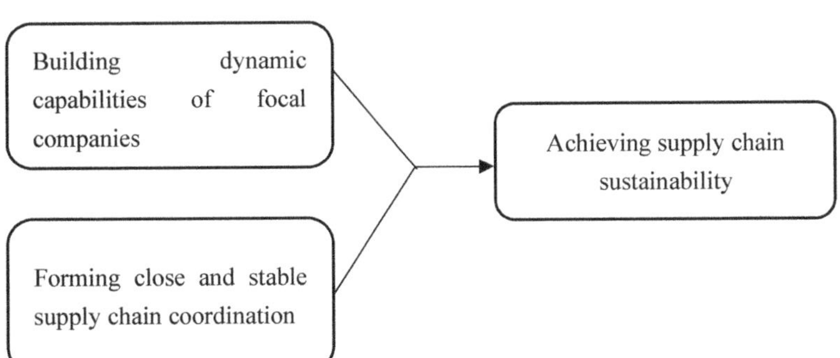

Figure 1. Pathway to achieve supply chain sustainability.

7. Conclusions

This study reached several conclusions for the sustainable development of the livestock sector in China.

First, focal companies in the Chinese livestock supply chains are important organizations in changing the sustainable practices of stakeholders in the supply chain. Companies use their resources and capabilities to innovate and drive the sustainability of the supply chain. Some notable sustainable practices that the focal companies adopted are as follows: (i) Strictly controlling the production pollution and implementation of recycling, (ii) managing epidemics of disease according to the disease environment of the farms, and (iii) recyclable disposal of the pigs that die from illness.

Second, using supply chain coordination, focal companies transfer their good sustainable practices to other stakeholders, especially to upstream farmers with whom they collaborate. Effective coordination includes vertical integration, "focal company + family-owned commercial farms," and "focal company + cooperatives + farmers." The coordination serves as a learning environment for farmers to gain sustainable development capabilities, and thus the entire supply chain becomes more sustainable.

Based on these conclusions, we present some implications of this study:

First, to encourage the focal companies to enhance their sustainable practices, core companies need to tentatively follow advanced sustainable practices in the industry. Learning and innovation are also important for companies to stay alert to new sustainable production technologies. The government can provide financial and technological support to core companies to nurture their capabilities in sustainable development.

Second, core companies in the pork supply chain in China can continue to increase and innovate supply chain coordination to improve supply chain sustainability. The coordination methods examined in this study can serve as examples for core companies which do not have supply chain coordination to adopt and test. To encourage close coordination through the pork supply chain, the government may provide training on how to coordinate through supply chain to farmers so that the willingness of farmers to join farmers' organizations increase. Core companies need to make efforts to coordinate and to constantly make this coordination efficient and effective.

Author Contributions: Conceptualization: C.J.; methodology: C.J. and N.Z.; formal analysis: N.Z. and C.J.; investigation: C.J. and N.Z.; data curation: N.Z. and C.J.; writing—original draft presentation: C.J. and N.Z.; writing—review and editing: N.Z. and C.J.; funding acquisition: C.J.

Funding: This research was funded by National Science Foundation of China Project, grant number 71403243.

Acknowledgments: The authors would like to acknowledge their appreciation to comments from the three anonymous reviewers to help improve the manuscript.

Conflicts of Interest: The authors declare no conflict of interest. The sponsors had no role in the design, execution, interpretation, or writing of the study.

References

1. Pagell, M.; Wu, Z. Building a more complete theory of sustainable supply chain management using case studies of ten exemplars. *J. Supply. Chain. Manag.* **2009**, *45*, 37–56. [CrossRef]
2. Amui, L.B.L.; Jabbour, C.J.C.; de Sousa Jabbour, A.B.L.; Kannan, D. Sustainability as a dynamic organizational capability: A systematic review and a future agenda toward a sustainable transition. *J. Clean. Prod.* **2017**, *142*, 308–322. [CrossRef]
3. Beske, P.; Land, A.; Seuring, S. Sustainable supply chain management practices and dynamic capabilities in the food industry: A critical analysis of the literature. *Int. J. Prod. Econ.* **2014**, *152*, 131–143. [CrossRef]
4. Miranda-Ackerman, M.A.; Azzaro-Pantel, C.; Aguilar-Lasserre, A.A. A green supply chain network design framework for the processed food industry: Application to the orange juice agrofood cluster. *Comput. Ind. Eng.* **2017**, *109*, 369–389. [CrossRef]
5. Vermeulen, W.J.V.; Seuring, S. Sustainability through the market: The impacts of sustainable supply chain management, introduction. *Sustain. Dev.* **2009**, *17*, 269–273. [CrossRef]
6. Wu, K.-J.; Liao, C.-J.; Tseng, M.-L.; Lim, M.K.; Hu, J.; Tan, K. Toward sustainability: Using big data to explore the decisive attributes of supply chain risks and uncertainties. *J. Clean. Prod.* **2017**, *142*, 663–676. [CrossRef]

7. Glover, J.L.; Champion, D.; Daniels, K.J.; Dainty, A.J.D. An Institutional Theory perspective on sustainable practices across the dairy supply chain. *Int. J. Prod. Econ.* **2014**, *152*, 102–111. [CrossRef]
8. Martins, F.; Trienekens, J.; Omta, O. Governance structures and coordination mechanisms in the Brazilian pork chain: Diversity of arrangements to support the supply of piglets. *Int. Food. Agribus. Man.* **2017**, *20*, 511–532. [CrossRef]
9. Trienekens, J.H.; Wognum, P.M.; Beulens, A.J.M.; van der Vorst, J.G.A.J. Transparency in complex dynamic food supply chains. *Adv. Eng. Inform.* **2012**, *26*, 55–65. [CrossRef]
10. Benton, W.C.; Maloni, M. The influence of power driven buyer/seller relationships on supply chain satisfaction. *J. Oper. Manag.* **2005**, *23*, 1–22. [CrossRef]
11. Han, J.; Trienekens, J.H.; Omta, S.W.F. Relationship and quality management in the Chinese pork supply chain. *Int. J. Prod. Econ.* **2011**, *134*, 312–321. [CrossRef]
12. Zhu, Q.; Sarkis, J.; Lai, K. Examining the effects of green supply chain management practices and their mediations on performance improvements. *Int. J. Prod. Res.* **2012**, *50*, 1377–1394. [CrossRef]
13. Chen, H.; Liu, S.; Oderanti, F. A knowledge network and mobilisation framework for lean supply chain decisions in agri-food industry. *Int. J. Decis. Support. Syst. Technol.* **2017**, *9*, 37–48. [CrossRef]
14. Banasik, A.; Kanellopoulos, A.; Bloemhof-Ruwaard, J.M.; Claassen, G.D.H. Accounting for uncertainty in eco-efficient agri-food supply chains: A case study for mushroom production planning. *J. Clean. Prod.* **2019**, *216*, 249–256. [CrossRef]
15. Crowder, D.W.; Reganold, J. Financial competitiveness of organic agriculture on a global scale. *PNAS* **2015**, *112*, 7611–7616. [CrossRef] [PubMed]
16. Ji, C.; Jia, F.; Xu, X. Agricultural co-operative sustainability: Evidence from four Chinese pig production co-operatives. *J. Clean. Prod.* **2018**, *197*, 1095–1107. [CrossRef]
17. Nadvi, K. Global standards, global governance and the organization of global value chains. *J. Econ. Geogr.* **2008**, *8*, 323–343. [CrossRef]
18. Naseer, A.M.; Ashfaq, M.; Hassan, S.; Abbas, A.; Razzaq, A.; Mehdi, M.; Ariyawardana, A.; Anwar, M. Critical issues at the upstream level in sustainable supply chain management of agri-food industries: Evidence from Pakistan's citrus industry. *Sustainability* **2019**, *11*, 1326. [CrossRef]
19. Andree, P.; Dibden, J.; Higgins, V.; Cocklin, C. Competitive productivism and Australia's Emerging 'alternative' agri-food networks: Producing for farmers' markets in Victoria and beyond. *Aust. Geogr.* **2010**, *41*, 307–322. [CrossRef]
20. Akhtar, P.; Tse, Y.K.; Khan, Z.; Rao-Nicholson, R. Data-driven and adaptive leadership contributing to sustainability: Global agri-food supply chains connected with emerging markets. *Int. J. Prod. Econ.* **2016**, *181*, 392–401. [CrossRef]
21. Bremmers, H.; Omta, O.; Kemp, R.; Haverkamp, D.J. Do stakeholder groups influence environmental management system development in the Dutch agri-food sector? *Bus. Strateg. Environ.* **2007**, *16*, 214–231. [CrossRef]
22. Aramyan, L.; Hoste, R.; van den Broek, W.; Groot, J.; Soethoudt, H.; Lan, T.; Nguyen, T.; Hermansen, J.; Van der Vorst, J. Towards sustainable food production: A scenario study of the European pork sector. *J. Ch. Netw. Sci.* **2011**, *11*, 177–189. [CrossRef]
23. Notarnicola, B.; Hayashi, K.; Curran, M.A.; Huisingh, D. Progress in working towards a more sustainable agri-food industry. *J. Clean. Prod.* **2012**, *28*, 1–8. [CrossRef]
24. Del Borghi, A.; Gallo, M.; Strazza, C.; Del Borghi, M. An evaluation of environmental sustainability in the food industry through life cycle assessment: The case study of tomato products supply chain. *J. Clean. Prod.* **2014**, *78*, 121–130. [CrossRef]
25. Peano, C.; Tecco, N.; Dansero, E.; Girgenti, V.; Sottile, F. Evaluating the sustainability in complex agri-food systems: The SAEMETH framework. *Sustainability* **2015**, *7*, 6721–6741. [CrossRef]
26. Rueda, X.; Garrett, R.D.; Lambin, E.F. Corporate investments in supply chain sustainability: Selecting instruments in the agri-food industry. *J. Clean. Prod.* **2017**, *142*, 2480–2492. [CrossRef]
27. Ahi, P.; Searcy, C. Measuring social issues in sustainable supply chains. *Meas. Bus. Excell.* **2015**, *19*, 33–45. [CrossRef]
28. Mota, B.; Gomes, M.I.; Carvalho, A.; Barbosa-Povoa, A.P. Towards supply chain sustainability: Economic, environmental and social design and planning. *J. Clean. Prod.* **2015**, *105*, 14–27. [CrossRef]

29. Saitone, T.L.; Sexton, R.J. Agri-food supply chain: Evolution and performance with conflicting consumer and societal demands. *Eur. Rev. Agric. Econ.* **2017**, *44*, 634–657. [CrossRef]
30. Aramyan, L.; Oude Lansink, A.; Van der Vorst, J.; Kooten, O. Performance measurement in agri-food supply chains: A case study. *Suppl. Chain. Manag.* **2007**, *12*, 304–315. [CrossRef]
31. Nyaga, G.N.; Whipple, J.M.; Lynch, D.F. Examining supply chain relationships: Do buyer and supplier perspectives on collaborative relationships differ? *J. Oper. Manag.* **2010**, *28*, 101–114. [CrossRef]
32. Van der Vorst, J. Effective food supply chains: Generating, modelling and evaluating supply chain scenarios. Ph.D. Thesis, Wageningen University, Wageningen, The Netherlands, January 2000.
33. Seuring, S.; Müller, M. From a literature review to a conceptual framework for sustainable supply chain management. *J. Clean. Prod.* **2008**, *16*, 1699–1710. [CrossRef]
34. Boström, M.; Jönsson, A.M.; Lockie, S.; Mol, A.P.J.; Oosterveer, P. Sustainable and responsible supply chain governance: Challenges and opportunities. *J. Clean. Prod.* **2015**, *107*, 1–7. [CrossRef]
35. Hamprecht, J.; Corsten, D.; Noll, M.; Meier, E. Controlling the sustainability of food supply chains. *Supply. Chain. Manag.* **2005**, *10*, 7–10. [CrossRef]
36. Rogers, D.S.; Carter, C.R. A framework of sustainable supply chain management: Moving toward new theory. *Int. J. Phy. Dist. Log. Manag.* **2008**, *38*, 360–387. [CrossRef]
37. Helfat, C.E.; Winter, S.G. Untangling Dynamic and Operational Capabilities: Strategy for the (N)ever-Changing World. *Strateg. Manag. J.* **2011**, *32*, 1243–1250. [CrossRef]
38. Silvestre, B.; Dalcol, P.R.T. Geographical proximity and innovation: Evidences from the Campos Basin oil & gas industrial agglomeration, Brazil. *Technovation* **2009**, *29*, 546–561. [CrossRef]
39. Rauer, J.; Kaufmann, L. Mitigating external barriers to implementing green supply chain management: A grounded theory investigation of green-tech companies' rare earth metals supply chains. *J. Supply. Chain. Manag.* **2014**, *51*. [CrossRef]
40. Tsai, M.-S.; Tsai, M.-C.; Chang, C.-C. The direct and indirect factors on affecting organizational sustainability. *J. Manag. Sustain.* **2013**, *3*, 67–77. [CrossRef]
41. Mangla, S.K.; Luthra, S.; Rich, N.; Kumar, D.; Rana, N.P.; Dwivedi, Y.K. Enablers to implement sustainable initiatives in agri-food supply chains. *Int. J. Prod. Econ.* **2018**, *203*, 379–393. [CrossRef]
42. Cooper, M.; Ellram, L. Characteristics of supply chain management & the implications for purchasing & logistics strategy. *Int. J. Logist. Manag.* **1993**, *4*, 13–24. [CrossRef]
43. Soylu, A.; Oruç, C.; Turkay, M.; Fujita, K.; Asakura, T. Synergy analysis of collaborative supply chain management in energy systems using multi-period MILP. *Eur. J. Oper. Res.* **2006**, *174*, 387–403. [CrossRef]
44. Azevedo, S.; Silva, M.; Matias, J.; Dias, G. The Influence of collaboration initiatives on the sustainability of the cashew supply chain. *Sustainability* **2018**, *10*. [CrossRef]
45. Yin, R.K. *Case Study Research: Design and Methods*, 3rd ed.; Sage Publications: London, UK, 2003; pp. 109–140.

International Journal of
Environmental Research and Public Health

Article

Moderating Effect of Dynamic Environment in the Relationship between Guanxi, Trust, and Repurchase Intention of Agricultural Materials

Lan Li [1], Gang Li [2,*], Xiaoling Feng [2], Zhigao Liu [3] and Fu-Sheng Tsai [4,5,6,*]

1 Research Center of Henan Economy, School of Business Administration, Henan University of Economics and Law, Zhengzhou 450046, China; lilan@huel.edu.cn
2 School of Management and Economics, North China University of Water Resources and Electric Power, Zhengzhou 450045, China; 201510751237@stu.ncwu.edu.cn
3 Jiangsu FoxLink Green Energy Technology Co. LTD., Lianyungang 222000, China; bob_liu@foxlinkpower.com
4 Department of Business Administration, Cheng Shiu University, Kaohsiung 83347, Taiwan
5 Center for Environmental Toxin and Emerging-Contaminant Research, Cheng Shiu University, Kaohsiung 83347, Taiwan
6 Super Micro Mass Research and Technology Center, Cheng Shiu University, Kaohsiung 83347, Taiwan
* Correspondence: ligang0840907@ncwu.edu.cn (G.L.); tsaifs@gcloud.csu.edu.tw (F.-S.T.)

Received: 6 July 2019; Accepted: 24 September 2019; Published: 8 October 2019

Abstract: Repurchasing intention of agricultural materials is a key to a sustainable food business system. The novel contribution of this study is that we go beyond technical aspect and look into human capital dynamics in a general context, by examining how different dimensions of 'guanxi' (i.e., personal relations and instrumentality) between farmers and agricultural retailers affect trust between the two and, in turn, repeated purchase intention of agricultural materials by farmers in China. To further generate implications for food system as a whole, we also examined how dynamic environment moderates the effects mentioned above. Adopting survey method and multivariate analyses, this study tests the hypotheses with a collected data set of 578 farmers from representative rural areas of China. The results show that guanxi between farmers and agricultural retailers has a positive effect on trust between them and on repeated purchase intentions of farmers. While instrumentality has a negative effect on trust between them and on repeated purchase intentions of farmers. The trust between farmers and agricultural retailers promotes farmers' repeated purchase intentions. The intensity of competition negatively moderates the positive relation between trust and repeated purchases. Demand uncertainty does not moderate the positive effect of trust on repeated purchases. The results and discussion shed light on agricultural food system sustainability from a dynamic environment embedded business relationship perspective.

Keywords: guanxi; trust; dynamic environment; repeated purchase intention; agricultural food system

---

1. Introduction

The competition in the agricultural material market of China is becoming more intense in recent years as the market economy in this sector is developing very fast. A saying that is popular among marketing practitioners is that "the cost of attracting new customers is five times that of maintaining old ones". Agricultural retailers must focus on cultivating farmers' repeated purchase intentions to generate satisfactory sales. Thus, exploring the factors that stimulate farmers' repeated purchase intentions is of great importance for agricultural retailers, and this question also deserves more research attention from academia.

Previous studies on customer repeated purchase intentions show that factors such as perceived value [1,2], customer satisfaction [3,4], customer trust [5,6], and guanxi [7–9] are important antecedents. As the business culture and environment is very unique in China, a group of scholars have explored the catalysts of repeated purchase intention in China. They found that the guanxi has positively promoted customer purchase intention [8,9]. Trust plays an important role in customer repeat purchases by reducing information asymmetry, and customer perceived risk, thus forming a positive intention of purchasing [10,11].

In a business-to-business (B2B) context, trust is also important in establishing a stable, continuous, and good supplier relationship, thereby enhancing the role of corporate customer share and customer retention [12]. Moreover, studies also find that, unlike the trust established in the legal provisions and written contracts in Western society [13], guanxi in Chinese society are the foundation of trust establishment [14,15]. It facilitates the establishment and development of interpersonal trust between enterprises [16].

However, this research on guanxi, trust, and repeated purchase intentions is mostly based on B2B or online marketing scenarios. The purchase of agricultural materials in rural area of China is business-to-customer (B2C) and mostly offline transactions. In addition, the agricultural materials market in China is very unique. Most farmers in China have very limited knowledge of agricultural materials such as pesticides, seeds, and fertilizers. They do not know the components or usage of them [17]. Therefore, farmers are heavily relied on the retailers' recommendation, and their purchasing behavior is restrained within the network with retailers [18]. In addition, fake pesticides, seeds, and fertilizers have made farmers more dependent on retailers who they have built trust with so as to get a sense of security. Thus, the importance of guanxi and trust is even more significant in the agricultural material market of China. The conclusions drawn by the extant research in the context of B2B and online marketing are unlikely to be able to explain the agricultural materials transactions in China. An independent investigation on the antecedents of repeated purchase intention of agricultural materials by farmers in China is necessary.

Different from western countries, research shows that the relationship between the business parties in China is of great importance in business transactions. It is a stepping stone of making a deal successfully. The level of trust between the buyers and sellers determines the way in which business is done [14]. More specifically, customer purchasing in China begins with guanxi, and trust is about whether the two parties can cooperate for a long time. This situation makes guanxi and trust a natural focus of research on trading relationships in China, especially in the rural area where the Chinese traditional culture of relationship-based society is dominant.

In addition, given the importance of environment in determining business activities, it is necessary to ask the question that will the relationship between guanxi, trust, and repeated purchase intentions be affected by the boundary conditions, and 'what are the implications for the food system?'. Besides being embedded deeply with Chinese traditional culture of relationship-based "acquaintance" society, the agricultural materials market in China is emerging and developing. The number of agricultural retailers is growing rapidly, and farmers increasingly enjoy more choices. In a dynamic environment such as this, can trust, as a marketing tool, still significantly increase repeat purchase intentions? This is also a question that needs to be addressed.

In view of the above questions, based on relationship marketing and social exchange theory, from a perspective of Chinese 'relationships', this paper investigates firstly, the impact of the two dimensions of guanxi (i.e., personal relations and instrumentality) on trust. An emotional relationship is a relationship based on human emotions and established by a person to satisfy the emotional needs of both parties [15]. The instrumental relationship refers to the social relationship established by individuals with others in order to achieve some goals [15]. Secondly, the effect of trust on repeated purchase intentions. Thirdly, the moderating effects of demand uncertainty and competition intensity in the dynamic environment on the relationship between trust and repeated purchase intentions.

The following section will review the relevant literature and theories, develop the conceptual framework, and propose the hypotheses. The research methods and results are given in the third section. The conclusions, limitation of the research, and the direction of further studies are discussed in the last section.

2. Theoretical Base and Hypotheses

2.1. Theoretical Base

2.1.1. Guanxi

'Guanxi' in China refers to the interrelationship between people that is based on personal emotions and interests, and established in the process of social activities in the context of Chinese culture [8]. China is a country with a relationship-based culture. People are concerned about the establishment, maintenance, and expansion of relationships. Guanxi is considered very important in everyday life, as well as economic and political activities [16]. Unlike the self-independence consciousness promoted by the West, China's guanxi shows a "difference sequence pattern" structure that is "self-centered" [17]. That is to say, Chinese culture emphasizes that individuals need to know where they are in the network of guanxi and should have different strategies for dealing with different people [18].

Guanxi, as a culturally embedded variable, is complex and contains multiple dimensions [19]. Some studies equal guanxi to personal relations, and test its impact on speculation [20,21]. Others categorized cross organizational guanxi into emotional and instrumental groups [22,23]. Given the purpose of relationship and communication, a few studies have categorized the guanxi in China into an emotional relationship, an instrumental relationship, and a mixed relationship [15]. In a study of inter-enterprise transaction purchase behavior, guanxi was divided into four dimensions, i.e., personal relations, human condition, face, and instrumentality [8].

A combination of the traditional 'relationship-based' and a modernized 'commercial economy' type of purchasing behavior is spearing in the rural area in China in recent years. The purchase behavior of farmers is increasingly instrumental. The purchase decision of agricultural materials is no longer simply based on the emotional relationship of blood, kinship, or geography, but increasingly reflects the instrumental orientation of pursuing economic and utilitarian ends [24]. It is this interweaving of the traditional pattern of difference sequence and the modern commercial economy that leads to the complexity of the formation of purchase decision of farmers and reflects a mixed guanxi. Therefore, this paper divides the guanxi between farmers and agricultural retailers into emotional and instrumental relationships.

An emotional relationship is a relationship based on human emotions and established by a person to satisfy the emotional needs of both parties [15]. In the 'acquaintance' society in rural areas of China, the relatively closed and homogeneous guanxi network means that the emotional relationship between people is mostly based on kinship or geographical relationship. While guanxi refers to blood, kinship, or 'symbolic' kinship including fellow villagers, friends, classmates, etc. [25]. Therefore, in most cases, emotional relationship equals to personal relations. In agricultural marketing, farmers' trust in agricultural retailers is based on frequent interactions between the two parties. It is a kind of interpersonal trust, reflecting farmers' confidence in the reliability and integrity of agricultural retailers.

The instrumental relationship refers to the social relationship established by individuals with others in order to achieve some goals [15]. In the instrumental relationship, the orientation of people's communication is very obvious, that is, the relationship is a means to achieve the goal. The instrumental relationship is different from the personal relationship. Its emotional component is negligible. The parties to the transaction have always adhered to the fairness principle to ensure their own interests are protected. The long-term orientation of the relationship is usually very low. Yang examined the interpersonal relationship to the market economy and proposed that the contractualization of interpersonal relationship would replace the emotionalization [26]. Liu also believes that the interpersonal relationship under the conditions of market economy is generally weakening in emotional components and strengthening in economic interests [27]. In agricultural materials transactions, farmers want to establish good interpersonal relationships with agricultural retailers in order to purchase pesticides, fertilizers, and seeds with low costs and better benefits. Instrumental relationships are a means to achieve a goal, so they are short-lived and unstable [8], which is not conducive to the enhancement of customer trust and the promotion of repeated purchase intentions.

2.1.2. Trust

Trust refers to the confidence of one party in the transaction relationship to the reliability and integrity of the other party [5]. Trust is an important part of relationship marketing theory. It is a key variable that determines the success or failure of relationship marketing and can promote the establishment of a complete partner-type trading relationship. In Chinese rural areas, the production and living space of farmers is relatively limited, the social network has a high degree of closeness and homogeneity, which makes the role of trust more indivisible [28]. In addition, farmers often lack sufficient professional knowledge of agricultural materials, thus agricultural retailers are often their main source of information. Therefore, farmers rely heavily on the trust of agricultural retailers. Furthermore, farmers face a high risk of buying fake and shoddy agricultural materials, and thus are more inclined to choose an agricultural retailer whom they trust. Therefore, in order to purchase high-quality agricultural materials with reasonable price, it is prominent for farmers to establish trustworthy relationship with retailers. Trust sets up a bridge between farmers and agricultural retailers, which improves the efficiency of interaction, reduces unnecessary conflicts, and increases the possibility of cooperation, and promotes repeated purchase intentions.

2.1.3. Repeat Purchase Intention

The influencing factors of repeated purchase intentions have been a hot topic in the research of marketing. In the mid-1980s, customer perceived value was considered to be the most important factor affecting customers' repeated purchase intentions, and it was also one of the important methods to measure whether enterprises have competitive advantage in service [1,2]. Customer perceived value has a direct impact on the way in which repeated purchase intentions work, so as indirect effects through customer satisfaction. In the early 1990s, relationship marketing theory attracted more attention. Scholars used this theory as a starting point to argue that customer satisfaction is a major driving factor for customers' repeated purchases [3,4].

In the mid-to-late 1990s, customer trust stood out among the influencing factors of repeated purchase intentions and was highly valued by research scholars. It believed that trust had a significant effect on the generation of repeated purchase intentions [5,6]. Since the beginning of this century, guanxi has become the focus of attention of the research on factors affecting repeated purchase intentions [7,9]. Unfortunately, this research is either conducted in Western countries or based on B2B or online marketing scenarios [7,11,14].

While agricultural marketing has its uniqueness compared to B2B or online marketing. Firstly, agricultural materials are distinguished from consumer products and living materials by their high aftereffect—i.e., the cost of purchasing fake and shoddy products is very high—and the effect of the use of agricultural materials can only be identified in the middle, late, or end of the agricultural production process. Secondly, the main body of purchase of agricultural materials in China is thousands of individual farmers with small farms. The expertise to purchase agricultural materials is poor. Most of them make the purchase according to previous farming and purchasing experience, and are easily affected by the surrounding voices.

Thirdly, different from other trading networks, farmers are at the edge of the rural social network; their ability to effectively obtain information and resources is weak. Agricultural retailers often deal with supply and marketing companies, and have many channels to obtain information. Therefore, they occupy a central position in the agricultural social network, and have control over information and resources. Thus, farmers depend heavily on agricultural retailers for information and suggestions [17].

Therefore, whether the relevant conclusions that are drawn from B2B or online marketing scenarios are applicable in the context of agricultural materials needs further verification.

Social exchange theory believes that interpersonal interaction is a process that both parties follow the principle of reciprocity [29,30]. The value and benefits obtained from the interaction directly promotes the individual's willingness to identify, attach, and maintain relationships with each other. In order to continue to benefit from the interaction between the two parties, the individual is motivated to perform actions beneficial to the other party, so as to construct and maintain relationship. The establishment of the farmer–agricultural retailer relationship means that the agricultural retailers provide the resources and services that meet the farmers' needs in exchange for the farmers' repeated purchase intentions. Farmers' repeated purchases make the agricultural retailers achieve better sales performance. This behavior falls into the category of social exchange.

In agricultural trading, repeated purchase intentions reflect the desire and inclination of farmers to maintain a relationship with a certain agricultural retailer. Due to the cyclical nature of agricultural production and the characteristics of dedicated asset investment, farmers will have a strong motivation in cooperation with agricultural retailers to ensure a desirable result of farming. The relatively closed and acquaintance-driven society gives guanxi a key position in influencing the repeated purchase intention of farmers. The trust in agricultural retailers is the guarantee for farmers to deal with the purchase risk. Guanxi positively influences farmers' repeated purchase intention through establishing trust between agricultural retailers and farmers.

2.1.4. Dynamic Environment

A dynamic environment is the most prominent feature of an enterprise's external environment. The economic behavior of both seller and purchaser is always embedded in a certain social environment, and is bound to be affected by environmental factors [17]. A dynamic environment makes agricultural retailers get into a state of dynamic change and uncertainty, which will have a major impact on their marketing decisions.

Farmers and other retailers are the two most important market players in the agricultural market for agricultural retailers. The rapid and ever-changing demand of farmers has put pressure on agricultural retailers. The huge group of competitors and the fierce competition in the industry has clearly formed a big obstacle for the retailers to achieve competitive advantages and business expansion. Therefore, this study considers that demand and competition are the two most important aspects of environmental factors that influence the repeated purchase intentions of farmers [31,32]. Demand uncertainty refers to the extent that customers seek new and different products and services, changes in the demand, and the difficulty of predicting of the changes by retailers. Competitive intensity refers to the degree of similarity, renewal speed, competitive strategy, and competitive incentives of the services or products offered by competitors [31–33].

The demand of agricultural materials is vulnerable to fluctuations in market conditions during the sales process. Therefore, compared with general consumer goods and industrial products, agricultural product marketing faces higher demand uncertainty and changes in competition intensity [21]. The dynamic environment will cause uncertainty in the transactions between farmers and agricultural retailers, which in turn will affect the cooperative relationship between the two parties. In an environment with high demand uncertainty and strong competition intensity, the operation of agricultural retailers faces greater risks and pressures and farmers will also be indecisive in trusting in agricultural retailers in this environment. Therefore, this paper further explores the moderation effect of demand uncertainty and competition intensity on the relationship between trust and the repeated purchase intention of farmers.

In summary, we propose a conceptual model in which guanxi between farmers and agricultural retailers improves the trust between them and the repeated purchase intentions of farmers, trust promotes the repeated purchase intention of farmers, and this effect is moderated by environmental dynamics. The conceptual framework is depicted in Figure 1.

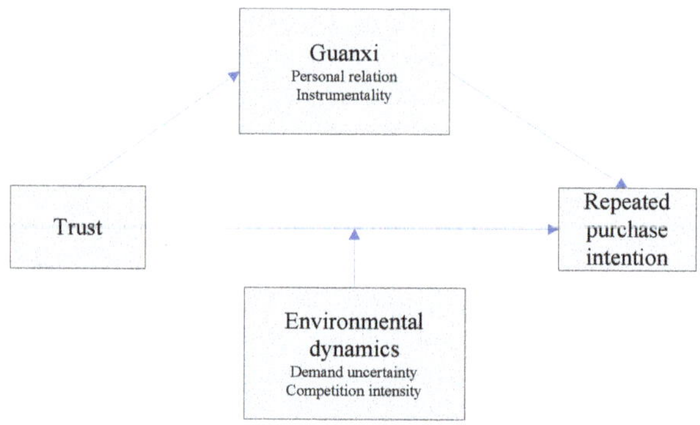

Figure 1. Conceptual model.

2.2. Research Hypothesis

2.2.1. The Impact of Guanxi on Trust

- Guanxi and Trust

The close guanxi between farmers and retailers indicates that the relationship has been tested for a long time and the two sides believe that the other party is a friend of their own. Agricultural retailers will not forge the information on quality and other aspects of agricultural materials. Instead, they will really care about the agricultural harvest and benefits of farmers and provide quality products and services that meet the needs of farmers. This will increase farmers' confidence of agricultural retailers in fulfilling their responsibilities and maintaining credibility. Therefore, farmers believe that agricultural retailers with good guanxi are worthy of trust and have the ability to recommend suitable products for them. In addition, in the acquaintance society of Chinese rural area, the close guanxi indicating that the two parties will not only be friendly and mutually supportive at work, but also extend the friendship to non-work issues, and the two sides will maintain contact. Frequent communication and interaction play a role of mutual benefit and eliminates contradictions. Thus, farmers have higher trust in agricultural retailers. As Huang (2015) argued that in the context of micro-marketing, guanxi can promote the generation of trust [18]. Therefore, it is proposed that all else being equal,

Hypothesis 1. *The closer the personal relations (i.e., higher guanxi) between farmers and agricultural retailers are, the higher the trust between them.*

- Instrumentality and Trust

On the contrary, if farmers only contact agricultural retailers when they want to buy agricultural materials, the relationship is established by the farmers solely to meet their goals, to achieve possible benefits, to do things smoothly and obtain internal information. In other words, both the farmers and the agricultural retailers treat the relationship as an instrumentality and will not trust each other.

Based on this short-term trading relationship, it is difficult for agricultural retailers to truly care about the agricultural harvest and interests of farmers. Naturally, the reliability of agricultural retailers perceived by farmers will be reduced. In addition, the trust established within the instrumental relationship is short-lived and unstable, or misplaced. Once other agricultural retailers can offer more favorable prices, more considerate services or more adequate agricultural information, the trust relationship is likely to collapse. Therefore, it is proposed that, all else being equal,

Hypothesis 2. *The more instrumental the relationship between farmers and agricultural retailers is, the less possible the trust between them will be established.*

2.2.2. Impact of Guanxi on Repeated Purchase Intentions

- Personal Relations and Repeated Purchase Intention

The emotional elements of guanxi established a foundation of cooperation between farmers and agricultural retailers. The better the personal relations, the more likely they form an 'inside circle'. In this circumstance, even if they do not consider doing business with the other side, the two parties will care about each other's private affairs and help each other, showing the responsibility and support between the members of the 'inside circle'. This kind of responsibility and affection not only weakens the motivation of agricultural retailers to pursue profit at the expense of morality, but also enables farmers to generate more business contacts with agricultural retailers in the future, which in turn will stabilize the personal relations between the two. The long-term stable relationship plays an important role in retaining customers and improving customers' repeated purchase intentions. In addition, better personal relations also mean that farmers will get more price concessions, credit sales services, and agricultural materials information in the actual purchase process. Thus, farmers are willing to buy again because of the perception of 'returning the favor'. Therefore, it is proposed that all else being equal,

Hypothesis 3. *The better the personal relations between farmers and agricultural retailers, the stronger the intention to repeat purchases.*

- Instrumentality and Repeated Purchase Intention

In the rural society of China, the role of instrumental rationality is apparent in the purchase decision of farmers choosing agricultural retailers [34]. The continuous deepening of reform and opening of the market has enabled more agricultural materials to enter the business. Farmers have more autonomy and initiative in the choice of agricultural materials, which makes the purchase of agricultural materials more utilitarian. Some farmers only contact agricultural retailers for buying agricultural materials, and this instrumental relationship is purely of interest. The more a farmer thinks that the relationship between him and the agricultural retailer is based on the instrumentality, the more likely it is that he will only consider which agricultural retailer can provide the maximum benefit the next time of the agricultural purchase, and tlower repeated purchase intentions is guaranteed. Therefore, it is proposed that all else being equal,

Hypothesis 4. *The more instrumental that farmers are in their relationship with agricultural retailers, the lower their intention to repeat purchases.*

2.2.3. Impact of Trust on Repeated Purchase Intentions

Individual consumers purchasing decisions are closely related to their trust in the seller [18]. The trust of farmers in agricultural retailers means, firstly, farmers believe that agricultural retailers will keep their promises during the transaction, guarantee the quality of agricultural materials, and provide services at preferential prices and with satisfactory technology, i.e., no false information about the quality, price, or service of agricultural materials will be provided to induce farmers' purchase. Secondly, farmers believe that agricultural retailers are always concerned about their interests, and really care about the agricultural harvest of farmers. Thirdly, farmers' ability to freely obtain effective information is limited, and the asymmetry of information makes farmers always in a weak position in the process of business transactions. In this case, trust acts as a compensation mechanism to effectively avoid the speculation of agricultural retailers. Farmers believe that agricultural retailers are trustworthy, because these retailers also believe that doing so will increase farmers' intentions of repeat purchases and maintain long-term and stable trading relationships benefit both sides. Therefore, it is proposed that all else being equal,

Hypothesis 5. *Farmers' trust in agricultural retailers positively affects farmers' repeated purchase intentions.*

2.2.4. Moderation Role of Dynamic Environment Played in the Relationship between Trust and Repeated Purchase Intentions

- Moderation Effect of Demand Uncertainty

In an environment with high demand uncertainty, farmers' trust in agricultural retailers will be more effective in promoting repurchase intention. In the condition of high demand uncertainty, farmers' demand preferences for agricultural materials appear inconstant and always seek something new and different, which indicates some risks in transferring from one brand or type to another brand or type of fertilizer, pesticides, and seeds. Therefore, farmers will choose a trusted agricultural retailer to avoid potential risks and provide protection for their replacement options. Firstly, farmers believe that the agricultural retailers they trust will guarantee the quality of newly purchased agricultural materials and provide corresponding technical services. Secondly, they believe that retailers will not exaggerate the quality of new agricultural materials to induce farmers to buy. Thirdly, farmers believe that retailers will really care about the agricultural harvest after farmers use new agricultural materials. This will further enhance farmers' intention to repeatedly purchase agricultural materials. Therefore, it is proposed that all else being equal,

Hypothesis 6. *Demand uncertainty positively moderates the relationship between trust and repeated purchase intentions.*

- The Moderation of Competition Intensity

A highly competitive environment reduces the incentives for trust to promote repeat purchase intentions. On the one hand, the increase in competition intensity indicates that many agricultural retail stores are in the market which providing similar agricultural materials, and the change of agricultural materials is very fast. In this situation, in order to be competitive, agricultural retailers will try their best to improve service, e.g., keep their promises in agricultural materials transactions, provide preferential prices and professional technical services, and go deep into the fields to actively support farmers of the use of agricultural materials and actual harvest. Farmers will thus not worry about the provision of false information or inferior agricultural materials by a certain agricultural retailer, and they can compare and replace agricultural retailers more easily. On the other hand, in order to cope with fierce competition, agricultural retailers will adopt strategies such as low-cost concessions, home delivery, and technical services. These concessions and convenience services will attract farmers to switch from formerly trusted agricultural retailers to new agricultural retailers. Therefore, it is proposed that all else being equal,

Hypothesis 7. *The intensity of competition will negatively moderate the relationship between trust and the repeated purchase intention of agricultural materials.*

3. Materials and Methods

3.1. Sampling and Data

A questionnaire approach is adopted by the study to collect data. The questionnaire collectors are mainly university students who were born and grow up in rural areas of China, so that they were familiar with the production and life in rural areas. They were trained on the skills required for the survey before going to do the data collection. In order to ensure the recovery rate and efficiency of the questionnaire, from July to September 2016, a stratified sampling method was adopted to select the samples. Firstly, a province with developed agriculture was randomly selected from the provinces and autonomous regions in the eastern, central, and western regions of China. Hebei province was drawn from the eastern region; the central region was drawn to Henan Province, and the western region was drawn to Guizhou Province. Because agriculture is a pillar industry in the western region, in order to make the research questions of this paper more thorough, Ningxia Province was selected from the west. The agricultural development conditions of the four provinces are as follows: Most of Hebei Province belongs to the North China, with more arable land and better sunshine. It is one of the national grain and cotton oil production areas and one of the country's 13 commodity grain production bases. Located in the middle and lower reaches of the Yellow River in central China, Henan Province has a humid climate, abundant rainfall and long sunshine hours. It is an important agricultural production base and agricultural province in China. The climate of Guizhou Province is mild, humid, rich in heat, and the conditions for agricultural development are good. Ningxia belongs to the plain area, with long sunshine hours, high effective accumulated temperature, sufficient precipitation and long frost-free period, and the agricultural foundation is unique.

Secondly, from the four provinces of Henan, Hebei, Ningxia and Guizhou, 10 cities with relatively developed agriculture—such as Luohe City, Baoding City, Yinchuan City, and Zunyi City—were randomly selected. Among them, Luohe City is located in Henan Province. In the south, warm and humid monsoon climate, four distinct seasons, more precipitation, better agricultural development foundation; Baoding City has four distinct seasons, sufficient sunshine, heat, and rain in the same season; Yinchuan City is located in Hetao Plain, an important irrigated agricultural area. Zunyi City has a large cultivated land and distinctive industries. Thirdly, 13 counties were selected with strong agricultural development advantages such as Linyi County, Tang County, Xixia District, and Wuchuan County from these 10 cities. Finally, in these counties, select some representative villages and farmers were selected in these counties.

Door-to-door visits were used to issue and collect questionnaires. The respondents selected are the farmers who are mainly engaged in agricultural production in each family and have experience in purchasing agricultural materials. A total of 605 households were selected for the survey, and 605 questionnaires were collected. After eliminating 27 invalid questionnaires, the effective questionnaire was 578. The effective recovery rate is 95.5%. The main reason for the invalidity is that it is not completed and/or there are extensive default values in the purification process of the questionnaire. A profile of the samples is shown in Table 1.

Table 1. Sample profile (%)

Variables	Profile
Gender	
Male	66.78
Female	33.22
Farming Experience	
≤10 years	23.9
11–30 years	38
≥31 years	38.1
Education	
Less literacy	10.38
Elementary school	31.66
Junior high school	45.67
High school/technical school	11.07
Diploma and above	1.21

Table 1. *Cont.*

Variables	Profile
Agri. Income/Total Income	
≤25%	50.9
26–50%	27.8
≥51%	21.3
Age	
≤36	10.38
37–46	26.12
47–54	27.85
55–65	23.70
Family Income (Yuan)	
≤10,000	25.78
10,000–30,000	47.58
30,000–60,000	19.55
60,000–90,000	3.80
>90,000	3.29

A pilot study was firstly conducted in a typical agricultural village in Henan Province in June 2016. 80 questionnaires were collected, of which 75 were valid, and the effective recovery rate is 93.75%. According to the information feedback of the interviewees and the data analysis results of the 75 answers of the pilot study, the questionnaire was further revised, and a formal questionnaire was finally formulated. To ensure the authenticity and validity of the survey data, all data collected in the pilot survey is not used as a final sample.

Harman's single factor test was conducted to ensure that common methods variance is not a threat to the study. All items in the questionnaire were put together for un-rotated factor analysis and a total of six principal component factors were extracted. The first factor explains 25.2% of the total variance, much lower than the threshold value of 40%, which means that the common method variance is not significant and will not affect the conclusion of the study.

3.2. Measurement

The measurement of the variables is based on extant literature [8,16,31,32,35], and adjusted and optimized according to the pilot study. For indicator purification, the items with the lower values of the factor loadings or the corrected item total correlation (CITC) is deleted by SPSS (version 22, IBM, New York, NY, USA), and were conceptually confirmed by all co-authors. Generally, if CITC is less than 0.4, the item is suggested to be deleted, because this means this item could not be well combined with other items to reflect the measured construct (variable) [36]. Guanxi is measured based on Wang et al. [16] and Li [8]. It is divided into two dimensions of personal relations and instrumentality, in total seven items. Trust mainly draws on the scales of Doney and Cannon et al. [35], a total of six items. Repeated purchase intention is measured by the scales developed by Paolo and Laurent [37], a total of three items. The dynamic environment includes two dimensions of demand uncertainty and competition intensity. Based on the scales developed by Yang et al. [31] and Jaworski and Kohli [32], they are measured by three items respectively. Table 2 shows the details. See Appendix A for the full questionnaire.

Table 2. Reliability and validity

Scale	Factor Loadings
Personal relations (CR = 0.837, AVE = 0.507, α = 0.752, Percentage of variance explained = 50.74%)	
When local farmers interact with sales staff in agricultural retail stores, they think each other is a friend of their own.	0.725
Local farmers are willing to help each other on non-work issues when they interact with sales staff at agricultural retail stores.	0.712
Local farmers often talk about some personal issues when they interact with salespeople in agricultural retail stores.	0.627
When local farmers interact with the sales staff of the agricultural retail store, even if the current buying and selling relationship is over, they will keep in constant contact with each other.	0.736
Local farmers think of each other as a circle when they interact with sales staff at agricultural retail stores.	0.688
Instrumentality (CR = 0.919, AVE = 0.850, α = 0.825, Percentage of variance explained = 85.08%)	
If local farmers are not buying agricultural materials (pesticide, fertilizer, seeds), they are not willing to contact agricultural retail stores.	0.922
I believe that if there is not a demand for agricultural materials, local farmers will not be willing to contact agricultural retail stores.	0.922
Trust (CR = 0.876, AVE = 0.542, α = 0.826, Percentage of variance explained = 54.16%)	
Agricultural retail stores are committed to us.	0.657
We believe that the information provided by agricultural retail stores.	0.775
Agricultural retail stores really care about our agricultural harvest.	0.753
When making important decisions, the agricultural retail store will consider giving both-sides benefits.	0.764
We believe that agricultural retail stores are always concerned about our interests.	0.696
Agricultural retail stores are worthy of trust.	0.764
Uncertainty in demand (CR = 0.870, AVE = 0.695, α = 0.775, Percentage of variance explained = 69.45%)	
The demand of local farmers in the agricultural retail industry is difficult to predict.	0.754
Local farmers in the agricultural retail industry always seek new differences.	0.864
The preferences of local farmers in the agricultural retail industry are always changing.	0.877
Competition intensity (CR = 0.793, AVE = 0.560, α = 0.607, Percentage of variance explained = 56.07%)	
There are many agricultural retail stores in the agricultural retail market that provide similar agricultural materials (pesticide, fertilizer, seeds).	0.704
Agricultural materials (pesticide, fertilizer, seeds) in the agricultural retail industry are changing rapidly.	0.677
The market competition of agricultural materials retail industry is very fierce.	0.752
Repeat purchase intention (CR = 0.878, AVE = 0.707, α = 0.789, Percentage of variance explained = 70.70%)	
Local farmers will have more business dealings with agricultural retail stores in the future.	0.763
Local farmers will purchase new agricultural materials (pesticide, fertilizer, seeds) or new services provided by frequent agricultural retail stores.	0.881
Local farmers will buy more agricultural materials (pesticide, fertilizer, seeds) or services from frequent agricultural retail stores.	0.873

Note. CR = construct reliability; AVE = Average Variance Extracted.

3.3. Reliability and Validity

To ensure the reliability and accuracy of the measurement results, this study firstly tested the reliability and validity of the data. Using internal consistency as a measure of reliability, Table 1 shows that the Cronbach's α of each scale reached 0.6 or more. In addition, the composite reliability values (CR) of each variable are greater than 0.7, and even close to or higher than the 0.8, which indicates that the internal consistency of the data is acceptable, and the variables of the questionnaire have good reliability.

As shown in Table 3, the Kaiser-Meyer-Olkin (KMO) values of each variable are greater than 0.5, and the Bartlett sphericity test has a significant probability of 0.000, indicating that each variable is suitable for factor analysis. The results of the factor analysis are shown in Table 1. The factor loadings of each variable are greater than 0.6, which is much higher than the standard of 0.4. The average variance extraction value (AVE) of all variables exceeds 0.5, indicating that the scale has good convergence validity. In addition, as shown in Table 3 below, the AVE square root of each variable is greater than the correlation coefficient of other variables, which meets the requirements of the discriminant validity test, indicating that the questionnaire has good validity. Furthermore, the percentage of variance explained of the factors is greater than 50%. It can be seen that the designed metrics have a higher degree of interpretation of the research variables and can truly measure the research variables. In summary, the scales used in this paper have good reliability and validity.

Table 3. KMO and Bartlett sphericity test.

KMO and Bartlett Sphericity Test		Personal Relations	Instrumentality	Trust	Demand Uncertainty	Competition Intensity	Repeated Purchase Intention
Kaiser–Meyer–Olkin		0.756	0.500	0.830	0.664	0.637	0.669
Bartlett sphericity test	Approximate chi-square distribution	663.836	390.141	1189.210	530.125	183.003	571.656
	Freedom	10	1	15	3	3	3
	Significant probability	0.000	0.000	0.000	0.000	0.000	0.000

Except for the AVE of all variables, the mean, standard deviation, and correlation coefficient of each variable are shown in Table 4.

Table 4. Correlation coefficient between statistical description and each research variable.

Variables	Means	Standard Deviation	1	2	3	4	5	6
1. Repeat purchase intention	5.641	0.996	(0.841)					
2. Personal relations	5.012	1.088	0.392 **	(0.712)				
3. Instrumentality	4.001	1.665	−0.137 **	0.095	(0.922)			
4. Trust	5.352	0.960	0.539 **	0.431 **	−0.022	(0.736)		
5. Demand uncertainty	4.737	1.372	0.105 *	0.236 **	0.160 **	0.144 **	(0.834)	
6. Competition intensity	5.555	0.986	0.211 **	0.197 **	0.017	0.199 **	0.228 **	(0.748)

Note: The number of samples is $N = 578$; the value in parentheses on the diagonal is the square root of the mean variation extraction (AVE). The non-diagonal is the correlation coefficient of each variable, ** $p < 0.01$ means significant at 99% confidence, and * $p < 0.05$ means significant at 95% confidence.

4. Results

4.1. Results of Regression Equation Modeling

In order to ensure the accuracy of the empirical results, the variables are sequentially placed into the regression equation model by the hierarchical regression method, and the influence of the predicted independent variables on the dependent variables is analyzed by comparing the changes of the regression coefficients. The results are shown in Table 5. In order to minimize the collinearity problem between interaction variables, the related variables were averaged before regression analysis. The independent and moderation variables are centralized and then multiply for the moderation analysis.

Table 5. Results of hierarchical regression.

Variables	Trust (TR)	Repeat Purchase Intention (RI)				
	Model 1	Model 2	Model 3	Model 4	Model 5	Model 6
Personal relation (PR)	0.437 ***	0.215 ***	0.203 ***	0.203 ***	0.205 ***	0.204 ***
Instrumentality (IV)	−0.063 *	−0.148 ***	−0.148 ***	−0.149 ***	−0.153 ***	−0.153 ***
Trust (TR)		0.443 ***	0.431 ***	0.430 ***	0.810 ***	0.820 ***
Uncertainty in demand (DU)			−0.001	0.002	0.003	0.002
Competition intensity (CI)			0.088 **	0.088 **	0.475 ***	0.486 ***
Uncertainty in demand × Trust (DU × TR)				−0.017		0.008
Competition intensity × Trust (CI × TR)					−0.601 **	−0.616 **
R	0.436	0.586	0.592	0.592	0.598	0.598
R^2	0.190	0.343	0.350	0.351	0.358	0.358
ΔR^2	0.187	0.340	0.345	0.344	0.351	0.350
F	67.465	99.950	61.734	51.418	52.960	45.325

Note: ***, $p < 0.01$; **, $p < 0.05$; *, $p < 0.1$.

For the influence of personal relations and instrumentality on trust, model 1 of Table 4 shows that the model fits the data well ($R^2 = 0.190$). In the regression results, the personal relations are significantly positively related with trust ($\beta = 0.437$, $p < 0.01$), thus Hypothesis 1 is supported. Instrumentality is significantly negatively related with trust ($\beta = -0.063$, $p < 0.1$), thus Hypothesis 2 is supported. For the effect of personal relations and instrumentality on repeated purchase intention, it is known from model 2 that the fit is better ($R^2 = 0.343$). In the regression results, the personal relations are significantly and positively related with the repeated purchase intention ($\beta = 0.215$, $p < 0.01$), thus the Hypothesis 3 is supported. The instrumentality is significantly and negatively related with the repeated purchase intention ($\beta = -0.148$, $p < 0.01$), thus Hypothesis 4 is supported. The effect of trust on repeated purchase intentions is also shown in model 2. In the regression results, trust is significantly and positively related with the repeated purchase intention ($\beta = 0.443$, $p < 0.01$), thus Hypothesis 5 is supported.

The three-step test of hierarchical moderating regression method is adopted to examine the moderation effect. The detailed approach is as follows. Firstly, the impact of trust on the intention of repeated purchase is tested. It can be seen from model 2 that trust has a significant positive impact on the repeated purchase intention. Secondly, the influence of trust and demand uncertainty on the repeated purchase intention is examined. The model 3 fits the data well ($R^2 = 0.350$). Trust has a significant impact on repeated purchase intentions ($\beta = 0.431$, $p < 0.01$), but the effect of demand uncertainty on repeated purchase intentions is not significant. Finally, the interaction between trust and demand uncertainty was added to the model 3, this leads to model 4. The model 4 fits the data well ($R^2 = 0.351$). However, it is found that the coefficient of the interaction term is not significant. Therefore, the uncertainty of demand does not moderate the relationship between trust and repeated purchase intention of agricultural materials, thus Hypothesis 6 is not supported.

The same method is used to examine the moderation effect of competition intensity. Firstly, model 2 has verified the significant positive impact of trust on repeated purchase intentions. Secondly, the impact of trust and competition intensity on repeated purchase intentions is examined. As shown by model 3, both have significant effects ($\beta = 0.431$, $p < 0.01$; $\beta = 0.088$, $p < 0.05$). Finally, the interaction term of trust and competition intensity is added to the model 3, this leads to model 5. The fit of model 5 is also good ($R^2 = 0.358$). The coefficient of the interaction term is significant ($\beta = -0.601$, $p < 0.05$). Therefore, the moderation of the intensity of competition is verified, and the Hypothesis 7 is supported.

To ensure the robustness of the empirical results, the two moderation variables were put in the full model (i.e., model 6) and examined again. The results are completely consistent with the models 4 and 5, indicating the results for the test of the moderation effect of dynamic environment are stable and reliable. Therefore, Hypothesis 6 is not supported, and Hypothesis 7 is supported.

In order to more intuitively reveal whether demand uncertainty and competitive intensity have a moderation effect on the relationship between trust and repeated purchase intentions, an interaction diagram of the two variables is shown in Figures 2 and 3. It can be seen from Figure 2 that the high and low demand uncertainty has no significant influence on the relationship between trust and repeated purchase intention, and the moderating effect is not established. As can be seen from Figure 3, the competition intensity negatively moderates the effect of trust on repeat purchase intention. Specifically, in a market environment with high competition intensity, trust has a weaker effect on the repeated purchase intention of agricultural materials; while in a market environment with low competition intensity, there is a strong positive relationship between trust and repeated purchase intentions of agricultural materials.

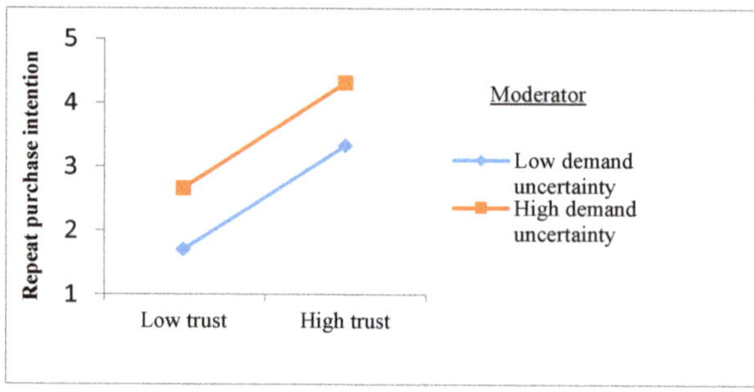

Figure 2. Moderation effect of demand uncertainty on the relationship between trust and repeated purchase intention.

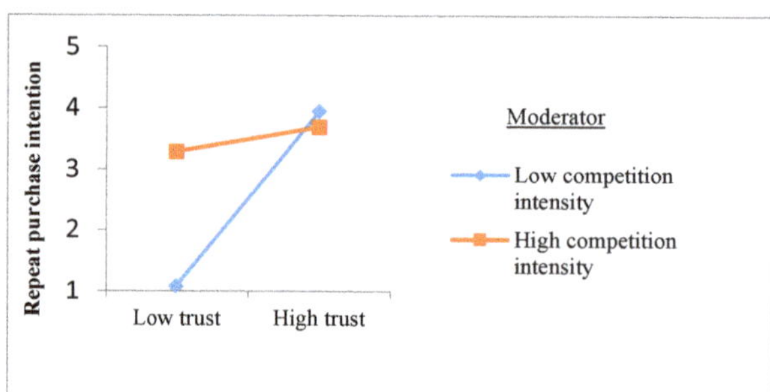

Figure 3. Moderation effect of competition intensity on the relationship between trust and repeated purchase intentions.

4.2. Robustness Test

In order to verify that there is no random trend or determined trend in the empirical results, 300 questionnaires were selected by the EXCEL random sampling method from 578 questionnaires for robustness test, and the hierarchical regression analysis was performed again. The results are shown in Table 6.

Table 6. Results of hierarchical regression analysis (after selecting 300 samples)

Variables	Trust (TR)	Repeated Purchase Intention (RI)				
	Model 1	**Model 2**	**Model 3**	**Model 4**	**Model 5**	**Model 6**
Personal relations (PR)	0.441 ***	0.239 ***	0.236 ***	0.235 ***	0.245 ***	0.244 ***
Instrumentality (IV)	−0.126 **	−0.210 ***	−0.207 ***	−0.205 ***	−0.213 ***	−0.210 ***
Trust (TR)		0.452 ***	0.447 ***	0.449 ***	−0.021 ***	0.433 ***
Uncertainty in demand (DU)			−0.026	−0.033	0.068	−0.031
Competition intensity (CI)			0.057	0.055	0.433	0.066
Uncertainty in demand × Trust (DU × TR)				0.038		0.059
Competition intensity × Trust (CI × TR)					−0.084 *	−0.096 **
R	0.444	0.632	0.634	0.635	0.639	0.642
R^2	0.197	0.399	0.402	0.403	0.409	0.412
ΔR^2	0.192	0.393	0.392	0.391	0.397	0.398
F	36.530	65.474	39.529	33.022	33.756	29.210

Note: ***, $p < 0.01$; **, $p < 0.05$; *, $p < 0.1$.

4.2.1. Comparative Analysis of Models

The R values of the six models vary between 0.008 and 0.046, the values of R^2 vary from 0.007 to 0.056, and the values of adjusted R^2 vary between 0.005 and 0.053. Thus, the variance is relatively small and negligible. Therefore, it can be concluded that there is no big difference in the model fitting between 578 questionnaires and 300 questionnaires.

4.2.2. Comparison of the Hypotheses Test Results

Regarding Hypothesis 1, in which the impact of personal relations on trust is to be examined. Both of the regression results of 578 questionnaires in model 1 ($\beta = 0.437$, $p < 0.01$) and the regression results of 300 questionnaires ($\beta = 0.441$, $p < 0.01$) indicate the relationship is significant and positive, with subtle differences in β values.

Hypothesis 2 proposes the impact of instrumentality on trust. Again, the regression results of 578 questionnaires in model 1 ($\beta = -0.063$, $p < 0.1$) and the regression results of 300 questionnaires ($\beta = -0.126$, $p < 0.05$) indicate the same effect, i.e., instrumentality is significantly but negatively associated with trust, only that the result for 300 questionnaires were slightly more significant.

Hypothesis 3 proposes that the impact of personal relations on repeated purchase intentions. The regression results of 578 questionnaires in model 2 ($\beta = 0.215$, $p < 0.01$) and the regression results of 300 questionnaires ($\beta = 0.239$, $p < 0.01$) indicate the same result, i.e., there is a significant positive relation between personal relations and repetitive purchase intentions, only with subtle differences in β values.

Hypothesis 4 proposes the impact of instrumentality on repeated purchase intentions. Both of the regression results of 578 questionnaires in model 2 ($\beta = -0.148$, $p < 0.01$) and the regression results of 300 questionnaires ($\beta = -0.210$, $p < 0.01$) show the same effect, i.e., instrumentality significantly and negatively associated with repeated purchase intentions, with subtle differences in beta values.

Hypothesis 5 proposes the impact of trust on repeated purchase intentions. Again, the regression results of 578 questionnaires in model 2 ($\beta = 0.443$, $p < 0.01$) and the regression results of 300 questionnaires ($\beta = 0.452$, $p < 0.01$) shows the same effect that trust is significantly and positively correlated with repeated purchase intentions, with only subtle differences in beta values.

Hypothesis 6 proposes the moderation effect of uncertainty of demand. Neither the results of 578 questionnaires nor the results of 300 questionnaires support the hypothesis. The results were stable and consistent.

Hypothesis 7 proposes the moderation effect of competition intensity on the relationship between trust and repeated purchase intentions. Both of the regression results of 578 questionnaires in model 5 ($\beta = -0.601$, $p < 0.05$) and the regression results of 300 questionnaires ($\beta = -0.084$, $p < 0.1$) indicate a significant negative moderation effect of competition intensity, except that the significance level of the 578 questionnaires was slightly higher.

In summary, the regression results in Table 5 are consistent with that in Table 4. After extracting about half of the total sample, the relationship between other variables in the model passed the significance test except that the demand uncertainty still did not moderate the relationship between trust and repeated purchase intentions. This confirms that personal relations positively affect trust and repeated purchase intentions; instrumentality negatively and significantly affects trust and repeated purchase intentions; trust positively affects repeated purchase intentions, and the competition intensity negatively moderates the relationship between trust and repeated purchase intentions. It demonstrates that the regression results of the above-mentioned whole sample are robust.

5. Discussions

The results of the study show that firstly, the two dimensions of guanxi have different effects on trust. Personal relations play a positive role in generating the trust of farmers in agricultural retailers, while instrumentality plays a negative role in this relationship. Good personal relations mean that the two sides of the transactions have frequent communication and interaction, thus establishing a bridge of mutual trust between them. This result is consistent with the extant research, e.g., Wang [16] and Li [8]. The higher instrumentality means that the transactions between the two parties are short-lived and utilitarian, and the farmers' perception of the reliability of the agricultural retailers is lower. However, this is inconsistent with the results of Li (2010) in the context of B2B marketing, in which it is argued that instrumentality positively affects interpersonal trust among enterprises. It is analyzed that the salespersons in the B2B situation are professional businessmen, where the most important consideration in dealing with the other party is how much benefits that the other party could bring to them. Therefore, the two parties could predict accurately the other party's actions based on economic rational. This is also a type of trust, but an 'instrumental trust'. To obtain expected benefits, e.g., repeated purchase, the seller needs only provide what the buyer wants. Thus, instrumental relationship positively affects instrumental trust between the buyers and sellers in the B2B market.

While in the agricultural materials market, farmers are not professional businessmen. Their purchasing decisions are very often influenced by factors other than pure economic interests. Especially in Chinese rural areas, due to the information asymmetry and the relatively closed networks, farmers often choose retailers they know well. Thus, making the deal more embedded in emotional elements. In this context, the stronger the instrumentality in the relationship, the less likely trust will be built between the two parties.

Based on the above discussion, conclusions drawn by previous research, e.g., Peng [15] and Kriz and Fang [14], that guanxi is the basis of trust establishment, should be applied cautiously. This study shows that, in the context of Chinese agricultural materials marketing, only the dimension of personal relations in guanxi is the basis of trust establishment. In this context, when the emotional component is greater than the instrumental component, that is, when the role of personal relations is greater than the instrumentality, the trust between people will be established.

The results also show that good personal relations lead to the cooperation between farmers and agricultural retailers, thus encourage farmers to choose the same agricultural retailers for new purchases. While the more instrumental the relationship between the two parties is, the more likely they will consider economic interests than other aspects during the business transactions, repeated purchase is thus more unpredictable. In addition, trust lays the foundation for long-term relationships between farmers and agricultural retailers and promotes repeated purchase intentions [38]. Therefore, guanxi between farmers and agricultural retailers could influence repeated purchase intentions of farmers through establishing trust between them. Good personal relations between farmers and agricultural retailers help building trust between the two parties, which in turn leads to more close cooperation between them, and repeated purchase is more likely to happen. While relationship based on pure economic interests does not help trust-building between farmers and agricultural retailers and will unlikely to encourage repeated business transactions between them.

This is inconsistent with previous research on the relationship between personal relationship and the repeated purchase intention in the B2B and e-commerce market, e.g., Li et al., (2010) [8] and Lin et al., [9], in which while both personal relationship and instrumentality improve the trust between the buyers and sellers, they do not necessarily encourage repeated purchase intention. This is understandable, in those markets, customers are more professional and well informed by product and market information, thus appears more rational and less relied on personal relationship to make repurchase decisions.

Furthermore, demand uncertainty does not have a moderation effect on the relationship between trust and repeat purchase intentions. The reason may be that, due to the relatively fixed planting area, the annual demand for agricultural materials of farmers is basically stable. In addition, farmers have always chosen agricultural materials based on past experience. Therefore, the choice of agricultural brands has not changed much. In this situation, relying on trust to increase farmers' intention to repeat purchases is more significant than increasing the product variety or adjusting the inventory by agricultural retailers. The intensity of competition moderates the relationship between trust and repeated purchase intentions, which indicates the convenience of transforming agricultural retailers brought about by the intensity of competition negatively affect the role played by trust in the intention of repeated purchases.

Finally, the interaction between guanxi, trust, competition intensity, and repeat purchase intention in the agricultural material market has important implications for food system. Global food security and sustainability is subject to food productivity and the way that food is produced. The application of circularity economy and reuse of food waste to improve the sustainability and security of food supply is gaining a ground in recent years [39,40]. This process is characterized by new agricultural materials to be used, and these might include nutrient inputs produced with emerging technologies for ammonia production, chemical recovery of phosphorus from digested food in sewage, and genetically modified seeds of crops [39]. For Chinese farmers whose knowledge of agricultural material is mainly learned from agricultural retailers, and the utilization of these new materials is largely dependent on the extent to which farmers believe that the retailers' suggestion would serve their best interests. Therefore, building the trust between farmers and retailers is especially critical to not only the retailers, but also to the circularity and reuse of food waste in food system, and in turn to food security and sustainability.

Even when some agricultural retailers are not keen to sell materials that are produced from food waste or by new technologies, given that circular economy is increasingly applied to food systems [39,40], intense competition in the market that provides better price, quality, and service would weaken the impact of trust on repeat purchase intention, and attract farmers to switch their purchase to these new retailers. This situation will also promote the security and sustainability of food system.

It is also worth noting that the unique characteristics of Chinese agricultural material market may be changing. The education of the new generation of farmers is improved significantly in recent years. The young generation are better educated and increasingly appear to be more reliant on science and technology to make growing and purchasing decisions. They are more interested in understanding market demands and customer preferences, and more capable of responding to the government's concern of food security and long-term strategy in developing industrial agriculture [41,42]. Given this situation, it is likely that the role played by personal relationship in Chinese agricultural material market will be weakened, and the importance of instrumentality will be increased. This tendency deserves research attention in the future, and it will be interesting to find out if, and to what extent, the changes in young generations of Chinese farmers will affect the interaction between guanxi, trust, and repeated purchase intention in Chinese agricultural material market.

5.1. Theoretical Contributions

This study makes several important theoretical contributions. Firstly, it extends the theory of relationship marketing by showing that the impact of guanxi on the trust varies crossing different contexts. The empirical examination of this study proves that instrumentality has a negative impact on trust in the context of agricultural material marketing in China, while the extant research shows the contrary in the B2B marketing. Secondly, this study enriches scholarly understanding of the impact of guanxi on repeated purchase intentions by showing that the former may have a different effect on the latter in different marketing contexts. Previous studies have confirmed the positive role played by guanxi in repetitive purchase intentions [8,9,11,38], the result of this study shows that the effect might not be positive in the context of Chinese agricultural marketing. Again, as a dimension of guanxi, instrumentality affects repeated purchase intentions of Chinese farmers negatively. This conclusion is obviously different from the extant research, indicating that the marketing context plays an important role in the effect of guanxi on repeated purchase intentions.

Thirdly, this study extends scholarly understanding of the effect of guanxi on repeated purchase intentions by showing that this effect may be transferred through trust, i.e., trust can be an influencing mechanism between the two. The integration of the three variables—i.e., guanxi, trust, and repeated purchase intentions—in the conceptual model shows that the interaction between them in a special context, such as Chinese agricultural material marketing, could effectively explain the repeated purchase decisions of buyers. In a special context of Chinese rural area, where Chinese traditional culture plays an important role in the social life, relationship marketing practice may well be different from that is in other contexts.

Finally, this study enriches the scholarly understanding of repeated purchase intentions by showing that the effects of some important antecedents, such as guanxi and trust, could be influenced by environmental conditions, such as competition intensity in the marketplace. Different from the previous literature of agricultural marketing, in which the external marketing and consumers' own characteristics are the focus of discussion, this study examines the relationship between trust and repeated purchase intentions based on the dynamic environment. The research results show that the more intense the competition between agricultural retailers is, the less the role of trust can play, and the dynamic environment will affect the effect of trust on repeated purchase intentions.

5.2. Implications for Management

The results of this study have important implication for practice. Firstly, it shows that the role of personal relations played in agricultural materials market is significant. This suggests that agricultural retailers should properly manage guanxi with farmers, paying particular attention to the establishment and maintenance of personal relations with them, and weaken farmers' instrumental choices, thus increasing their repeated purchase intentions.

Secondly, trust plays a key role in the purchase of agricultural materials. This suggests that agricultural retailers should emphasize cultivating farmers' trust in themselves. Agricultural retailers ought to take actions such as improving the quality of agricultural materials, keeping promises to farmers, providing satisfactory after-sales services, and upgrading their professional skills. So that trust can be built and repeated purchase intentions can be obtained subsequently.

Finally, the highly competitive environment has a negative impact on the economic interests of agricultural retailers. This suggests that agricultural retailers should aware the competition intensity in their industries and adopt appropriate marketing strategies. In the case of high competition, agricultural retailers ought to take extra efforts, such as offering preferential prices, credit sales services and door-to-door delivery to maintain customers. Therefore, farmers received more 'real' benefits, and perceive the deals as 'big bargains'. On the contrary, if the competition in the market is relatively weak, agricultural retailers should pay more attention to cultivating farmers' trust to maintain customers and achieve more sales.

5.3. Limitations and Further Research

Although the study has important contributions to the literature, it is not without limitations. Firstly, the agricultural materials, the purchase of agricultural materials and the rural social network in Chinese rural area compose a special context of this study. Therefore, caution is advised when applying the results of this research to other contexts.

Secondly, this study focuses on the two dimensions of guanxi—i.e., personal relations and instrumentality—other dimensions such as human feelings and face have not been considered. Therefore, more elements of guanxi should be explored in the future, thus a deeper and more comprehensive understanding of the effect of guanxi on repeated purchase intentions could be achieved.

Thirdly, the data collected only from four provinces of the eastern, central, and western regions of China, but did not reach a wider region, thus the generalization of the research results needs further investigation. In addition, this paper uses only horizontal data but not longitude data, it is more conducive to discover the change of data along time span. Therefore, longitudinal research design can be considered in the future to verify the causal relationship in the conceptual model.

Finally, as discussed above, the Chinese agricultural material market is changing, especially the education of young generation of farmers is improved significantly in recent years, combined with new technology, and the development of industrial agriculture, and the long-term strategy of government in food security, the interaction between guanxi, trust, and repeated purchase intention in Chinese agricultural material market is very likely to change. Thus, a dynamic perspective must be applied to understand the conclusions of this study. At the same time, it will be interesting to conduct a research in a few years time to find out if and to what extent the expected change would happen.

6. Conclusions

This study explores the influencing factors of repeated purchase intentions of farmers in the agricultural resource market in China. In particular, it examines the impact of guanxi and trust on repeated purchase intention, and the moderating effect of dynamic environment on the relationship between trust and repeated purchase intentions. A data set of 578 samples was used and hierarchical regression analysis was conducted to examine the conceptual models. The results generally support the hypotheses that based on the conceptual model. The results show that guanxi between farmers and agricultural retailers has a positive effect on trust between them and on repeated purchase intentions of farmers. While instrumentality has a negative effect on trust between them and on repeated purchase intentions of farmers. The trust between farmers and agricultural retailers promotes farmers' repeated purchase intentions. The intensity of competition negatively moderates the positive relation between trust and repeated purchases. Demand uncertainty does not moderate the positive effect of trust on repeated purchases. The results and discussion shed light on the agricultural food system sustainability from a dynamic environment embedded business relationship perspective. They also suggest that conclusions drawn by previous research that based on B2B and e-commerce market may not be applicable to Chinese agricultural material market, where customers' background and their interaction with suppliers are of unique characteristics.

Author Contributions: L.L. conceptualized and wrote the original draft of the paper; G.L. was a major co-writer of the first draft; X.F. conducted data curation and formal analysis; Z.L. validated the research and results. F.-S.T. reviewed and edited the manuscript, and would be responsible for revisions and re-submissions.

Funding: The authors gratefully acknowledge the National Social Science Fund of China (15BGL087), Henan Soft Science Foundation (172400410353), Program for Science & Technology Innovation Talents in Universities of Henan Province (2016-cx-003), Henan University Philosophy and Social Sciences Outstanding Scholar Funding Project (2019-YXXZ-14), and National Natural Science Foundation of China (Project 71672164) for financial support.

Conflicts of Interest: The authors declare no conflicts of interest.

Appendix A

Appendix Questionnaire

Completion Instructions:

Dear Sir/Madam:

Thank you very much for taking the time to participate in the research activities carried out by the research group of "Trust Transfer Mode and Marketing Strategy Research in Agricultural Sales".

In order to obtain accurate data, please answer or fill in the relevant questions according to the actual situation. We will treat all information you provide in absolute confidence. Scientific research, policy analysis, and opinion review are published in a large number of questionnaires, rather than the case information of your individual, family, and village, and will not cause leakage of your personal, family, and village information. We apologize for any inconvenience caused to your life and agricultural production, and thank you for your understanding and assistance in our academic research.

Please fill in the relevant information or mark "√" on the number you think is appropriate. Sincerely thank you for your support and help in the development of China's business management discipline.

Basic Information:

01.The type of respondent: 1) ordinary farmers; 2) farming expert; 3) village cadres; 4) agricultural technicians
02. Address of the village: ________ City _________ County _________ Township _______ Village Your name: ____________ Contact number: ___________
03. Your gender: 0—male; 1—female
04. Your age: 1) under 36 years old; 2) 37~46 years old; 3) 47~54 years old; 4) 55~65 years old; 5) over 66 years old
05. Education level: 1) less literacy; 2) elementary school; 3) junior high school; 4) high school or technical school; 5) college and above
06. Farming experience: 1) 10 years and below; 2) 11 to 30 years; 3) 30 years or more
07. Size of your farm: 1) 5 mu and below; 2) 6 ~ 15 mu; 3) more than 15 mu
08. Type of land: 1) cultivated land; 2) mountainous land; 3) forest fruit land; 4) water surface; 5) pasture
10. The average annual income of your family: 1) less than 10,000 (inclusive) yuan; 2) 10,000 to 30,000 (inclusive) yuan; 3) 40,000 to 60,000 (inclusive) yuan; 4) 70,000 to 90,000 (inclusive) yuan; 5) 100,000 yuan or more
11. Agricultural income as a percentage of total income: 1) 25% and below; 2) 26% ~ 50%; 3) 51% or more

Please choose according to the actual situation: 1—completely disagree; 2—basically disagree; 3—do not agree; 4—does not matter; 5—partially agree; 6—basically agree; 7—fully agree **Guanxi**								
Personal relations								
S1.1	When local farmers interact with sales staff in agricultural retail stores, they think each other is a friend of their own.	1	2	3	4	5	6	7
S1.2	Local farmers are willing to help each other on non-work issues when they interact with sales staff at agricultural retail stores.	1	2	3	4	5	6	7
S1.3	Local farmers often talk about some personal issues when they interact with salespeople in agricultural retail stores.	1	2	3	4	5	6	7
S1.4	When local farmers interact with the sales staff of the agricultural retail store, even if the current buying and selling relationship is over, they will keep in constant contact with each other.	1	2	3	4	5	6	7
S1.5	Local farmers think of each other as a circle when they interact with sales staff at agricultural retail stores.	1	2	3	4	5	6	7
S1.6	When the local farmers interact with the sales staff of the agricultural retail store, the relationship between the two parties is tested for a long time.	1	2	3	4	5	6	7

Instrumentality								
S2.1	Local farmers maintain their relationship with agricultural retail stores and contribute to the choice of agricultural products (pesticide, fertilizer, seeds)	1	2	3	4	5	6	7
S2.2	Local farmers maintain relationships with agricultural retail stores and help to obtain information or resources on agricultural products (pesticide, fertilizer, seeds)	1	2	3	4	5	6	7
S2.3	Local farmers maintain their relationship with agricultural retail stores, helping to reduce the cost of agricultural products (pesticide, fertilizer, seeds)	1	2	3	4	5	6	7
S2.4	If local farmers are not buying agricultural materials (pesticide, fertilizer, seeds), they are not willing to contact agricultural retail stores.	1	2	3	4	5	6	7
S2.5	I believe that if there is not a demand for agricultural materials, local farmers will not be willing to contact agricultural retail stores.	1	2	3	4	5	6	7
Dynamic environment								
Demand uncertainty								
T1.1	The demand of local farmers in the agricultural retail industry is difficult to predict.	1	2	3	4	5	6	7
T1.2	Local farmers in the agricultural retail industry always seek new differences.	1	2	3	4	5	6	7
T1.3	The preferences of local farmers in the agricultural retail industry are always changing.	1	2	3	4	5	6	7
Competition intensity								
T2.1	There are many agricultural retail stores in the agricultural retail market that provide similar agricultural materials (pesticide, fertilizer, seeds).	1	2	3	4	5	6	7
T2.2	Agricultural materials (pesticide, fertilizer, seeds) in the agricultural retail industry are changing rapidly.	1	2	3	4	5	6	7
T2.3	In the agricultural retail industry, the strategy of price competition is often adopted between agricultural retail stores.	1	2	3	4	5	6	7
T2.4	The market competition of the agricultural materials retail industry is very fierce.	1	2	3	4	5	6	7
Trust								
X2.1	Agricultural retail stores are committed to us.	1	2	3	4	5	6	7
X2.2	Agricultural retail stores do not always treat us honestly (R).	1	2	3	4	5	6	7
X2.3	We believe the information provided by agricultural retail stores.	1	2	3	4	5	6	7
X2.4	Agricultural retail stores really care about our agricultural harvest.	1	2	3	4	5	6	7
X2.5	When making important decisions, the agricultural retail store will consider giving both-sides benefits.	1	2	3	4	5	6	7
X2.6	We believe that agricultural retail stores are always concerned about our interests.	1	2	3	4	5	6	7
X2.7	Agricultural retail stores are worthy of trust.	1	2	3	4	5	6	7
X2.8	We found it necessary to be cautious about agricultural retail stores (R).	1	2	3	4	5	6	7
Repeat purchase intention								
Z2.1	Local farmers will have more business dealings with agricultural retail stores in the future.	1	2	3	4	5	6	7
Z2.2	Local farmers will purchase new agricultural materials (pesticide, fertilizer, seeds) or new services provided by frequent agricultural retail stores.	1	2	3	4	5	6	7
Z2.3	Local farmers will buy more agricultural materials (pesticide, fertilizer, seeds) or services from frequent agricultural retail stores.	1	2	3	4	5	6	7

References

1. Dodds, W.; Monroe, K. The effect of brand and price information on subjective product evaluations. *Adv. Consum. Res.* **1985**, *12*, 85–90.
2. Wu, L.Y.; Chen, K.Y.; Chen, P.Y.; Cheng, S.L. Perceived value, transaction cost, and repurchase-intention in online shopping: A relational exchange perspective. *J. Bus. Res.* **2014**, *67*, 2768–2776. [CrossRef]
3. Rust, R.T.; Zahorik, A.J. Customer satisfaction, customer retention, and market share. *J. Retail.* **1993**, *69*, 193–215. [CrossRef]
4. Yi, Y.; La, S. What influences the relationship between customer satisfaction and repurchase intention? Investigating the effects of adjusted expectations and customer loyalty. *Psychol. Mark.* **2010**, *21*, 351–373. [CrossRef]
5. Morgan, R.M.; Hunt, S.D. The commitment-trust theory of relationship marketing. *J. Mark.* **1994**, *58*, 20–38. [CrossRef]
6. Upamannyu, N.K.; Gulati, C.; Chack, A.; Kaur, G. The effect of customer trust on customer loyalty and repurchase intention: The moderating influence of perceived CSR. *Int. J. Res. IT Manag. Eng.* **2015**, *5*, 1–31.
7. Lang, B.; Colgate, M. Switching barriers in consumer markets: An investigation of the financial services industry. *J. Consum. Mark.* **2001**, *18*, 332–347. [CrossRef]
8. Li, G.; Lu, H.; Liu, H. An empirical study of the influence of interpersonal relationship on enterprise purchase intention—Based on china's cultural background. *J. Shanxi Univ. Financ. Econ.* **2010**, 60–67. [CrossRef]
9. Lin, J.; Yan, Y.; Chen, S. Understanding the impact of social commerce website technical features on repurchase intention: A Chinese guanxi perspective. *J. Electron. Commer. Res.* **2017**, *18*, 225–244.
10. Gustafsson, A.; Johnson, M.D.; Roos, I. The effects of customer satisfaction, relationship commitment dimensions, and triggers on customer retention. *J. Mark.* **2005**, *69*, 210–218. [CrossRef]
11. Kim, D.J.; Ferrin, D.L.; Rao, H.R. Trust and satisfaction, two stepping stones for successful e-commerce relationships: A longitudinal exploration. *Inf. Syst. Res.* **2009**, *20*, 237–257. [CrossRef]
12. Yan, X.; Zhou, T.; Li, Y. Trust, commitment, relationship behavior and relationship performance: buyer's perspective. *Manag. Rev.* **2011**, *23*, 71–81. [CrossRef]
13. Leung, T.K.P.; Chan, R.Y.K.; Lai, K.; Ngai, E.W. An examination of the influence of guanxi and xinyong (utilization of personal trust) on negotiation outcome in China: An old friend approach. *Ind. Mark. Manag.* **2011**, *40*, 1193–1205. [CrossRef]
14. Kriz, A.; Fang, T. Interpersonal trust in Chinese relational networks: Moving from guanxi to xinren. In *Proceedings of the IMP 19th Annual International Conference*; IMP Group: Lugano, Switzerland, 2003.
15. Peng, S. The establishment of trust: Relationship operation and legal means. *Sociol. Res.* **1999**, *2*, 53–66.
16. Wang, X.; Yan, G. The impact of personal relations and contractual relationship on two-level trust between enterprises—An empirical analysis from the distribution channel of household appliances. *Shanghai Manag. Sci.* **2006**, *28*, 5–8.
17. Fei, X. *Native China*; Beijing People's Publishing House: Beijing, China, 2008; ISBN 9787208118164.
18. Sun, J.; Li, Y. Willingly or helplessly? The preliminary study on the driving factors of farmers' lock-in purchasing behavior of agricultural materials. *Econ. Manag. J.* **2014**, 81–93. Available online: http://en.cnki.com.cn/Article_en/CJFDTotal-JJGU201411011.htm (accessed on 25 September 2019).
19. Xin, K.R.; Pearce, J.L. Guanxi: Connections as subtitutes for formal institutional support. *Acad. Manag. J.* **1996**, *39*, 1641–1658. [CrossRef]
20. Tian, M.; Zhang, W.; Xia, C. Market uncertainty and the speculative behavior of agricultural product buyers: The role of guanxi. *J. Bus. Econ.* **2013**, 11–20. Available online: http://en.cnki.com.cn/Article_en/CJFDTOTAL-SYJG201312003.htm (accessed on 25 September 2019).
21. Zhang, W.; Xu, J.; Du, N.; Zhou, N. Research on the influence of personal relations on speculation in marketing channels based on local culture. *Chin. J. Manag.* **2016**, *13*, 958–971.
22. Lee, D.Y.; Dawes, P.L. Guanxi, trust, and long-term orientation in Chinese business markets. *J. Int. Mark.* **2005**, *13*, 28–56. [CrossRef]
23. Zhuang, G.; Xi, Y.; Tsang, A. Power, conflict, and cooperation: The impact of guanxi in Chinese marketing channels. *Ind. Mark. Manag.* **2010**, *39*, 137–149. [CrossRef]

24. Yan, Y. Difference pattern and hierarchical view of Chinese culture. *Sociol. Res.* **2006**, 201–213. Available online: http://kns.cnki.net/kcms/detail/detail.aspx?dbcode=CJFD&filename=SHXJ200604011&dbname=CJFD2006&uid=WEEvREdxOWJmbC9oM1NjYkZCbDdrdXJKMFptQjdpaStkamhnUy83U3lMQ3g%3D%24R1yZ0H6jyaa0en3RxVUd8df-oHi7XMMDo7mtKT6mSmEvTuk11l2gFA!! (accessed on 25 September 2019).
25. Weber, M. translated by Hong, T. ; *Confucianism and Taoism*; Jiangsu People's Publishing House: Nanjing, Jiangsu, China, 1995; ISBN 9787214049520.
26. Yang, F. Investigating interpersonal relationships in the period of market economy. *J. Jiangxi Soc. Sci.* **1994**, 63–65. Available online: http://kns.cnki.net/kcms/detail/detail.aspx?dbcode=CJFD&filename=JXSH406.027&dbname=CJFD9495&uid=WEEvREcwSlJHSldRa1FhdkJkVG1ERERDdzB1clMrUnA5VnhMWnd3ZDE2bz0%3D%249A4hF_YAuvQ5obgVAqNKPCYcEjKensW4IQMovwHtwkF4VYPoHbKxJw!! (accessed on 25 September 2019).
27. Liu, H. An Empirical Study of the Influence of Interpersonal Relationships on Business Intention in Chinese Culture Background. Master's Dissertation, Nankai University, Tianjin, China, 2009.
28. Zhuang, G.; Xi, Y. Cultural basis of relationship marketing in China. *Manag. World* **2003**, 98–109. [CrossRef]
29. Liu, X. The formation process of employee organizational commitment: Internal mechanism and external influence—An empirical study based on social exchange theory. *Manag. World* **2011**, 92–104. [CrossRef]
30. Li, L.; Li, G.; Chan, S. Corporate responsibility for employees and service innovation performance in manufacturing transformation: The mediation role of employee innovative behavior. *Career Dev. Int.* **2019**. [CrossRef]
31. Yang, Z.; Deng, L.; Fang, E. Market orientation, strategic flexibility and firm performance: The regulating effect of environmental uncertainty. *China Soft Sci.* **2010**, 130–139. Available online: http://kns.cnki.net/kcms/detail/detail.aspx?dbcode=CJFD&filename=ZGRK201009015&dbname=CJFD2010&uid=WEEvREdxOWJmbC9oM1NjYkZCbDdrdXJKMFptQjdpaStkamhnUy83U3lMQ3g%3D%24R1yZ0H6jyaa0en3RxVUd8df-oHi7XMMDo7mtKT6mSmEvTuk11l2gFA!! (accessed on 25 September 2019).
32. Jaworski, B.J.; Kohli, A.K. Market orientation: Antecedents and consequences. *J. Mark.* **1993**, *57*, 53–71. [CrossRef]
33. Li, L.; Li, G.; Yang, X.; Yang, Z. Pursuing superior performance of service innovation through improved corporate social responsibility. *Asia Pac. J. Mark. Logist.* **2019**, *31*, 925–943. [CrossRef]
34. Li, T.; Li, Y.; Li, W. Research on trust chain model in farmers' purchase decision based on psychological distance situation. *Forecast* **2016**, *35*, 35–42.
35. Doney, P.M.; Cannon, J.P. An examination of the nature of trust in buyer-seller relationships. *J. Mark.* **1997**, *61*, 35–51. [CrossRef]
36. Churchill, G.A. A paradigm for developing better measures of marketing constructs. *J. Mark. Res.* **1979**, *16*, 64–73. [CrossRef]
37. Paolo, G.; Laurent, G. Interpersonal trust in commercial relationships. *Eur. J. Mark.* **2010**, *44*, 114–138. [CrossRef]
38. Yan, X.; Zhou, T.; Li, Y. Trust, commitment, relationship behavior and relationship performance: A vendor perspective. *Chin. J. Manag.* **2010**, *7*, 1032–1038.
39. Tseng, M.L.; Chiu, A.S.; Chien, C.F.; Tan, R.R. Pathways and barriers to circularity in food systems. *Resour. Conserv. Recycl.* **2019**, *143*, 236–237. [CrossRef]
40. Piergiuseppe, M.; Pasquale, M.F.; Valentina, E.T. Food waste valorisation: Assessing the effectiveness of collaborative research networks through the lenses of a COST action. *J. Clean. Prod.* **2019**, *238*, 117868. [CrossRef]
41. Falcone, P.M. Tourism-Based Circular Economy in Salento (South Italy): A SWOT-ANP Analysis. *Soc. Sci.* **2019**, *8*, 216. [CrossRef]
42. David, J.H.B.; Poon, T.; Pongtip, T.; Richard, M.F.; Bob, D.; Samarthia, T. Looking at complex agri-food systems from an actor perspective: The case of Northern Thailand. In *Advances in Food Security and Sustainability*; Elsevier: Amsterdam, The Netherlands, 2019. [CrossRef]

International Journal of
Environmental Research and Public Health

Article

Evaluation of Policies on Inappropriate Treatment of Dead Hogs from the Perspective of Loss Aversion

Chenchen Yang [1] **and Jianhua Wang** [1,2,*]

[1] School of Business, Jiangnan University, Wuxi 214122, China
[2] Food Safety Research Base of Jiangsu Province, Jiangnan University, Wuxi 214122, China
* Correspondence: Jianhua.w@jiangnan.edu.cn; Tel.: +86-182-5154-9736

Received: 26 June 2019; Accepted: 10 August 2019; Published: 15 August 2019

Abstract: Punishment policies on the inappropriate treatment of dead hogs play a key role in safeguarding public health and environmental protection. These policies aim to regulate the behavior of farmers and promote the development of sustainable agriculture. Farmers' evaluation of a policy can be used to measure its effectiveness, and loss aversion is a factor that has been little studied. This study surveyed 404 hog farmers in China, and analyzed the factors that influenced their evaluation of the penalties for the inappropriate treatment of dead hogs during 2016 and 2017. We used three indicators for the evaluation of the penalties: the degree of necessity, implementation, and effectiveness. Special attention was paid to farmers' aversion to financial penalties and police detention time, which was elicited using economic experiments. The results show that farmers are more likely to be averse to police detention time than financial penalties, and suggest that the level of each indicator needs to be increased. The results from an ordered Probit model show that there are both similarities and differences between the formation paths of the three indicators. An aversion to financial penalties will help to improve the degree of implementation. An aversion to police detention time will lead to a negative trend in the degree of effectiveness. An in-depth analysis of the factors that influence farmers' evaluation of policies to punish inappropriate treatment of dead hogs may provide a basis for the design of government policies to improve environmental protection performance.

Keywords: loss aversion; dead hogs; policy evaluation; hog production

---

1. Introduction

Inappropriate treatment of dead hogs pollutes the environment and threatens human health. To ensure that hog products are safe and of high quality, China's government has been attempting to reduce the inappropriate treatment of dead hogs. For example, the government has promulgated a series of policies and regulations on hog production, including the Animal Epidemic Prevention Law in 2007, the Interpretation of Several Issues Concerning the Application of Laws in Handling Criminal Cases Hazardous in 2013, and the Opinions on Establishing a Mechanism for Harmless Treatment of Dead Livestock and Poultry in 2014. These policies prohibit farmers from inappropriately handling dead hogs. The relevant provisions of the Law of the People's Republic of China on Animal Epidemic Prevention state that persons who dispose of dead animals (including dead hogs) for an unknown cause will be punished. The relevant agencies shall order the offender to conduct harmless treatment of animals, require the offender to bear the handling costs, and impose a fine on the offender of not more than 3000 yuan. If the circumstances are serious, they shall be addressed based on the specific situation. The acts of producing or selling pork products from sick animals are deemed to be illegal and persons who perform them will receive punishment. Persons who sell dead hogs no longer only receive a fine, and those who cause a serious food poisoning incident or other serious incident involving a foodborne disease shall be sentenced to a fixed-term imprisonment of not more than three years. If the consequences are serious, the offender will be sentenced to more than seven years in prison or life imprisonment. The Opinions on Establishing a Mechanism for the Harmless Treatment of Dead Livestock and Poultry stipulates that in cases where animals (including dead hogs) are slaughtered, sold, or have an unknown cause of death, the relevant department must order the perpetrator to take remedial measures, confiscate the products, and impose a fine of more than five times the value of the goods. The penalties for the inappropriate handling of dead hogs in the existing law and regulations are primarily fines, with detention as a supplement. Therefore, this paper considered both fines and detention.

What is the effect of a punishment policy on the implementation process? How does a farmer evaluate the policy? Follow-up investigations on a policy's implementation are one of the means through which to ensure its effectiveness. Understanding farmers' evaluation of a policy on the inappropriate treatment of dead hogs is a premise of and the basis for its effective implementation and continuous improvement. Therefore, analyzing the formation process of a policy's evaluation based on punishment policy evaluation indicators is of theoretical and practical importance. Investigating the role played by farmers' characteristics, animal breeding characteristics and breeding conditions, and other factors in policy evaluation may provide a specific direction for improving farmers' policy evaluations and the optimization of policies. With the deepening of prospect theory in the field of policy evaluation, psychological factors have gradually become important indicators in policy evaluation research. The value function in prospect theory consists of a risk preference coefficient, a loss aversion measure, and a nonlinear probability weighting coefficient [1]. The loss aversion measure refers to the ratio of the absolute value of the negative utility produced by losing one unit of wealth to the absolute value of the positive utility of giving one unit of wealth when a person makes a future decision; that is to say, losses can have a greater psychological effect than returns [2,3]. Therefore, losses are often more difficult to accept than the same amount of earnings [4]. Provisions on losses in public policy can fully amplify the loss that people suffer when they behave inappropriately, and lead people to generate an aversion to loss to avoid the occurrence of inappropriate behavior.

The objective of this study was to understand the characteristics of farmers and their evaluation of a punishment policy on the inappropriate handling of dead hogs through an in-depth investigation. In other countries, food safety regulation is more stringent. In the United States, the listing of genetically modified (GM) foods is subject to review by three departments. Several years are needed before any GM food enters the market, and an outlay of $10 to $15 million is required to collect the data to complete the approval process, which considerably reduces the number of incidents involving unsafe food. In Germany, food safety issues can be traced back to the country of origin. Once food safety problems are discovered, the German government quickly isolates pig and poultry farms and forces the slaughter of affected livestock. The Japanese government comprehensively intervenes in and supervises food safety, and the compensation awarded by the court in food safety cases is generally reasonable. The government has established a liability insurance fund to provide compensation to victims in product liability cases. Based on the perspective of an aversion to loss, this paper explored the formation mechanism of farmers' evaluation of policies on the inappropriate treatment of dead hogs. Finally, suggestions were provided for improving the level of farmers' satisfaction of these policies and using farmers' aversion to loss to construct efficient and reasonable policies.

The rest of this paper is structured as follows. Section 2 presents a literature review. Section 3 describes the data source and presents descriptive statistics. Section 4 outlines the experimental design; the results are provided in Section 5. Our conclusions and discuss policy implications are presented in the final section.

2. Literature Review

The research on policies has mainly focused on the status quo of policies, the transmission mechanism of policy evaluation, and the formulation of policy evaluation indicators. With regard to policy status quo, farmers say that the government's policy on inappropriate treatment of dead hogs is enforceable, and the degree of influence of supervision and punishment policies on farmers' behavior is obvious [5]. However, some scholars have reported that the punishment policies on the inappropriate handling of dead hogs are insufficient to deter farmers from this behavior. In the long term, non-standard treatment of dead hogs will occur. Due to the arbitrariness of a farmer's inappropriate handling of dead hogs, the probability of being discovered and punished is low, which limits the policy's effectiveness [6]. With regard to the transmission mechanism of policy evaluation, Crabbé stated that it is crucial to identify the path to a policy's implementation in a policy evaluation study [7]. As agriculture is the industry with the longest development history, the government has issued many policies to ensure its sustainable development. Clarifying the problems in the process of a policy's implementation by evaluating agricultural policies is a research topic in the field of agriculture. For example, Tang et al. assessed the level of performance of support policies through provincial and agricultural commodity levels and suggested that China's agricultural product policy needs to be further improved [8]. Specifically, an important way of improving the level of performance of an agricultural product policy is to improve the technical efficiency of agricultural products. Moser et al. used framed experiments and games to provide respondents with different stimulations (rewards and penalties) to obtain their preferences from a producer's perspective [9]. The results showed that the probability of the implemented policy and the direction of the stimulus are the key indicators that influence the policy's effectiveness, and a high return fertilizer policy with a low probability is the most motivating policy. In a study of the sustainable development of food safety in Ireland, Lynch showed that the refinement, expansion, and transformation of sustainable development indicators can promote the evaluation and development of agricultural policies [10]. In an analysis of the effect of an agricultural policy's implementation, breeding characteristics, cultivation period, breeding conditions [11], education level, and income level are generally the influencing factors [12,13].

In our exploration of the indicators for measuring a policy's evaluation and the development of evaluation mechanisms, this work found that not many studies have been conducted on the policy indicators for managing dead hogs. Therefore, the work drew on the research conducted in other fields and divided the policy evaluation index into three indicators: the degree of necessity, the degree of implementation, and the degree of effectiveness. This is similar to the conclusions of Barbiroli on evaluation indicators for China's urban green transformation policy [14]. The author pointed out that the three core factors that ultimately identify and refine the evaluation index system of urban green transformation are policy attributes, the implementation process, and the implementation effects. Zhang used policy formulation, policy implementation, and policy results as the evaluation criteria to measure the overall implementation effect in Hubei province in an evaluation of the effectiveness of high-tech industrial policies [15].

This work incorporated psychological factors into the variable system to expand our understanding of the formation mechanism of a policy's evaluation, and the relationship between the degree of aversion to loss of policy evaluators and a policy's evaluation to analyze the impact of loss aversion was selected. In our study of relevant policies, our work did not find many studies that considered psychological factors, and few penalties were found regarding policies on the inappropriate handling of dead hogs. Drawing on policy research in other fields, Huang and Du evaluated China's housing security policy from the perspective of public satisfaction [16]. The results showed that when residents' level of trust and fairness is higher, their satisfaction with housing security is higher. Xia et al. introduced the free patriarchal system into the design of public policy when discussing the implementation of public policy on land use control [17], which involves considering the characteristics of people's aversion to loss to enhance the acceptance of public policy and optimize existing policies. The results of their analysis show that an increase in the use of compensation as a method of control, a reduction in law-abiding losses, an increase in the use of penalties for illegal transfers, and an increase in the prohibitions in the law can comprehensively improve a policy's effectiveness. Kibeta et al. found that risk aversion promoted the recognition of GAP (Good Agricultural Practices) certification among bean growers, and loss avoidance was not conducive to the popularity of the certification [18].

After reviewing the literature, two gaps were found to exist in the research. The first is that the research has not included the degree of aversion to loss in the analysis of the factors that influence a policy's evaluation, and ignores the guiding role that social psychological factors play in an evaluation of a policy on the inappropriate treatment of dead hogs. The existing research on farmers' policy evaluation and the transmission mechanism of policy evaluation does not include many psychological factors. Therefore, analyzing the effects of psychological factors of farmers on their evaluation of a policy and identifying the mechanism of the implicit factors is crucial to optimizing the policy's design and improving farmers' recognition of and satisfaction with the policy. Therefore, the contribution of this paper is a policy evaluation that includes a loss aversion factor. The second gap is that the existing data on psychological factors were mostly obtained directly using a Likert scale or a risk equivalent. These data have a large subjective component and are consequently of low value. A few studies on farmer policies introduced experimental design methods, such as answer selection. The experiments by Tanaka et al. [19] and Holt and Laury [20] were used to provide the basis for our research methods.

3. Theoretical and Research Framework

3.1. Theoretical Framework

The theoretical framework was constructed based on prospect theory (PT) and assumed that farmers are bounded rational decision-makers [21]. Kahneman and Tversky proposed PT and found that individuals are less sensitive to returns [3]. When facing a loss, the economic behavior that individuals exhibit is an appetite for risk, whereas they are considered risk averse in the case of gains. In particular, loss aversion is the degree of an individual's evasion of loss, and, in cumulative PT, the value function is based on the return and loss, that is, the degree of positive and negative deviation from the decision point of the decision-maker. The degree of negative deviation is expressed by the loss. The loss is weighted by a loss aversion measure, and the weight that is assigned to the loss is larger, so that the benefit is treated differently from the loss. To explain the existence of loss aversion, some scholars have used evolutionary theory, which suggests that people prefer more conservative and cautious behavior because aversion to loss in decision-making can improve their survival and reproduction success rate; so, to improve survival and evolution, people prefer loss aversion [22]. In the field of behavioral economics, the multiple price list (MPL) method is a common economic experimental method for measuring risk preference or loss aversion. This method originated from research that Binswanger conducted on Indian farmers [23]. More recently, domestic scholars have used this method to measure farmers' risk aversion and study its impact on agricultural production factors [24] and pesticide application behavior [25]. The MPL method is easy to implement, its subject matter is easy to understand, and it more realistically represents risk appetite. Based on PT, Tanaka et al. constructed an MPL experimental design that was used to measure the time preference coefficient [19], the risk preference coefficient, and a loss aversion measure. The authors also designed a lottery decision experiment that consisted of three parts and 35 pairs, including risky returns, risky losses, and combinations of different probabilities of these gains or losses. This work used the MPL Tanaka et al. [19] and Holt and Laury's [20] methods to measure loss aversion in pig farmers.

3.2. Research Framework

Based on the literature review and theoretical framework, the developed research framework is shown in Figure 1.

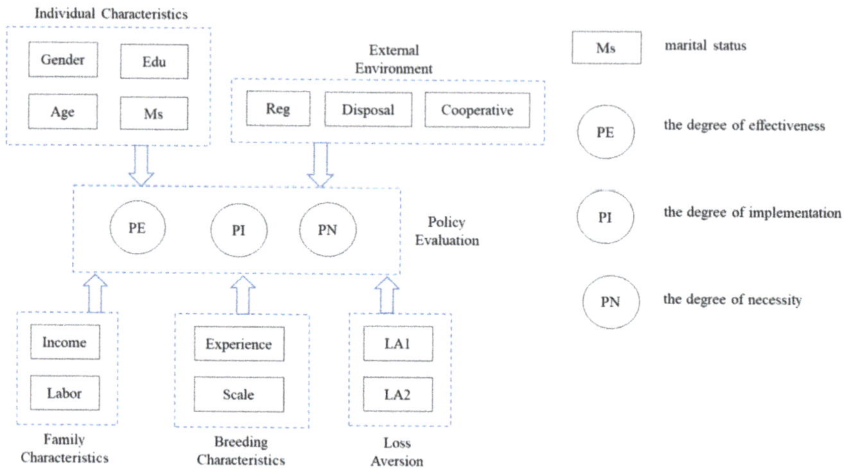

Figure 1. Research framework for factors that influence a punishment policy's evaluation.

4. Data Source and Sample Characteristics

4.1. Data Source

The data used in this study were obtained from a questionnaire survey of hog farmers in 12 provinces in China. The survey was administered using a layered design and random sampling technique and was conducted by Jiangnan University from January to March 2018. Due to the agglomeration and difference in scale of hog production among provinces, the intensity of hog breeding was divided into three categories: large, medium, and small. This paper used the number of live pigs in each province in 2016 as the standard for dividing the large, medium, and small categories. In the survey's first stage, 12 provinces were selected according to the geographical distribution and hog production scale of the sample provinces. The survey area's distribution is shown in Figure 2. In the survey's second stage, two to three sample areas (counties) were selected in provinces with a large amount of hog production, and one to two sample areas (counties) were selected in provinces with a medium or small volume. In the survey's third stage, this paper randomly selected 15–20 farmers for investigation in each county. Since the investigators are local residents, we can obtain information on all pig farmers through them and their related personnel as well as local village committees. On this basis, we randomly selected farmers. To ensure the validity of the survey and the reliability of the data, experts in the relevant field were invited to review and optimize the questionnaire and a pre-survey was conducted in Anhui province and Jiangsu province. Finally, a total of 450 questionnaires were distributed and 404 valid questionnaires were received after data cleaning.

Face-to-face interviews were conducted by well-trained interviewers who were hired from local universities in each province. Before the interview, the farmers received questionnaires issued by the investigators. Our survey was structured, but the investigators needed to provide assistance during the process of filling out the questionnaire to solve the confusion of farmers. The survey collected information on household-level and farm-level characteristics (e.g., age, sex, education, and hog livestock status) and external environmental characteristics for the loss aversion measure.

Figure 2. A map of the survey area. Note: The color change from dark to light indicates a larger to smaller number of surveyed people, respectively.

4.2. Variable Description and Sample Characteristics

Table 1 presents the definitions of and descriptive statistics for the variables that were used in this study. Individual characteristics, family characteristics, and animal breeding characteristics were included as control variables that affect the treatment of dead hogs in the system of factors influencing a punishment policy's evaluation. External conditions and environmental variables are expressed in terms of farmers' production conditions and the organizational environment. What needs special explanation is that "External environmental characteristics" mean the support and help that the government and society can provide during the breeding process. As an individual's aversion to loss is considerably affected by long-term growth in their environment, the socio-economic differences between the surveyed areas are significant. Therefore, regional dummy variables (Reg) were added to control for regional fixed effects. According to the data from many Internet companies on cities, the comprehensive strength of the 338 prefecture-level cities in China in 2017 was ranked Second-tier cities and cities above the second tier in the survey area were assigned a value of 1, and cities below the second-tier were assigned a value of 0.

The results of Table 1 show that 70.3% of the respondents were male, with few female farmers, the average age was 49.33 years old, the average education level was 7.34 years, and 96% of the farmers were married. Regarding family characteristics, the 21%–40% proportion of income from pig farming was the largest, and the number of people in the other four sections was roughly the same. The internal differentiation of pig farmers in China was relatively large. In terms of the proportion of people engaging in hog production labor, the number of people in the range 21%–40% was the highest (38.1%), and the number of people in the range of 81%–100% only accounted for 1.5%. Most farmers were engaged in hog breeding work, and family members constituted the main labor force.

The average time spent as a breeder was 12.72 years, and most of the farmers had rich experience in pig farming. Regarding the breeding scale variable, the majority of people fell into the interval between 0–50 and 51–500 pigs, accounting for 53.5% and 41.8% of the respondents, respectively. Only 4.7% of the respondents had a breeding scale of more than 500 pigs, and a considerable number of farmers conducted free-range pig farming. A total of 84.9% of the surveyed areas were cities below the second tier, with a general level of development. A total of 44.8% of the farmers indicated that there were no diseased pigs in the breeding area. Of the respondents, 68.8% indicated that they had not joined a cooperative, and the level of popularity of pig breeding cooperatives was low.

Table 1. Variable selection and feature analysis.

Variable		Variable Assignment	Frequency (Mean)	Frequency (Standard Deviation)
Individual Characteristics (IC)	Sex	Female = 0	120	29.7%
		Male = 1	284	70.3%
	Age	Years	(49.33)	(8.68)
	Education	Schooling years	(7.34)	(4.39)
	Marital Status	Unmarried = 0	16	4%
		Married = 1	388	96%
Family Characteristics (FC)	Income	0%–20% = 1	73	18.1%
		21%–40% = 2	110	27.2%
		41%–60% = 3	86	21.3%
		61%–80% = 4	72	17.8%
		81%–100% = 5	63	15.6%
	Labor	0%–20% = 1	102	25.2%
		21%–40% = 2	154	38.1%
		41%–60% = 3	98	24.3%
		61%–80% = 4	44	10.9%
		81%–100% = 5	6	1.5%
Breeding Characteristics (BC)	Experience	Numerical Value	(12.72)	(9.22)
	Scale	0–51 = 1	216	53.5%
		51–500 = 2	169	41.8%
		501–3000 = 3	14	3.5%
		3001–10,000 = 4	4	1.0%
		10,001 and above = 5	1	0.2%
Dummy Variable (DV)	Reg	Cities below the second tier = 0	343	84.9%
		Second-tier cities and above = 1	61	15.1%
External Environment (EE)	Disposal	No = 0	181	44.8%
		Yes = 1	223	55.2%
	Cooperative	No = 0	278	68.8%
		Yes = 1	126	31.2%

Note: Reg = regional dummy variables.

4.3. Ordered Probit Model

A farmer's evaluation of a punishment policy on the inappropriate treatment of dead hogs is a multi-ordered variable. The ordered Probit model is suitable for managing cases where the dependent variable is an ordered multi-class discrete variable. In the dependent variable that was selected in this study, the degree of necessity indicator was divided according to the five-point Likert scale as follows: Y1 = 1, which means that the policy is totally unnecessary; Y2 = 2, which means that the policy is unnecessary; Y3 = 3, which means that the policy is generally necessary; Y4 = 4, which means that the policy is necessary; and Y5 = 5, which means that the policy is very necessary. The degree of effectiveness and the degree of implementation indicators were similarly divided according to the five-point Likert scale.

The general form of the model can be expressed as follows:

$$\begin{aligned} Y^* &= \beta_0 + \beta_1 Gender + \beta_2 Age + \beta_3 Edu + \beta_4 Ms + \beta_5 Income \\ &+ \beta_6 Lab + \beta_7 Exp + \beta_8 Scale + \beta_9 \mathrm{Reg} + \beta_{10} Disposal \\ &+ \beta_{11} Cooperative + \beta_{12} LA1/LA2 + \xi \end{aligned}, \tag{1}$$

where is a latent variable, is a residual term that is subject to a normal distribution with a variance of σ, β_0, β_1, β_2, ..., β_{12} are the coefficients to be estimated. Let Y denote the farmers' evaluation of the punishment policy. The greater the value, the more positive the farmers' evaluation of the policy. Assuming there are tangent points k_1, k_2, k_3, the relationship between Y and Y^* can be expressed as follows:

$$Y = \begin{cases} 1 \text{ if } Y^* \leq k_1 \\ 2 \text{ if } k_1 < Y^* \leq k_2 \\ 3 \text{ if } k_2 < Y^* \leq k_3 \\ 4 \text{ if } k_3 < Y^* \leq k_4 \\ 5 \text{ if } Y^* > k_4 \end{cases}. \tag{2}$$

Writing Equation (1) in its vector form: this paper can derive:

$$\begin{aligned} &\Pr(Y = 1) = \Pr(Y^* \leq k_1) = \Pr(X\beta + \xi \leq k_1) \\ &= \Pr(\xi \leq k_1 - X\beta) = \Phi(k_1 - X\beta) \end{aligned}. \tag{3}$$

As such, this paper can obtain:

$$\Pr(Y = 2) = \Phi(k_2 - X\beta) - \Phi(k_1 - X\beta), \tag{4}$$

$$\Pr(Y = 3) = \Phi(k_3 - X\beta) - \Phi(k_2 - X\beta), \tag{5}$$

$$\Pr(Y = 4) = \Phi(k_4 - X\beta) - \Phi(k_3 - X\beta), \tag{6}$$

$$\Pr(Y = 5) = 1 - \Phi(k_4 - X\beta), \tag{7}$$

where $\Phi(\bullet)$ represents the cumulative distribution function of the general normal distribution. Unlike in the general least squares estimation method, the explanatory variables in the multi-ordered Probit model describe a probability problem, whose solution can be estimated by the maximum likelihood method [26].

5. Experimental Design

To obtain more realistic micro-scale data on each respondent farmer's degree of aversion to loss, this paper measured the degree of aversion to loss using an experimental economics method. The authors Kemel et al., using a loss aversion measure [27], found a difference between the degree of aversion to a loss of personal freedom and the degree of aversion to a loss of money. When a public policy involves detention and fines, an individual will exhibit a different degree of aversion to loss. With respect to detention, people have a higher degree of aversion to loss [27]. Therefore, two items, detention and fines, were set to obtain and compared different loss aversion measures. Based on the MPL proposed by Tanaka et al. [20] and Holt and Laury [21], this paper inputs the punishments of fines and detention for the inappropriate treatment of dead hogs into the MPL, which enabled the farmers to understand and perceive the experimental design more realistically to improve the experiment's authenticity and effectiveness. In the experiment's first stage, to check whether the respondents truly understood the meaning of the questions that were being asked, at the beginning of the experiment, the investigators introduced the rules of the experiment to the respondents and helped the respondents to follow them.

The investigators asked the respondents: "If the local government department was to impose penalties on pig production farmers who randomly dispose of or sell dead hogs, which option would you choose in the following two situations?" Table 2 lists the options.

Table 2. Options in the experimental exercise.

Question Number	The First Case	The Second Case
1	50% may be found and a fine 1000 yuan, and 50% may not be found	100% found and a fine of 500 yuan

According to each farmer's response, the investigators recorded the farmer's choice between the two situations. Since the first case has a certain probability of generating a fine of 1000 yuan but there is also the possibility of not being fined, farmers with a lower degree of loss aversion will choose this option. In contrast, although the second case will result in a fine (500 yuan) that is lower than that in the first, it is an inevitable event, so those with higher degree of loss aversion will prefer this option to avoid high fines. If the respondent was able to make a choice and provide a reasonable explanation, then this paper considered the responder to understand the question thoroughly and conducted a formal experiment. When the respondent indicated that they did not understand the question or arbitrarily answered it, then the investigator was required to explain the question in more detail until the respondent truly understood, and then conducted the formal experiment.

In the experiment's second stage, after the respondents were familiarized with the experiment's rules, the investigators provided two sets of 20 multiple-choice questions. In the MPL's first group, 10 groups of fines for the inappropriate handling of dead hogs were set as the multiple-choice questions for the study sites (Table 3). In the MPL's second group, 10 groups of detention periods for the inappropriate treatment of dead hogs were set as the multiple-choice questions (Table 4). In both sets of experimental protocols, the probability of being fined was 50%, which was designed to exclude farmers from being affected by probability weighting. The first option in the MPL's first group was 100% probability of being fined 50 yuan or 100–500 yuan, and the first group of the MPL's second group was 100% probability of being detained for 1 or 2 h. At 10 h, the same period of detention and the amount of fines corresponded one-to-one. According to the principle of high risk and high return, the first case in each MPL group is a low risk option, and the second case is a high risk option, allowing respondents to choose between the two schemes. In the experiment's third stage, based on the final choices of the respondents, this paper was able to provide the option to jump from the first case to the second case. According to each farmer's actual selection, this paper calculated their loss aversion measure: LA1 = (the selection of the first case or the second case in the fine question)/10, LA2 = (the option from the first case to the second case in the detention question)/10. If a loss aversion measure is 1, the respondent's loss aversion is high. If a loss aversion measure is 0, the respondent's loss aversion is low. The investigators asked the respondents: "The local government department imposes fines on pig farmers who randomly dispose of or sell dead hogs. Will you choose the first or the second case?" Table 3 shows the options. Investigators were required to emphasize that the investigation would not affect the respondent's farm's breeding qualifications and would not affect their reputation and information was collected for research purposes only. Given farmers' perception of the punishment policy, and to avoid unrealistic answers due to farmers' resistance, the questionnaire indicated that "this paper only knows your opinion about the fine and will not report you". This paper required the investigators to make this clear to the farmers during the investigation.

Table 3. Punishment policy penalty amount options (Group 1).

Question Number	The First Case	The Second Case
1	100% found and a fine of 50 yuan	50% may be found and a fine of 500 yuan
2	100% found and a fine of 100 yuan	50% may be found and a fine of 500 yuan
3	100% found and a fine of 150 yuan	50% may be found and a fine of 500 yuan
4	100% found and a fine of 200 yuan	50% may be found and a fine of 500 yuan
5	100% found and a fine of 250 yuan	50% may be found and a fine of 500 yuan
6	100% found and a fine of 300 yuan	50% may be found and a fine of 500 yuan
7	100% found and a fine of 350 yuan	50% may be found and a fine of 500 yuan
8	100% found and a fine of 400 yuan	50% may be found and a fine of 500 yuan
9	100% found and a fine of 450 yuan	50% may be found and a fine of 500 yuan
10	100% found and a fine of 500 yuan	50% may be found and a fine of 500 yuan

Table 4. Punishment policy detention time options (Group 2).

Question Number	The First Case	The Second Case
1	100% found and detention for 1 h	50% may be found and detention for 10 h
2	100% found and detention for 2 h	50% may be found and detention for 10 h
3	100% found and detention for 3 h	50% may be found and detention for 10 h
4	100% found and detention for 4 h	50% may be found and detention for 10 h
5	100% found and detention for 5 h	50% may be found and detention for 10 h
6	100% found and detention for 6 h	50% may be found and detention for 10 h
7	100% found and detention for 7 h	50% may be found and detention for 10 h
8	100% found and detention for 8 h	50% may be found and detention for 10 h
9	100% found and detention for 9 h	50% may be found and detention for 10 h
10	100% found and detention for 10 h	50% may be found and detention for 10 h

The investigators asked the respondents: "The local government department will detain pig farmers who randomly dispose of dead hogs. Would you choose the first case or the second case?" Table 4 shows the options. Investigators were required to emphasize that the investigation would not affect the respondent's farm's breeding qualifications and would not affect their reputation.

Figure 3 shows the frequency distribution of farmers' aversion index for fines and detention losses. The penalty and detention loss aversion index is the highest in 0.5–0.6. The number of farmers with a loss aversion index greater than 0.5 is higher than the number of farmers with a loss aversion index less than 0.5. This result shows that most farmers had a higher degree of aversion to loss, whether facing fines or detention. In the first experimental protocol, the average loss aversion index of the farmers was 0.59. In the second experimental protocol, the average loss aversion index of the farmers was 0.61. These results show that the loss aversion index for detention was slightly larger than that for fines.

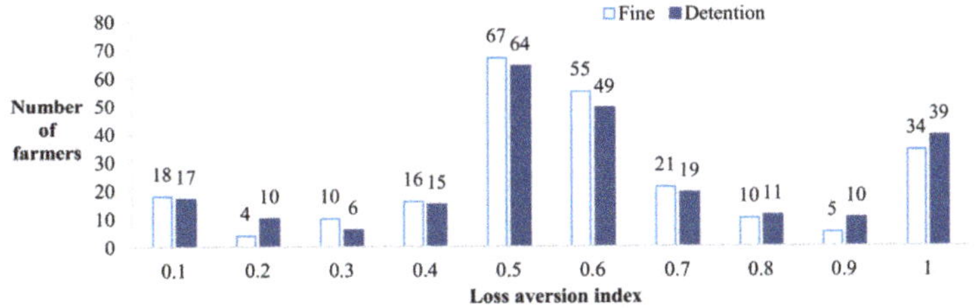

Figure 3. Frequency distribution of the aversion coefficient of loss to the amount of the fine and the detention time of farmers.

6. Empirical Analysis

6.1. Current Status of Farmers' Evaluation of Punishment Policies for the Inappropriate Treatment of Dead Hogs

In this study, farmers' evaluation of the punishment policy for the inappropriate treatment of dead hogs was divided into three indicators according to the five-point Likert scale—An evaluation of the degree of necessity (PN), the degree of implementation (PI), and the degree of effectiveness (PE) of the punishment policy. This paper then collected data on the farmers' evaluation of the punishment policy in the survey area. These data are shown in Table 5.

Table 5. Frequency distribution of farmers' evaluation of punishment policies for the inappropriate treatment of dead hogs.

	Variable	Variable Assignment	Frequency	Rate of Recurrence
Policy evaluation	How necessary do you think the punishment policy is for the inappropriate treatment of dead hogs? (PN)	Totally unnecessary = 1	12	3.00%
		Unnecessary = 2	31	7.70%
		General = 3	73	18.10%
		Necessary = 4	144	35.60%
		Very necessary = 5	144	35.60%
	How well do you think the local government is implementing the punishment policy for the inappropriate treatment of dead hogs? (PI)	Not executed at all = 1	27	6.70%
		Basically not Executed = 2	72	17.80%
		General = 3	108	26.70%
		Mostly executed = 4	109	27.00%
		Fully executed = 5	88	21.80%
	How effective do you think the current policy is for the inappropriate treatment of dead hogs? (PE)	Totally ineffective = 1	21	5.20%
		No effect = 2	38	9.40%
		General = 3	103	25.50%
		Mostly effective = 4	139	34.40%
		Very effective = 5	103	25.50%

The data shown in Table 5 represent the current status of the farmers' evaluation of the punishment policy. Firstly, in the evaluation of the degree of necessity of the punishment policy (PN), the "necessary" and "very necessary" options accounted for 35.6% of the farmers, and only 10.7% of the farmers think that the punishment policy is unnecessary. That is, most farmers had a positive evaluation of the policy in terms of its degree of necessity. For the degree of implementation (PI), 51.2% of the farmers provided a score of three or below, indicating that most farmers believed that the degree of implementation of the punishment policy is low. Finally, regarding the evaluation of the degree of effectiveness (PE) of the punishment policy, 59.9% of the farmers believed that the punishment policy is mostly effective in regulating the behavior of the farmers, and 40.1% of the farmers provided scores of three or below. The above analysis shows that problems remain with the current status of farmers' evaluation of the punishment policy, and the overall evaluation is not positive, especially in terms of implementation and effectiveness.

6.2. Results and Discussion

In this study, this paper investigated the factors that influence farmers' evaluation of a punishment policy for the inappropriate treatment of dead hogs. The dependent variable in Models I and II is the evaluation of the degree of necessity of the punishment policy. Model I adds a loss aversion measure of a pecuniary penalty and Model II adds a loss aversion measure of detention time. The dependent variable in Models III and IV is the evaluation of the degree of implementation of the punishment policy. Model III adds a loss aversion measure of a pecuniary penalty and Model IV adds a loss aversion measure of detention time. The dependent variable in Models V and VI is the evaluation of the degree of effectiveness of the punishment policy. Model V adds a loss aversion measure of a pecuniary penalty and Model VI adds a loss aversion measure of detention time. The analysis was performed using Stata version 15.1 (STATA, North Carolina, USA). Tables 6–11 shows the results. This paper estimates the effect on the probability of choosing each category (e.g., the effect on the probability of choosing "Unnecessary = 1" to "Unnecessary = 5" in the necessary for the punishment policy), separately, and we use "Margin (1)" to "Margin (5)" to express the probability.

Table 6. Estimation results of the formation mechanism of policy evaluation (Model I).

Variable	Model I					
	Coefficients (DV = PN)	**Margin (1)**	**Margin (2)**	**Margin (3)**	**Margin (4)**	**Margin (5)**
Sex	−0.010 (0.118)	0.002	0.003	0.001	−0.002	−0.004
Age	0.005 (0.007)	−0.001	−0.001	−0.000	−0.000	0.001
Education	0.016 (0.276)	0.001	0.001	−0.000	−0.001	−0.002
Marital Status	0.015 (0.019)	−0.002	−0.002	−0.000	0.001	0.003
Income	0.156 *** (0.054)	−0.024	−0.028	−0.006	0.017	0.041
Labor	0.048 (0.062)	−0.006	−0.007	−0.001	0.004	0.010
Experience	−0.016 ** (0.007)	0.002	0.003	0.001	−0.002	−0.004
Scale	0.204 * (0.106)	−0.037	−0.042	−0.009	0.026	0.062
Reg	0.197 (0.173)	−0.026	−0.030	−0.007	0.019	0.044
Disposal	0.391 *** (0.117)	−0.060	−0.069	−0.015	0.043	0.101
Cooperative	0.241 ** (0.116)	−0.036	−0.041	−0.009	0.025	0.060
LA1	0.001 (0.021)	−0.009	−0.010	−0.002	0.006	0.015
LA2						
Observations	404					
Log likelihood	−203.751					
Prob > chi^2	0.000					
Pseudo R^2	0.159					

Note: *, **, and *** denote significance at the statistical levels of 10%, 5%, and 1%, respectively. The numbers in parentheses are standard errors. DV means dependent variable.

Table 7. Estimation results of the formation mechanism of policy evaluation (Model II).

Variable	Model II					
	Coefficients (DV = PN)	**Margin (1)**	**Margin (2)**	**Margin (3)**	**Margin (4)**	**Margin (5)**
Sex	−0.015 (0.118)	0.002	0.003	0.001	−0.002	−0.004
Age	0.004 (0.007)	−0.001	−0.001	0.000	0.000	0.001
Education	0.008 (0.275)	0.001	0.001	0.000	−0.001	−0.002
Marital Status	0.012 (0.019)	−0.002	−0.002	0.000	0.001	0.003
Income	0.164 *** (0.053)	−0.024	−0.028	−0.006	0.017	0.041
Labor	0.039 (0.062)	−0.006	−0.007	−0.001	0.004	0.010
Experience	−0.016 ** (0.007)	0.002	0.003	0.001	−0.002	−0.004
Scale	0.251 ** (0.107)	−0.037	−0.042	−0.009	0.026	0.062
Reg	0.177 (0.173)	−0.026	−0.030	−0.007	0.019	0.044
Disposal	0.408 *** (0.117)	−0.060	−0.069	−0.015	0.043	0.101
Cooperative	0.244 ** (0.116)	−0.036	−0.041	−0.009	0.025	0.060
LA1						
LA2	0.060 *** (0.023)	−0.009	−0.010	−0.002	0.006	0.015
Observations	404					
Log likelihood	−203.982					
Prob > chi^2	0.000					
Pseudo R^2	0.158					

Note: *, **, and *** indicate significance at the statistical levels of 10%, 5%, and 1%, respectively. The numbers in parentheses are standard errors.

Table 8. Estimation results on the formation mechanism of policy evaluation (Model III).

Variable	Model III					
	Coefficients (DV = PI)	**Margin (1)**	**Margin (2)**	**Margin (3)**	**Margin (4)**	**Margin (5)**
Sex	−0.161 (0.119)	0.039	0.013	−0.011	−0.023	−0.018
Age	0.011 (0.007)	−0.003	−0.001	0.001	0.002	0.001
Education	0.005 (0.019)	−0.021	−0.007	0.006	0.012	0.010
Marital Status	0.086 (0.288)	−0.001	0.000	0.000	0.001	0.001
Income	0.050 (0.054)	−0.012	−0.004	0.004	0.007	0.006

Table 8. *Cont.*

Variable	Model III					
	Coefficients (DV = PI)	Margin (1)	Margin (2)	Margin (3)	Margin (4)	Margin (5)
Labor	0.255 *** (0.064)	−0.062	−0.021	0.018	0.037	0.029
Experience	−0.018 *** (0.007)	0.004	0.002	−0.001	−0.003	−0.002
Scale	0.296 *** (0.108)	−0.072	−0.025	0.021	0.042	0.034
Reg	0.237 (0.173)	−0.058	−0.020	0.017	0.034	0.027
Disposal	0.738 *** (0.118)	−0.180	−0.062	0.052	0.106	0.084
Cooperative	0.119 (0.118)	−0.029	−0.010	0.008	0.017	0.014
LA1	0.044 ** (0.021)	−0.011	−0.004	0.003	0.006	0.005
LA2						
Observations	404					
Log likelihood	-550.397					
Prob > chi^2	0.000					
Pseudo R^2	0.107					

Note: *, **, and *** denote significance at the statistical levels of 10%, 5%, and 1%, respectively. The numbers in parentheses are standard errors.

Table 9. Estimation results on the formation mechanism of policy evaluation (Model IV).

Variable	Model IV					
	Coefficients (DV = PI)	Margin (1)	Margin (2)	Margin (3)	Margin (4)	Margin (5)
Sex	−0.163 (0.119)	0.040	0.014	−0.012	−0.024	−0.019
Age	0.010 (0.007)	−0.002	−0.001	0.001	0.001	0.001
Education	0.004 (0.019)	−0.006	−0.002	0.002	0.004	0.003
Marital Status	0.025 (0.286)	−0.001	0.000	0.000	0.001	0.000
Income	0.026 (0.053)	−0.006	−0.002	0.002	0.004	0.003
Labor	0.256 *** (0.064)	−0.063	−0.022	0.018	0.037	0.029
Experience	−0.019 *** (0.007)	0.005	0.002	−0.001	−0.003	−0.002
Scale	0.333 *** (0.109)	−0.081	−0.029	0.024	0.048	0.038
Reg	0.213 (0.172)	−0.052	−0.018	0.015	0.031	0.024
Disposal	0.761 *** (0.118)	−0.186	−0.065	0.054	0.110	0.087

Table 9. *Cont.*

Variable	Model IV					
	Coefficients (DV = PI)	Margin (1)	Margin (2)	Margin (3)	Margin (4)	Margin (5)
Cooperative	0.127 (0.118)	−0.031	−0.011	0.009	0.018	0.015
LA1						
LA2	0.016 (0.023)	−0.004	−0.001	0.001	0.002	0.002
Observations	404					
Log likelihood	−552.298					
Prob > chi^2	0.000					
Pseudo R^2	0.104					

Note: *, **, and *** denote significance at the statistical levels of 10%, 5%, and 1%, respectively. The numbers in parentheses are standard errors.

Table 10. Estimation results on the formation mechanism of policy evaluation (Model V).

Variable	Model V					
	Coefficients (DV = PE)	Margin (1)	Margin (2)	Margin (3)	Margin (4)	Margin (5)
Sex	−0.096 (0.119)	0.026	0.006	−0.013	−0.010	−0.009
Age	0.006 (0.007)	−0.002	0.000	0.001	0.001	0.001
Education	0.022 (0.019)	0.034	0.008	−0.017	−0.013	−0.012
Marital Status	−0.125 (0.282)	−0.006	−0.001	0.003	0.002	0.002
Income	0.002 (0.054)	0.000	0.000	0.000	0.000	0.000
Labor	0.253 *** (0.064)	−0.068	−0.016	0.034	0.027	0.024
Experience	−0.009 (0.007)	0.002	0.001	−0.001	−0.001	−0.001
Scale	0.559 *** (0.112)	−0.151	−0.035	0.075	0.059	0.052
Reg	0.169 (0.174)	−0.046	−0.011	0.023	0.018	0.016
Disposal	0.289 ** (0.118)	−0.078	−0.018	0.039	0.031	0.027
Cooperative	0.259 ** (0.119)	−0.070	−0.016	0.035	0.027	0.024
LA1	0.022 (0.022)	−0.006	−0.001	0.003	0.002	0.002
LA2						
Observations	404					
Log likelihood	−527.046					
Prob > chi^2	0.000					
Pseudo R^2	0.094					

Note: *, **, and *** denote significance at the statistical levels of 10%, 5%, and 1%, respectively. The numbers in parentheses are standard errors.

Table 11. Estimation results on the formation mechanism of policy evaluation (Model VI).

Variable	Model VI					
	Coefficients (DV = PE)	Margin (1)	Margin (2)	Margin (3)	Margin (4)	Margin (5)
Sex	−0.091 (0.120)	0.024	0.006	−0.012	−0.009	−0.008
Age	0.006 (0.007)	−0.002	0.000	0.001	0.001	0.001
Education	0.024 (0.019)	0.037	0.009	−0.018	−0.014	−0.013
Marital Status	−0.137 (0.282)	−0.006	−0.001	0.003	0.002	0.002
Income	−0.017 (0.053)	0.004	0.001	−0.002	−0.002	−0.002
Labor	0.263 *** (0.064)	−0.071	−0.017	0.035	0.028	0.025
Experience	−0.010 (0.007)	0.003	0.001	−0.001	−0.001	−0.001
Scale	0.549 *** (0.113)	−0.147	−0.035	0.073	0.057	0.051
Reg	0.179 (0.174)	−0.048	−0.011	0.024	0.019	0.017
Disposal	0.292 ** (0.118)	−0.078	−0.018	0.039	0.031	0.027
Cooperative	0. 264 ** (0.119)	−0.071	−0.017	0.035	0.028	0.025
LA1						
LA2	−0.042 * (0.023)	0.011	0.003	−0.006	−0.004	−0.004
Observations	404					
Log likelihood	−525.860					
Prob > chi^2	0.000					
Pseudo R^2	0.096					

Note: *, **, and *** denote significance at the statistical levels of 10%, 5%, and 1%, respectively. The numbers in parentheses are standard errors.

The pseudo coefficient of determination (R^2) values in Table 6 suggest that our estimations have a good degree of fit across the models. This paper first analyzed the effect of loss aversion. In Model II, the respondents' aversion to detention time has a positive effect on the evaluation of the degree of policy implementation at the 1% level. That means that the higher the loss aversion, the better the evaluation of the degree of necessity. In Model III, the respondents' aversion to the amount of the fine in the penalty policy has a positive effect on the evaluation of the degree of policy implementation at the 5% level: the greater the coefficient of the loss aversion at the level of the monetary sum, the smaller the amount of speculative psychology, and the better the evaluation of the degree of implementation. The loss aversion measure of detention time did not have a significant effect on the evaluation of the degree of implementation. From the perspective of loss aversion, when a farmer receives a fine as punishment for the inappropriate handling of dead hogs and this produces resistance, the better the evaluation of the degree of the policy's implementation. However, farmers' degree of aversion to loss in terms of detention does not significantly affect their evaluation of the degree of the policy's implementation. The reason for this may be that the amount of the fine in the punishment policy can be quantified, is easy to implement, and is more relevant to the interests of most farmers. Therefore, the degree of aversion to a monetary loss is more related to the evaluation of the degree of the policy's implementation. The particulars of a detention policy may be vague and difficult to implement. Policy-makers can use farmers' aversion to the amount of the fine and its impact on the level of implementation to develop policies that are more restrictive and better regulate the behavior of farmers. Notably, the loss aversion measure is different in Model VI. In Model VI, after adding an abandonment coefficient of the time limit for the detention policy, the loss aversion measure was found to be negatively affected by the degree of effectiveness of the punishment policy at the 10% significance level: the stronger the farmers' aversion to a loss of personal freedom caused by a punishment policy, the lower their evaluation of the effect of the punishment policy on the disposal of dead hogs. The possible reason for this is that in real life, the government penalties that farmers face in the study area or in the surrounding areas are mostly punishments involving money. Only when there is a major epidemic or a food poisoning incident will the persons involved be detained. This paper concludes that the detention time clause in the punishment policy has a poor effect on farmers' inappropriate handling of dead hogs. Therefore, the greater the aversion coefficient of the farmers' evaluation of the amount of the fine, the more the farmers believe that the current punishment policy is effectively regulating their treatment of dead hogs. The greater the aversion coefficient of the farmers' loss with respect to detention time, the more the farmers believe that the current punishment policy's detention time is not able to effectively regulate their treatment of dead hogs. Accordingly, policy-makers should provide a more detailed description of the detention duration when creating policies to penalize farmers for inappropriate treatment of dead hogs.

Regarding the six models, the animal breeding scale was found to have a positive impact on the evaluation of degree of necessity, the degree of implementation, and the degree of effectiveness of the punishment policy. The reasons for the impact of scale and labor were found to be similar. The more energy and financial resources that farmers invest in the breeding of hogs, the higher the degree of specialization in hog breeding. Farmers prefer the government to impose penalties. The policy regulates the hog breeding market, creates a good market environment, and prevents the occurrence of inappropriate disposal of dead hogs. In the external environment, disposal points for dead hogs were found to have a positive impact on the degree of necessity, the degree of implementation, and degree of effectiveness of the punishment policy at the 1% and 5% levels of significance. The construction of equipment to dispose of dead hogs is the measure that the government uses to regulate the disposal of dead hogs. The availability of disposal equipment makes farmers feel that the government is paying attention to problems associated with the disposal of dead hogs and raising awareness about the harmless disposal of dead hogs, thus improving the farmers' evaluation of the degree of necessity of a punishment policy. The existence of disposal equipment for dead hogs improves the efficiency of the disposal of dead hogs, and the government can transmit a message of safe disposal of disease-carrying hogs. Additionally, government-created disposal sites for dead hogs have a higher degree of regulation of the disposal of dead hogs, and are more strict in the implementation of their policies, which may increase farmers' evaluation of the degree of implementation and effectiveness of punishment policies.

Finally, the unique characteristics of farmers were found to be factors that influenced the different indicators of punishment policy evaluation. In Models I and II, the income level of the farmers had a positive impact on the perception of degree of necessity of the punishment policy at the 1% significance level. The higher the income level, the higher the degree of necessity. The higher the household's income, the more likely farmers are to protect their property and to require policies to protect them. The labor was found to have a positive impact on the degree of implementation and degree of effectiveness at the 1% significance level. The larger the workforce that farmers allocate to hog breeding, the more they hope that the hog market functions in a virtuous circle. Therefore, the higher their evaluation of the indicators of the punishment policy, the better their evaluation of the degree of implementation and the degree of effectiveness. In Models I–IV, breeding time was negatively affected by the degree of implementation at the 1% significance level. The descriptive statistics show that the farmers' evaluation of the implementation degree of the punishment policy is poor, indicating that the current punishment policy has not been well-implemented. Farmers with more years of experience in hog breeding would have been aware of this phenomenon for a longer period, so their evaluation of the degree of implementation would be worse. In Models I, II, V, and VI, at the external environment level, farmers who had joined a hog breeding cooperative were found to have a better evaluation of the degree of effectiveness. Farmers that participate in hog breeding cooperatives can be exposed to more normative management and systematic training, have a higher familiarity with and pay more attention to government policies, and better evaluate the effectiveness of the punishment policy.

The marginal effect from the Probit model enabled us to identify different degrees of change for each evaluation indicator. For example, in Model I, when the sex variable is changed from female to male, Margin (2) is 0.003 and Margin (4) is −0.002, which indicates that the evaluation of the degree of necessity first rises and then falls. As the trends are all different and there are many variables, this paper will not repeat them all here.

6.3. Policy Recommendations

Based on our results, this paper provides the following policy recommendations. First, the human capital structure of hog farmers should be optimized, and a high-quality farmer team should be built. In mass media, reports of highly educated individuals returning to the countryside to plant or breed are increasing, and highly educated farmers rely on higher education levels and large-scale farming to achieve greater economic benefits. Therefore, both the media and the government ought to provide more positive publicity to encourage college students to return to their hometowns to improve the quality and level of the entire group of farmers. Second, moderate-scale breeding should be promoted and the degree of specialization of farmers should be increased. If large-scale farms (households) invest more in farming and are more dependent on the generated income, they will face greater barriers to performing inappropriate actions. Publicity and training to improve the level of professional farming can encourage farmers to focus on advanced technology for handling diseased dead hogs rather than on dealing with penalties for violations. Third, more equipment for the disposal of dead hogs should be constructed and farmers should be encouraged to participate in hog breeding cooperatives. Strengthening the supervision of the construction of disposal sites for dead hogs, implementing the government's preferred policies for the disposal of dead hogs, and providing more convenient channels for the disposal of dead hogs are additional recommendations. A hog breeding cooperative should be constructed and implemented to create a standardized breeding environment for farmers and farmers should be encouraged to participate actively. Fourth, more stringent regulations and laws should be formulated to combat misconduct and strengthen the government's enforcement of inappropriate behavior. The penalties for fines and detention should be clearer, the costs associated with violations should be increased, the idea of farmers' taking risks with respect to violations needs to be dispelled, and legal means should be used to enhance the compliance of farmers with laws. The handling of dead hogs must be standardized. Fifth, the "boost" approach to normative behavior should be used. The purpose of boosts in behavioral economics is to influence people's choices and make them profitable on the basis of decision-makers' judgments. According to this, after an incident involving the inappropriate disposal of dead hogs and a legal ruling, the local government may require the hog farmers to report the progress of the incident and publicize it via SMS. A boost uses the psychology of people's aversion to loss to guide them to make correct choices in an unenforced way, and to subtly strengthen their awareness of legal norms. Sixth, economic methods should be used to determine the optimal amount of a fine and detention time. A method for specifying the amount of a fine and the time of detention through formulas is worth developing. A potential formula is penalty strength = the loss caused by misconduct/the probability of misconduct being discovered.

7. Conclusions

In this study, this paper investigated the loss aversion of hog production farmers using an experimental economics method, and examined the determinants of loss aversion using an ordered Probit model. This paper used questionnaire survey data collected from 404 hog production farmers from 12 provinces in China in 2018. The results show that farmers are more averse to detention time than to fines. Kemel et al. suggested that in public policy, individuals are more averse to personal security threats than losses from fines [27]. These two conclusions are similar. Regarding the three indicators of a policy's evaluation, farmers' evaluation of the effectiveness is low, and their evaluation of the degree of implementation level is the lowest. Breeding scale and disposal points for dead hogs were both found to be associated with each of the three indicators to a certain extent. The remaining variables had different effects on different dependent variables. The aversion coefficient of the time of detention in the punishment policy had a positive effect on the evaluation of the degree of necessity of the policy. The aversion coefficient of the amount of the fine in the punishment policy had a positive effect on the evaluation of the degree of implementation, and the aversion coefficient of the time of detention in the punishment policy had a negative effect on the evaluation of the degree of effectiveness of the policy. With respect to the problem of endogeneity, this paper selected as many control variables as possible that affect both the explanatory variables and the explained variables to address any endogenous problems, such as missing variables [28,29].

Author Contributions: C.Y. performed most of the work. The paper was revised finalized based on J.W.'s comments and suggestions, and J.W. also had a role in supervising the work.

Funding: The research was funded by the Qing Lan Project; the National Natural Science Foundation of China (grant numbers 71673115 and 71540008); the Jiangnan University Basic Research Plan—Social Science Major Projects (grant number. JUSRP1808ZD); and the Postgraduate Research & Practice Innovation Program of Jiangsu Province (grant number. KYCX19_1885).

Conflicts of Interest: The authors declare no conflicts of interest.

References

1. Gimpel, H. Loss Aversion and Reference-Dependent Preferences in Multi-Attribute Negotiations. *Group Decis. Negot.* **2007**, *16*, 303–319. [CrossRef]
2. Tversky, A.; Kahneman, D. Loss Aversion in Riskless Choice: A Reference-dependent Model. *Q. J. Econ.* **1991**, *106*, 1039–1061. [CrossRef]
3. Kahneman, D.; Tversky, A. Prospect theory: An Analysis of Decision under Risk. *Econ. J. Econ. Soc.* **1979**, *47*, 263–291. [CrossRef]
4. Kahneman, D.; Knetsch, J.L.; Thaler, R.H. Experimental Tests of the Endowment Effect and the Coase Theorem. *J. Political Econ.* **1990**, *98*, 1325–1348. [CrossRef]
5. Chen, X.; Qiu, G.; Wu, L.; Xu, G.; Wang, J.; Hu, W. Influential Impacts of Combined Government Policies for Safe Disposal of Dead Pigs on Farmer Behavior. *Env. Sci. Pollut. Res. Int.* **2017**, *24*, 3997–4007. [CrossRef] [PubMed]
6. Fei, W. Game Analysis on the Safety Problem of Recycling and Processing in Food Supply Chain—Taking the Case of "Abandoning Pig" as an Example. *Issues Agric. Econ.* **2015**, *36*, 94–101. (In Chinese)
7. Crabbé, A.; Leroy, P. *The Handbook of Environmental Policy Evaluation*; Earthscan: London, UK, 2008.
8. Xiaolin, T.; Jieqiong, W.; Biao, Z.; Lingxian, Z. Application of the DEA on the Performance Evaluation of the Agricultural Support Policy in China. *Agric. Econ. Zemed. Ekon.* **2017**, *63*, 510–523. [CrossRef]
9. Moser, S.; Mußhoff, O. Ex-ante Evaluation of Policy Measures: Effects of Reward and Punishment for Fertiliser Reduction in Palm Oil Production. *J. Agric. Econ.* **2016**, *67*, 84–104. [CrossRef]
10. Lynch, J.; Donnellan, T.; Finn, J.A.; Dillon, E.; Ryan, M. Potential Development of Irish Agricultural Sustainability Indicators for Current and Future Policy Evaluation Needs. *J. Env. Manag.* **2019**, *230*, 434–445. [CrossRef]

11. Borges, P.J.; Fragoso, R.; Garciagonzalo, J. Assessing Impacts of Common Agricultural Policy Changes on Regional Land Use Patterns with a Decision Support System: An Application in Southern Portugal. *Policy Econ.* **2010**, *12*, 111–120. [CrossRef]
12. Zhou, F.J.; Liu, X. Research on Satisfaction of Agricultural Subsidy Policy and Its Influencing Factors—Based on Probit Regression Model. *Price Theory Pr.* **2018**, *2*, 75–78. (In Chinese)
13. Pufahl, A.; Weiss, C.R. Evaluating the Effects of Farm Programmes: Results from Propensity Score Matching. *Eur. Assoc. Agric. Econ.* **2009**, *36*, 79–101. [CrossRef]
14. Barbiroli, G. Economic Consequences of the Transition Process toward Green and Sustainable Economies: Costs and Advantages. *Int. J. Sustain. Dev. World Ecol.* **2011**, *18*, 17–27. [CrossRef]
15. Zhan, L. Effectiveness Evaluation of High-Tech Industrial Policy of Hubei Province. *Forum Sci. Technol. China* **2015**, *5*, 89–95. (In Chinese)
16. Huang, Z.; Du, X. Assessment and Determinants of Residential Satisfaction with Public Housing in Hangzhou, China. *Habitat Int.* **2015**, *47*, 218–230. [CrossRef]
17. Zhou, F.Z.; Yang, Y.M.; Chen, H. Reform Exploration on the Use Control of Territory Space: An Application of Libertarian Paternalism. *China Land Sci.* **2018**, *32*, 23–29. (In Chinese)
18. Kibeta, N.; Obareb, G.A.; Lagatb, J.K. Risk Attitude Effects on Global-GAP Certification Decisions by Smallholder French Bean Farmers in Kenya. *J. Behav. Exp. Financ.* **2018**, *18*, 18–29. [CrossRef]
19. Tanaka, T.; Camerer, C.; Nguyen, Q. Risk and Time Preferences: Linking Experimental and Household Survey Data from Vietnam. *Am. Econ. Rev.* **2010**, *100*, 557–571. [CrossRef]
20. Holt, C.A.; Laury, S.K. Risk Aversion and Incentive Effects. *Am. Econ. Rev.* **2002**, *92*, 1644–1655. [CrossRef]
21. Campbell, J.Y.; Mankiw, N.G. Permanent Income, Current Income, and Consumption. *Soc. Sci. Electron. Publ.* **1990**, *8*, 265–279.
22. Nelson, R.R.; Winter, S.G. Evolutionary Theorizing in Economics. *J. Econ. Perspect.* **2002**, *16*, 23–46. [CrossRef]
23. Binswanger, H.P. Attitudes Toward Risk: Experimental Measurement in Rural India. *Am. J. Agric. Econ.* **1980**, *62*, 395–407. [CrossRef]
24. Gong, Y.; Baylis, K.; Kozak, R.; Bull, G. Farmers' Risk Preferences and Pesticide Use Decisions: Evidence From Field Experiments in China. *Agric. Econ.* **2016**, *47*, 411–421. [CrossRef]
25. Liu, E.M.; Huang, J.K. Risk Preferences and Pesticide Use by Cotton Farmers in China. *J. Dev. Econ.* **2013**, *103*, 202–215. [CrossRef]
26. Jayachandran, N. A Probit Latent Variable Model of Nutriton Information and Dietary Intake. *Am. J. Agric. Econ.* **1996**, *78*, 628–639.
27. Kemel, E.; Paraschiv, C. Deciding about Human Lives: An Experimental Measure of Risk Attitudes under Prospect Theory. *Soc. Choice Welf.* **2018**, *51*, 163–192. [CrossRef]
28. Schijven, M.; Hitt, M.A. The Vicarious Wisdom of Crowds: Toward a Behavioral Perspective on Investor Reactions to Acquisition Announcements. *Strat. Manag. J.* **2012**, *33*, 1247–1268. [CrossRef]
29. Foss, N.J.; Frederiksen, L.; Rullani, F. Problem-Formulation and Problem-Solving in Self-Organized Communities: How Modes of Communication Shape Project Behaviors in The Free Open-Source Software Community. *Strat. Manag. J.* **2016**, *37*, 22. [CrossRef]

MDPI
St. Alban-Anlage 66
4052 Basel
Switzerland
Tel. +41 61 683 77 34
Fax +41 61 302 89 18
www.mdpi.com

International Journal of Environmental Research and Public Health Editorial Office
E-mail: ijerph@mdpi.com
www.mdpi.com/journal/ijerph